T0325076

GLOBAL SUSTAINABLE CITIES

Global Sustainable Cities

City Governments and Our Environmental Future

Edited by
Danielle Spiegel-Feld,
Katrina Miriam Wyman, and
John J. Coughlin

NEW YORK UNIVERSITY PRESS
New York

NEW YORK UNIVERSITY PRESS
New York
www.nyupress.org

Please contact the Library of Congress for Cataloging-in-Publication data.
ISBN: 9781479805747 (hardback)
ISBN: 9781479805754 (paperback)
ISBN: 9781479805716 (library ebook)
ISBN: 9781479805730 (consumer ebook)

New York University Press books are printed on acid-free paper, and their binding materials are chosen for strength and durability. We strive to use environmentally responsible suppliers and materials to the greatest extent possible in publishing our books.

Manufactured in the United States of America

10 9 8 7 6 5 4 3 2 1

Also available as an ebook

CONTENTS

Introduction

DANIELLE SPIEGEL-FELD, KATRINA MIRIAM WYMAN, AND
JOHN J. COUGHLIN

In recent years, scholars and policy makers alike have become increasingly interested in the role that cities can play in building a more environmentally sustainable future. This interest makes sense: more than 50 percent of the world's population lives in cities, and cities generate more than 80 percent of global economic output.[1] Urban populations account for the lion's share of global energy consumption as well (Lifshitz, chapter 11) and account for 70 percent of global CO_2 emissions.[2] Cities have also expressed lofty environmental ambitions, which often outpace the corollary goals of the nation-states in which they are located. At one time or another in recent decades, Beijing, Berlin, London, New York City, and Shanghai have committed to reducing greenhouse gas (GHG) emissions faster or more aggressively than their nation-states.[3] And, as cultural hubs, cities play an outsized role in creating and disseminating norms, including norms regarding environmental protection (Coughlin, chapter 1). Thus, the choices that cities make about how to acquire drinking water, whether to encourage fossil-fuel-powered cars, low-density development, or renewable energy, or how to reduce air pollution can have far-reaching consequences, both for the billions of people who live in cities and for the global climate.

This book seeks to shed new light on the role that major cities around the world are playing in global efforts to protect the environment. Our overarching goal is to understand the extent to which cities, as opposed to higher levels of government, are devising the environmental policies that are needed to build a more sustainable future. Relatedly, we seek to understand the challenges that cities face in developing and implementing environmental policies given their legal, economic, and spatial situation within nation-states.

Our hypothesis in planning this volume was that scholars and policy makers should be cautiously optimistic about the potential for cities to address environmental issues. We were neither as bullish nor as bearish as some of the other commentators who have written about cities' abilities to address environmental issues such as climate change.[4] Our nuanced view of the potential for cities as environmental actors is consistent with a synthesis of two leading schools of thought about cities.

One school sees local governments positively because they are responsive to local political preferences. Ideally, this responsiveness bolsters democratic processes and leads to an efficient provision of goods and services that accords with the preferences of local residents.[5] The twenty-first-century scholarship extolling the potential for progressive, liberal cities to address climate change is consistent with this line of thinking; it portrays cities as responsive levels of government in which concerned citizens can exert meaningful influence, in contrast to more distant national governments that are prone to capture by powerful special interests.[6]

There is, however, a second, dimmer view of local government in local government law scholarship. It also sees local governments as responsive to the preferences of local residents but emphasizes that local governments' responsiveness to certain local constituencies—the very feature that makes them fonts of democratic participation—also makes them prone to neglect the interests of less powerful local constituencies and the broader public outside their borders.[7] According to this view, local governments focused on maximizing the interests of their residents will, if they can, export problems, such as garbage and water and air pollution, to people living beyond their borders and import resources from other jurisdictions with little concern for the problems that this resource extraction creates in the areas from which it comes.[8] The chapters in this book provide a number of examples of cities improving their local environments by externalizing harms onto others: Beijing and Delhi have helped improve their air quality by closing local coal-fired power plants and importing more electricity from coal-fired plants outside the city instead. (Ganguly, chapter 7; Lin, chapter 9; Kuldeep and Biswas, chapter 10). Abu Dhabi meets its needs for water partly by desalinating huge quantities of water from the Persian Gulf, even though desalination is carbon intensive (at least as it is performed today) and harms marine biodiversity in the Persian Gulf (Savarani, chapter 4). New York City dumps untreated sewage and rainwater into the water bodies surrounding it when it rains heavily and the capacity of the city's sewage system is exceeded. (Wyman, chapter 6).

Notwithstanding the potential for cities to export environmental problems, we believe that cities do have substantial potential to advance environmental protection; the key is to distinguish the areas in which they have greatest potential to act from those in which they are most constrained. Broadly speaking, we hypothesize that cities are most likely to seriously address environmental issues when they perceive that doing so will confer a concentrated local benefit and/or when the costs of inaction will be concentrated locally. If a key driver of local action is a perception that action will benefit the locality, it is logical to ask what gives rise to a perception of local benefit. As law professors, we are not well equipped to answer this question. However, John Coughlin's contribution to this volume on the papal encyclical *Laudato Si': On the Care for Our Common Home* suggests some possibilities, including human attitudes toward nature, which he underscores can change over time.

Although this volume does not rigorously test the hypothesis that cities address environmental issues in response to a perception of local benefits, the chapters provide some suggestive evidence to support it.[9] The chapters analyze the responses of cities to four environmental challenges—securing clean drinking water, reducing local air pollution, reducing greenhouse gas emissions, and adapting to climate change—focusing on seven case-study cities (Abu Dhabi, Beijing, Berlin, Delhi, London, New York, and Shanghai). Of the four challenges, the one that the cities discussed in this volume have arguably done the most to address is securing drinking water (Długosz-Stroetges and Anton, chapter 3; Savarani, chapter 4; Vijayaraghavan, chapter 5; Wyman, chapter 6). This is a challenge that cities clearly stand to benefit from addressing; and if they fail, their populations will suffer in a tangible way. It is a truism that water is necessary for human life; no city could exist without water, and providing clean drinking water has historically been an important prerequisite for urban life.[10] Reducing greenhouse gas emissions to mitigate climate change is a challenge with a different cost-benefit calculus for cities. While all cities would benefit from limiting climate change, at first glance, no city has a strong incentive to unilaterally reduce greenhouse gas emissions within its borders because it will be costly for local actors to reduce their emissions and the benefits of such reductions will accrue to people around the world. The benefits of any single city's actions could also be canceled out by the choices made by other jurisdictions to continue emitting.

Given that reducing GHG emissions would seem to impose costs on cities for few local benefits, there is an extensive literature seeking to explain why some cities and other subnational governments have demonstrated a commitment to reducing GHG emissions.[11] Essays in this volume suggest that some cities may be tackling GHG emissions because they derive local co-benefits from doing so, such as reductions in particulate matter and other forms of local pollution that affect the health of city residents (Kuldeep and Biswas, chapter 10; Mao et al., chapter 12). However, it seems unlikely that the local co-benefits of reducing GHG emissions will be sufficient to induce cities to fully decarbonize on their own, and policy interventions by higher levels of government, such as national governments, probably will be necessary to make the dramatic changes in urban life necessary to mitigate climate change. Indeed, the chapters in this volume suggest that the actions of cities such as Beijing and Berlin to reduce GHG emissions should be understood as consistent with national policy commitments to reducing GHG emissions, which lead local decision-makers to see reducing GHG emissions as having local benefits on their own terms.

At the peril of generalizing, the provision of clean air and adapting to climate change lie somewhere between securing drinking water and reducing GHG emissions in their cost-benefit structures from the local government perspective. Reducing emissions of local air pollutants such as particulate matter stands to benefit city residents by improving public health, but it also may be costly for individuals and businesses to change the behaviors that lead to these emissions, such as burning coal for electricity and residential heating and driving fossil-fuel-powered cars. Moreover, due to wind patterns, local air quality is often influenced by activities beyond a city's borders. Thus, local efforts to improve air quality may be undermined by pollution coming from elsewhere—and local efforts may in turn benefit residents of other cities, similar to efforts to reduce GHG emissions.

With respect to adaptation, it is often suggested that cities have strong incentives to adapt to the extreme heat, rising sea levels, flooding, and other changes associated with climate change.[12] However, the chapters in this volume indicate that many cities have only made limited investments in climate adaptation to date (Minelli, chapter 17; Zavadski, chapter 18). There are several potential explanations for this seeming underinvestment. For one thing, cities may not perceive a strong local benefit to investing heavily in adaptation because many of the most severe risks to the quality of local life may lie in the future or may seem extremely costly—or perhaps impossible—for cities to address on their own due to limited fiscal resources and constrained local authority in nation-states.[13] In addition, as Harry den Hartog makes clear in chapter 16, there is a tension between coastal cities' desire to maximize land values, which incentivizes development of wetlands, and the desire to maintain coastlines in a natural state for storm surge protection. In Shanghai's case, the pressures to develop seem to be preventing the city from fully achieving its conservation goals.

Notably, even when cities perceive a strong local benefit to addressing an environmental issue—such as securing clean drinking water—cities may not address that issue on their own. Of the four cities' whose approaches to securing clean drinking water are discussed in this volume—Abu Dhabi, Berlin, Delhi, and New York City—only Berlin obtains all of its drinking water from within its borders, and that self-sufficiency seems to be strongly rooted in the city's isolation during the Cold War, though Berlin did consciously reaffirm its commitment to remaining self-reliant for water in the decades after the Berlin Wall fell (Długosz-Stroetges and Anton, chapter 3). To varying degrees, the other three cities draw on sources outside their borders. As mentioned earlier, Abu Dhabi desalinates water from the Persian Gulf in addition to relying on groundwater (Savarani, chapter 4). New York City and Delhi use freshwater coming from outside their jurisdictions, which necessitates interstate agreements with other jurisdictions to meet the cities' water needs (Vijayaraghavan, chapter 5; Wyman, chapter 6). Thus, even when the local benefits of cities addressing an environmental issue are high, we see that cities may not be able to fully address the issue by themselves. Cities may be limited by a lack of physical resources within their borders to meet the needs of their location populations, by an unwillingness to impose costly policies on residents to conserve the limited in-city resources or by their legal inability to govern actors beyond their borders, among other constraints. Regardless of the reasons, the inability of leading cities such as Delhi and New York City to meet the basic need of their populations for drinking water on their own further underscores the need to be cautious in considering the potential for cities to be environmental saviors.

The chapters in this volume about how cities obtain drinking water and manage wastewater and rainfall also suggest that there is a relationship between the decisions that cities make about how to address one environmental issue and how they address other such issues. Historians of the development of urban water systems have argued that the development of urban water-supply systems contributed to the need to develop systems for managing wastewater, as the supply systems increased water consumption.[14] The sources of cities' drinking water also may affect their approaches to managing wastewater and rainfall. Berlin, which receives its drinking water from groundwater sources that are within its geographic boundaries, has pursued aggressive, and effective, action to conserve water and reduce pollution from stormwater (Długosz-Stroetges and Anton, chapter 3). By contrast, New York City, which receives its drinking water from surface-water sources outside the city's borders, has done relatively little to reduce stormwater pollution or reduce water usage. When it has acted, it has typically only done so following prodding from federal regulators (Wyman, chapter 6).

The history of the development of sewage treatment in New York City also underscores that even if cities act to address an environmental challenge, they may not act equitably. Required by the federal government to build a sewage-treatment plant to treat sewage deposited into the Hudson River, the city sited the plant close to an African American and Latino neighborhood.[15] As environmental justice scholars and advocates in the United States have emphasized, cities may perpetuate injustices not only through their actions but also through continued inaction that disproportionately burdens low-income communities and communities of color.[16] The differences between the environmental challenges facing cities in the Global South and the Global North, and their differential capacities to address them, underscore the unequal distribution of environmental harms and benefits at the international level.[17]

Overall, the chapters in this volume support the proposition that cities will seek to address environmental challenges when there is a perception that the local benefits of doing so exceed the local costs. The chapters also reinforce the idea that there are important constraints on the ability of cities to protect the environment, even when they have an incentive to do so. Chief among these constraints are the legal constraints imposed on cities by higher levels of government, the limited geographic territory that they control, and the path dependency that flows from choices made in the past about the type of urban infrastructure such as water supply, wastewater, and electricity systems to develop. Scholars, policy makers, and environmental advocates should bear in mind these constraints on cities in considering their potential to deal with climate change and other environmental issues.

The Organization of the Book

To help elucidate the contemporary role that leading cities around the globe are playing in advancing environmental protection, we held a series of virtual workshops during the fall of 2020 with experts from major cities around the globe to assess their cities' efforts to tackle environmental challenges. The chapters are grouped into five parts; within each part, the chapters are organized alphabetically based on the last name of the author (or first author when there are multiple authors). The first part features two chapters that address issues that cut across specific environmental challenges. John Coughlin's essay reflects on Pope Francis's encyclical *Laudato Si': On the Care for Our Common Home* and in so doing brings out the essential role of norms and values in spurring human interest in environmental protection and the potential role of cities in giving rise to these values. The transformation in human consciousness toward nature that Coughlin argues is needed to "resolve the environmental crisis" might alter people's perceptions of the actions that would be beneficial and potentially lead to more environmental protection if, as we suggested earlier, local governments act in response to perceptions of local benefit. Josephine van Zeben underscores the importance of cities' legal situation within nation-states for their ability to realize their "environmental ambitions." No city has complete legal autonomy from the state that houses it. Instead, higher levels of government carve out specific areas of jurisdictional competence for their cities and expressly remove some subjects from local competence. Often, cities share jurisdiction over a given subject with higher levels of government. Van Zeben's chapter foreshadows the exploration in a number of the chapters of the institutional framework in which cities operate and how legal constraints imposed on cities limit their space to act. Perhaps owing to the fact that relatively little scholarship in this area has been written by legal scholars, few existing works have systematically considered the ways in which legal hierarchies between cities and superior levels of government within the nation-state influence the environmental policy decisions that cities make.

The remaining four parts of the volume contain chapters about how cities are addressing one of four environmental challenges: securing clean drinking water, reducing local air pollution, mitigating climate change, and adapting to climate change. A considerable amount of the existing literature examining cities as environmental actors focuses on cities' role in combating climate change.[18] This is paradoxical if the goal is to understand local governments' contributions to environmental protection because, as already discussed, it is not self-evident that cities have a strong incentive to reduce GHG emissions given the limited benefits that would accrue to local residents from reducing these emissions.[19] We look beyond just what cities are doing to address climate change and also look to see how they are tackling more traditional environmental challenges such as provision of drinking water and clean air, where the benefits of local action are clearer and thus cities might be expected to be more aggressive.

In inquiring into the actions that large cities around the world are undertaking under the banner of sustainability, this book is asking what these cities are doing to protect the environment and what challenges they face in doing so. The Brundtland Commission's 1987 report *Our Common Future* is credited with popularizing the idea of sustainable development, which is often thought to mean that development should attend to environmental, social, and economic considerations.[20] In recent decades, governments, including city governments, and international institutions have embraced the idea that development should be sustainable, and indeed there is now a critique that sustainability has been co-opted by governments in the service of economic growth.[21] According to this critique, urban governmental sustainability policies often prioritize economic growth and embrace greening as a signal to promote growth rather than to meaningfully protect the environment or achieve social justice.[22] In this book, we use sustainability in a rather limited sense, to refer to environmental protection, albeit protection against long-standing environmental threats, such as polluted drinking water, and newer challenges, such as climate change.

Most of the chapters in this volume focus on seven case-study cities: Abu Dhabi, Beijing, Berlin, Delhi, London, New York, and Shanghai. A number of contemporary studies, including works by Sara Hughes, Stephen Jones, and Anél Du Plessis, present important comparisons of how different cities in the Global North or South are tackling climate change.[23] However, there appear to be fewer analyses that compare the cities in the Global North and South to see how their responses differ.[24]

The seven case-study cities studied in this volume share a number of commonalities. To begin with, each of the cities is its nation's most populous city, its capital, or both. Moreover, while the cities are at different stages of development, services, rather than heavy industry, dominate their economies.[25] These cities are hubs of the world economy where the knowledge economy flourishes and prosperity has been largely unmoored from local resource exploitation.[26] They are also all places that are experiencing some severe environmental challenges, from smog-filled skies to drinking-water shortages to rising seas. All of this would seem to make these cities fertile ground for bold sustainability policies to take hold.

But there are also some critical differences among the case-study cities. Four of the cities are in high-income countries (Abu Dhabi, Berlin, London, and New York), while the other three cities are in middle-income countries (Beijing, Delhi, and Shanghai). The cities run the gamut with regard to governance approaches too. Several of the cities are national capitals and have been given special powers that other cities in their nations have not been allocated. Each of the northern cities, as well as Delhi, is situated in a liberal democratic nation, and the municipal government in these cities is selected via local elections. These locally elected governments function with a degree of independence from higher levels of authority (although, as Josephine van Zeben explains in chapter 2, they are also prone to conflicts with these higher levels of government). By contrast, the Chinese cities, as well as Abu Dhabi, exist within autocracies. These city governments are not truly independent political entities. In Shanghai and Beijing, the central government directly appoints large swaths of the cities' leadership, and the central government provides the general policy direction for the cities.[27] Thus, when Beijing undertook efforts to dramatically reduce air pollution as part of its plans for the 2008 Summer Olympics, Beijing did so with the full support and coordination of the central government, including access to central fiscal resources (Lin, chapter 9).[28] As its nation's capital, Beijing, like Abu Dhabi, Delhi, Berlin, and London, also enjoys a special status (and perhaps a closeness with the national government).

Notably, this volume was planned before the COVID-19 pandemic; but many of the chapters were written and first presented in draft in the midst of the pandemic, and some of them refer to the effects of the pandemic on cities (Ganguly, chapter 7; Kuldeep and Biswas, chapter 10; Lifshitz, chapter 11; Ohlhorst and Schreurs, chapter 13; Spiegel-Feld, chapter 14; den Hartog, chapter 16). Casual observation suggests that the COVID-19 pandemic accelerated many cities' efforts to develop bicycle infrastructure, as well as pedestrian and green spaces, providing residents with room to safely commute and commune in open air. However, there is also a concern that the economic challenges that the pandemic has imposed on some cities has diminished their capacity to pursue new environmental regulations or fully enforce existing ones (Spiegel-Feld, chapter 14).

Directions for Future Research

The chapters that follow provide valuable insight into the extent to which leading cities in the Global North and South are advancing environmental protection. Still, many questions remain for future research; we identify three categories of questions based on our reading of the chapters.

One set of outstanding questions concerns the approaches that cities use to achieve their environmental goals. Sara Hughes argues that there is much less literature analyzing *how* cities are seeking to reduce

GHG emissions compared with the ample literature addressing why cities seek to do so in the first place.[29] The point might be generalized into an observation about the lack of research on how cities are satisfying their environmental goals generally, not merely their GHG emission reductions priorities. Some patterns are apparent from the case studies in this volume; it would be interesting to examine the extent to which these patterns hold across a wider range of cities. Using Harriet Bulkeley's typology of "modes of governing climate change in the city," all the cities whose approach to securing drinking water is described in this volume "provision" water services directly.[30] While there is some degree of private-sector involvement in various of the sample cities, the general trend is toward public provision of water.[31] However, the question of whether water services should be provided by the government or the private sector is an enduring one. In the nineteenth century, New York City developed a public water-supply system (with the assistance of New York State) after the private sector proved incapable of building a system (Wyman, chapter 6). More recently, there have been debates about privatizing water services in Delhi, where demand for water exceeds the amount that the Delhi Jal Board is able to supply (Vijayaraghavan, chapter 5). Notably, Berlin experimented with partial privatization of its water utility, only to fully remunicipalize it (Długosz-Stroetges and Anton, chapter 3).

In contrast to the tendency toward public provision of water services, the chapters in this volume suggest that local governments are seeking to regulate to improve local air quality and reduce GHG emissions.[32] Cities are using variations of traditional forms of "command-and-control" regulation that require particular technologies or behaviors or set mandatory performance standards. For example, Beijing prohibits public buildings from setting air conditioners to less than 78.8 degrees Fahrenheit (26 degrees Celsius) during summer (Mao et al., chapter 12). Delhi mandates the installation in existing and new buildings of solar water heaters for all industries, hotels, education institutions, and other commercial and residential buildings with an area of more than five hundred square meters (Kuldeep and Biswas, chapter 10). In 2019, New York capped GHG emissions for large new and existing buildings and both residential and commercial properties (Spiegel-Feld, chapter 14). Cities also are introducing market-based mechanisms that incentivize changes in behavior and technologies. To tackle local air pollution, the mayor of London introduced a congestion pricing scheme in 2003, becoming one of the first jurisdictions in the world to implement such a program (Kelly, chapter 8; Lifshitz, chapter 11). Beijing "established a [pilot] carbon emission trading program in December 2013" (Mao et al., chapter 12).[33] In a work published in 2013, Bulkeley observed that regulation was "perhaps the least used [mode of governing] in the climate change domain," but the chapters in this volume on Berlin, Beijing, London, and New York City suggest that the policy landscape is changing in at least some cities in major GHG-emitting countries in the world.[34] It would be useful to know whether this is an emerging trend in other cities and, if so, what regulatory approaches cities are using and which sectors they are targeting for regulation of GHG emissions.

In addition to raising questions about the types of strategies that cities are using to protect the environment, the chapters in this volume raise questions about the impacts of local environmental policies. While the case-study chapters present a picture of measures that the various cities are taking to address environmental challenges, they do not systematically assess the impacts of these actions. This is a subject that is ripe for future interdisciplinary analysis involving scholars from public policy, economics, sociology, anthropology, law, and other areas. Depending on the availability of resources for ex post evaluation, many different impacts of local policies could be examined, including the effects that policies are having on the physical environment (for example, are policies to reduce GHG and local pollutant emissions actually reducing these emissions and, if so, by how much?) and the impacts on human health (for example, are policies to reduce GHG and local pollutant emissions saving lives and reducing pollution-related disease?). The costs of policies on cities and actors within them might be analyzed and compared to the benefits that cities and actors receive from the policies. The distribution of these costs and benefits might be assessed to analyze whether disadvantaged populations in cities are bearing undue burdens or enjoying fewer benefits from policies than other populations. Rigorous retrospective analysis of local government environmental policies would help improve them and facilitate the development of new policies. It also would be a sign of the growing recognition of the importance of local governments in protecting the environment if the impacts of more innovative local environmental policies were comprehensively analyzed ex post.[35]

A third set of questions that the chapters in this volume highlight is the importance of better understanding how policies that are used at the local level diffuse to other jurisdictions. The diffusion may be horizontal, to other cities in the same country or abroad, or vertical, to higher levels of government within the same country or potentially even internationally.[36] While noting that Beijing still does not meet Chinese or World Health Organization (WHO) standards for particulate matter, Lin's chapter in this volume brings out the progress that Beijing, with the assistance of national authorities, has made in improving air quality in the city since "Beijing launched the first local government air pollution program in China in 1998" (Lin, chapter 9). Lin indicates that several air pollution control policies first adopted in Beijing "have become models for China's national approach to air pollution" (Lin, chapter 9).[37] Spiegel-Feld provides an example of policy diffusion in the United States when she refers to the way other cities adopted regulations to reduce GHG emissions from buildings after New York City did so (chapter 14). Indeed, at the time of the writing, the federal government in the United States is working to adopt a performance standard for federally owned buildings that would mimic standards developed in several leading US cities, including New York.[38] These examples from Beijing and New York suggest that cities play an important role in the ecosystem of environmental policy makers by piloting new approaches to environmental regulation that may later diffuse to other governments.[39] It would be useful to better understand how this diffusion occurs within and across nation-states and whether there are factors—such as in the design and the marketing of the policy or the identity of the city implementing it—that lead some local policy innovations to be emulated by other governments. It would also be useful to explore whether policy diffusion occurs primarily between large cities, such as the cities that are the focus of this book, or whether the policies developed by smaller cities also frequently percolate to other cities and/or higher levels of government. Indeed, we are mindful that many of the lessons that we draw from the case studies in this book may be primarily applicable to the world's capitals and other large cities.

In short, while the chapters in this volume help advance our understanding of how leading cities in the Global North and South are contributing to the environmental agenda, there is a substantial need for further research. With a clear-eyed understanding of cities' potential and shortcomings as environmental regulators, scholars can help local leaders to be more productive participants in the global push to build a more sustainable future.

NOTES

1. World Bank. n.d. "Urban Development." Accessed September 2, 2021, www.worldbank.org.
2. C40. n.d. "A Global Opportunity for Cities to Lead." Accessed September 2, 2021, www.c40.org.
3. Lifshitz, chapter 11; Mao et al., chapter 12; Ohlhorst and Schreurs, chapter 13; den Hartog, chapter 16.
4. For an example of the bullish view, *see* Barber, Benjamin. 2017. *Cool Cities: Urban Sovereignty and the Fix for Global Warming*. New Haven, CT: Yale University Press. For a more bearish perspective, *see* Melosi, Martin. 2008. *The Sanitary City: Environmental Service in Urban America from Colonial Times to the Present*. Pittsburgh: University of Pittsburgh Press.
5. *See, e g.*, Frug, Gerard E. 1980. "The City as a Legal Concept." *Harvard. Law Review* 93(6): 1057–1154; Tiebout, Charles. 1956. "A Pure Theory of Local Expenditures." *Journal of Political Economy* 64(5): 416–424.
6. *See, e.g.*, Barber 2017; *see also* Miller, David. 2020. *Solved: How the World's Great Cities Are Fixing the Climate Crisis*. Toronto: University of Toronto Press.
7. The debate over the desirability of local autonomy is related to a larger body of scholarship about how responsibilities should be allocated between different levels of government across a range of issue areas, not only the subject of this book, environmental protection. On the desirability of various allocative arrangements, *see generally* Inman, Robert, and Daniel L. Rubinfeld. 2020. *Democratic Federalism: The Economics, Politics, and Law of Federal Governance*. Princeton, NJ: Princeton University Press. There is an extensive literature going back decades about the appropriate allocation of authority over environmental law between national and subnational levels of government (such as states and provinces) in federal systems. *See, e.g.*, Revesz, Richard L. 1997. "The Race to the Bottom and Federal Environmental Regulation: A Response to Critics." *Minnesota Law School* 82: 535–564; Schoenbrod, David, Richard B. Stewart, and Katrina M. Wyman. 2010. *Breaking the Logjam: Environmental Protection that Will Work*. New Haven, CT: Yale University Press.
8. *See* Howell-Moroney, Michael. 2008. "The Tiebout Hypothesis 50 Years Later: Lessons and Lingering Challenges for Metropolitan Governance in the 21st Century." *Public Administration Review* 68(1): 97–109. To some extent, this drive

to export pollution outside major cities may be appropriate; because many of cities' worst urban environmental problems are caused by the heavy concentration of polluting sources in a relatively small area, spreading out the sources over a larger area can prevent the most severe pollution problems from materializing. It may also lead to fewer public health problems if polluting sources are reallocated to less populous areas. And yet shifting pollution around is hardly an ideal ecological outcome and resembles the broader tendency toward NIMBYism that have hamstrung other attempts at environmental progress. It also highlights the reasons to be skeptical of cities' ability to tackle a global environmental problem like climate change.

9. Spiegel-Feld and Wyman plan on further developing the idea that local governments act to protect the environment on the basis of perceptions of the local benefits of environmental perception in a subsequent book on cities and the environment.
10. Mumford, Lewis. 1961. *The City in History: Its Origins, Its Transformations, and Its Prospects*. New York: Harcourt, 142.
11. Hughes, Sara. 2019. *Repowering Cities: Governing Climate Change Mitigation in New York City, Los Angeles, and Toronto*. Ithaca, NY: Cornell University Press, 5; Farber, Daniel A., Yuichiro Tsuji, and Shiyuan Jing. Forthcoming. "Thinking Globally, Acting Locally: Lessons from the U.S., Japan and China." *Ohio State Law Journal*. https://doi.org/10.2139/ssrn.3792510; Stewart, Richard B. 2008. "States and Cities as Actors in Global Climate Regulation: Unitary vs. Plural Architectures." *Arizona Law Review* 50(3): 688–693.
12. Wyman, Katrina M., and Danielle Spiegel-Feld. 2020. "The Urban Environmental Renaissance." *California Law Review* 108(2): 305–377. https://doi.org/10.15779/Z38DZ0325P.
13. Kahn, Matthew E. 2021. *Adapting to Climate Change: Markets and the Management of an Uncertain Future*. New Haven, CT: Yale University Press.
14. Tarr, Joel A. 1996. *The Search for the Ultimate Sink: Urban Pollution in Historical Perspective*. Akron, OH: University of Akron Press, 133–134.
15. Miller, Vernice D. 1994. "Planning, Power and Politics: A Case Study of the Land Use and Siting History of the North River Water Pollution Control Plant." *Fordham Urban Law Journal* 21: 707–722.
16. Bullard, Robert D. 2001. "Environmental Justice in the 21st Century: Race Still Matters." *Phylon* 49(3–4): 151–171.
17. Natarajan, Usha. 2021. "Environmental Justice in the Global South." In *The Cambridge Handbook of Environmental Justice and Sustainable Development*, edited by Sumudu A. Atapattu, Carmen G. Gonzalez, and Sara L. Seck. Cambridge: Cambridge University Press, 44 (referring to "the divide between the Global South and Global North when it comes to environmental issues"). This chapter emphasizes the relevance of the idea of environmental justice in the Global South, even though the environmental justice movement initially developed in the United States.

18. *See, e.g.*, Bulkeley, Harriet. 2013. *Cities and Climate Change*. Abington, UK: Routledge; Fitzgerald, Joan. 2010. *Emerald Cities: Urban Sustainability and Economic Development*. Oxford: Oxford University Press; D. Miller 2020.
19. Stewart 2008, 689; *see also* Wyman and Spiegel-Feld 2020.
20. Greenberg, Miriam. 2015. "'The Sustainability Edge': Competition, Crisis, and The Rise of Green Urban Branding." In *Sustainability in the Global City: Myth and Practice*, edited by Cindy Isenhour, Gary McDonogh, and Melissa Checker. Cambridge: Cambridge University Press, 106. World Commission on Environment and Development. 1987. *Report of the World Commission on Environment and Development: Our Common Future*, para. 39 ("Sustainability requires views of human needs and well-being that incorporate such non-economic variables as education and health enjoyed for their own sake, clean air and water, and the protection of natural beauty. It must also work to remove disabilities from disadvantaged groups, many of whom live in ecologically vulnerable areas, such as many tribal groups in forests, desert nomads, groups in remote hill areas, and indigenous peoples of the Americas and Australasia."). *See also* Mensah, Justice. 2019. "Sustainable Development: Meaning, History, Principles, Pillars, and Implications for Human Action: Literature Review." *Cogent Social Sciences* 5(1).
21. For critical analysis of the history of the concept of sustainability and its meaning, *see* Isenhour, Cindy, Gary McDonogh, and Melissa Checker. 2015. "Introduction: Urban Sustainability as Myth and Practice." In *Sustainability in the Global City: Myth and Practice*, edited by Cindy Isenhour, Gary McDonogh, and Melissa Checker. Cambridge: Cambridge University Press, 1–26; Greenberg 2015; Checker, Melissa. 2020. *The Sustainability Myth: Environmental Gentrification and the Politics of Justice*. New York: New York University Press, 33–42.
22. *See, e.g.*, Greenberg 2015, 110–111; Isenhour, McDonogh, and Checker 2015; Checker 2020.
23. Hughes 2019; Jones, Stephen. 2018. "City Governments Measuring Their Response to Climate Change." *Regional Studies* 53(1): 146–155. https://doi.org/10.1080/00343404.2018.1463517; Du Plessis, Anel, and Louis Jacobus Kotze. 2014. "The Heat Is On: Local Government and Climate Governance in South Africa." *Journal of African Law* 58(1): 145–174.
24. Harriet Bulkeley's work is an important exception to this general rule. *See, e.g.*, Bulkeley, Harriet. 2010. "Cities and the Governing of Climate Change." *Annual Review of Environment and Resources* 35: 229–253. https://doi.org/10.1146/annurev-environ-072809-101747.
25. Abu Dhabi is an outlier in this regard. While oil exports have historically accounted for the majority of Abu Dhabi's gross domestic product (GDP), since 2014, the non-oil economy has contributed the largest share of GDP. In 2020, the non-oil economy accounted for 56 percent of Abu Dhabi's GDP. The government seeks to enlarge the relative contribution of non-oil sectors further and transition toward a more diversified "sustainable knowledge-based economy." *See* Government of Abu Dhabi. 2008. *The Abu Dhabi Economic Vision 2030*, 6, 11, www

.ecouncil.ae. But because the non-oil economy includes some heavy industries such as manufacturing, it cannot be said that services presently dominate the economy.

26. Note that we do not formally call the case-study cities "global cities," as that term has been famously used by Saskia Sassen, but these cities do evidence some of the same characteristics that Sassen used to identify global cities, including being "key locations for finance and for specialized service firms." Sassen, Saskia, 2001. *The Global City: New York, London, Tokyo*. 2nd ed. Princeton, NJ: Princeton University Press, 3. Indeed, Sassen dubbed two of our case-study cities—New York and London—global cities in her study.
27. *See, e.g.*, Xinhua. 2017. "Job Responsibilities Adjustments for Key Officials of Communist Party in 3 Provinces and Municipalities Including Shanghai." *XinhuaNet*, October 29, www.xinhuanet.com (in Chinese, announcing that the Chinese Community Party has decided to name Li Qiang as the new secretary of the Chinese Communist Party Shanghai Municipal Committee, which is the highest-ranking member of the Chinse Communist Party in Shanghai). *See also* Kostka, Genia, and Jonas Nahm. 2017. "Central-Local Relations: Recentralization and Environmental Governance in China." *China Quarterly* 231: 567–582 ("Within this decentralized governance structure, the central government has historically exerted top-down control over the appointment and promotion of sub-national officials."). Describing the relationship between the central state and local governments, Alex Wang has observed that, "in China, central authorities, in theory, can govern agency staff, governors of provinces, mayors, state-owned enterprise heads, and a variety of other state actors through cadre evaluation." Wang, Alex. 2013. "The Search for Sustainable Legitimacy: Environmental Law and Bureaucracy in China." *Harvard Environmental Law Review* 37: 383. However, Wang also notes that while the Chinese government retains "top-down control over major priorities" and set policy targets, the government "offers local actors tremendous flexibility in how to meet targets." Wang 2013, 379, 381. In a recent publication, Yifei Li and Judith Shapiro explain that under President Xi Jingping's leadership, power has become even more centralized than it was under prior leaders. Li, Yifei, and Judith Shapiro. 2020. *China Goes Green: Coercive Environmentalism for a Troubled Planet*. Cambridge, UK: Polity, 18. Under Xi's leadership, Li and Shapiro observe, "measures to tackle 'airpocalypses' in urban areas have followed a top-down model that excludes even local-level officials from the political process." Li and Shapiro 2020, 18.
28. On the governance structure of Abu Dhabi, *see* Savarani, chapter 4.
29. Hughes 2019, 5.
30. Bulkeley 2013, 92. Hughes also draws on Bulkeley's typology. Hughes 2019, 61–66. In Bulkeley's terms, "a mode of governing refers to a specific set of processes and techniques through which governing is pursued." She identifies four modes: "self-governing," "provision," "regulation," and "enabling." Bulkeley 2013, 92–97. Provision "involves the development of low-carbon or resilient infrastructure systems,

as well as the delivery of services and goods that have a lower-carbon footprint or seek to improve adaptive capacity." Bulkeley 2013, 93.

31. However, even where there are publicly owned water utilities, private actors might be engaged in supplying water, for example, from groundwater.
32. Bulkeley defines regulation as "approaches intended to oversee, guide and determine particular conditions, ways of operating, behaviour and standards." Bulkeley 2013, 93.
33. *See also* Farber et al., forthcoming (discussing Shenzhen's emissions trading system).
34. Bulkeley 2013, 93.
35. For existing examples of ex post evaluation of municipal environmental policies, *see* Kontokosta, Constantine E., Danielle Spiegel-Feld, and Sokratis Papadopoulos. 2020. "The Impact of Mandatory Energy Audits on Building Energy Use." *Nature Energy* 5: 309–316.
36. *See* Bulkeley 2013, 99–100 (discussing "vertical" and "horizontal" "relationships" and "interactions" and the relevance of networks such as C40 for policy diffusion in the climate change context). *See also* Chu et al., chapter 15.
37. *See also generally* Teets, Jessica C., and William Hurst, eds. 2014. *Local Governance Innovation in China: Experimentation, Diffusion, and Defiance*. Abingdon, UK: Routledge (describing how local-level policy innovations sometimes diffuse up to central government policy in China).
38. Majersik, Cliff, and Ryan Freed. 2021. "What the White House's Building Performance Standard Means and How to Do It Right." Institute for Market Transformation, May 20, www.imt.org.
39. These cities' ingenuity gives credence to the old American adage that the states are the "laboratories of democracy" in the United States. New State Ice Co. v. Liebmann, 285 U.S. 262 (1932) (Brandeis, dissenting). Just as the states in the United States provide a proving ground in which new ideas can be explored, cities throughout the world are playing an important part in testing out new approaches to environmental policy.

PART I

Cross-Cutting Issues

Norms and Laws

1

Global Sustainable Cities and *Laudato Si'*

JOHN J. COUGHLIN

In this volume discussing the sustainability of global cities, I would like to draw our attention to Pope Francis's encyclical *Laudato Si': On Care for Our Common Home.*[1] When he published the encyclical in 2015, the pope spoke of the "appeal, immensity and urgency" presented by the global ecological crisis.[2] As we gather five years later, the crisis's severity has only intensified, and I shall discuss four topics about the encyclical and the sustainability of global cities: (1) the idea of integral ecology; (2) the goodness of our common humanity; (3) the encyclical's trust in law from the critical legal studies perspective; and (4) the possibility of religious wisdom as a conversation partner in the ecological discussion. Each of these four topics is a short essay, and I hope that, taken together, they will count as a modest partial analysis of the complex phenomena of global sustainable cities. Prior to proceeding to my musings about each of the four topics, I offer some preliminary observations about the concept of the global city.

We are examining case studies of seven "mega-cities" (Abu Dhabi, Beijing, Berlin, London, New Delhi, New York, and Shanghai). Of these seven, Shanghai had a 2018 population of approximately 25.5 million people, as compared to Abu Dhabi, with a 2018 population of approximately 1.4 million persons.[3] New York City generated a 2019 gross domestic product (GDP) of approximately $1.75 trillion, while Berlin's 2019 economy yielded a GDP of approximately $179 billion (€153 billion).[4] Given the disparities, what, then, constitutes the global city? While I confess that I am a novice when it comes to the construct of global cities, it seems to me that the elements of population size and economic power are far from the only factors that go into global city status.[5] First, an adequate description should also acknowledge that global cities serve not only as international commercial hubs but also as centers of science, technology, law, medicine, politics, culture, language, university education, art, and history. Second, another characteristic of the global city seems to be the ethnic, linguistic, cultural, and religious diversity of its population mass. In this regard, it is not only the labor market that offers gainful employment but also national immigration laws that encourage the diversity of the citizens of the global city. Third, global cities are constituted by a complex series of interrelations. The global city is part of the national state, and it is also in relation to other urban and rural areas within the nation; while at the same, global cities are in relation to each other and to other nations. Although my preliminary observations are far from an adequate description of global cities, I shall suggest that each of these three characteristics is relevant to global sustainable cities and Pope Francis's encyclical.

Integral Ecology

The first major theme of *Laudato Si'* that I wish to explore is what the pope terms "integral ecology," an idea that flows from the basic principle that "everything is closely interrelated."[6] Integral ecology recognizes that an isolated fix to a particular environmental issue will be insufficient to achieve overall environmental sustainability. In the pope's words: "Recognizing the reasons why a given area is polluted requires a study of the workings of society, its economy, its behavior patterns, and the ways it grasps reality. Given the scale of change, it is no longer possible to find a specific, discrete answer for each part of the problem."[7] The idea that "everything is closely interrelated" is not merely theoretical but enjoys a scientific basis. In Chad's Sahara Desert, the geologist Moussa Abderamane researches how dust from six- to eight-thousand-year-old fossilized fish in the Bodélé Depression starts on a long trip in the wind over western Africa to the Atlantic Ocean.[8] In fact, there is so much dust from the Sahara (approximately 150 million tons annually) that astronauts in space can see it with the naked eye. NASA satellites record that Saharan dust mitigates the formation of hurricanes off the coast of western Africa. Dust-collecting buoys in the Atlantic provide data about how the dust nourishes microscopic plants on the seafloor, which in turn consume carbon dioxide and produce oxygen by photosynthesis, helping to alleviate greenhouses gases. The dust not only cedes life but also takes it. Some of the dust settles in the Florida Gulf Coast, where it feeds algae that produce the toxins of red tide that wreak havoc on marine life and on the ecosystem in general. The dust ultimately settles in the Amazon rain forest, where the meteorologist Stefan Wolf climbs the Amazon Tall Tower Observatory to collect samples that are traceable to those of Abderamane. Heavy rain and flooding in the Amazon washes away its soil, which is replenished by the Saharan dust supplying critical nutrients. Every tree in the Amazon probably contains some of these nutrients. Not only do the trees produce a cornucopia of nuts and berries, but they are also the lungs of the Earth, sustaining its life and biodiversity.[9] The Sahara is not the sole source of global dust. Dust from the Arabian Desert eventually falls as snow in the mountains of central Europe. The global journey of desert dust constitutes only one illustration of why integral ecology depends on the recognition that everything is closely interrelated.

How might integral ecology pertain to the sustainability of global cities? Think about the environmental goal of maintaining clean city streets. First, the goal is interrelated with other essential environmental concerns, such as toxic landfills, contaminated water, and polluted air. For example, the refuse collected from the city streets, which cannot be effectively recycled, may adversely impact the landfill, where it might be dumped and pollute the water supply through seepage into the ground. Effective recycling and safe disposal of refuse from the streets of the global city also entail ensuring that the trucks and other equipment employed in the street-cleaning efforts are designed to reduce the global city's air pollution. Recognizing the interrelatedness of various forms of environmental damage constitutes the first essential step in the integral ecology of global cities.

Second, the goal of clean city streets also raises a host of interrelated scientific, technological, financial, legal, and other social concerns. The scientific and technological research to develop the most effective method for recycling street refuse will require cost-effective approaches based on economic models. The financing for the implementation of a cost-efficient method of meeting the clean-street goal will depend on the political will to raise and budget the necessary capital, as well as on legal means to secure and enforce the arrangement. Public hearings prior to the promulgation of the recycling regulations will be vital to communicating an understanding of the democratic process and reasonableness of the policies to the citizenry. Artistically designed signage will be indispensable to educate and persuade citizens to comply with the regulations, and the communication will probably profit from presentation in a variety of languages so that the citizens can understand and appropriate the message. Facilitating a shift in consciousness among citizens about the attractiveness of integral ecology is another constitutive move in the sustainability of the global city.

Third, the nexus between the clean-streets goal and other concerns within the global city points to the interrelatedness between the global city and urban and rural areas within borders of the nation-state. Global cities, for example, sometimes transport their refuse to landfills and dumps in other geographic locations within the nation. At the same time, the ecological issues of global cities are not limited to national boundaries but impact other cities and areas in different nations. As an illustration of the international dimensions of integral ecology, one might consider how the air pollution of one large mega-city contributes to the global problem of greenhouse gases. The interrelatedness of global cities to national and international environmental issues also forms a crucial dimension of integral ecology.

Desert dust and clean city streets offer pragmatic examples of integral ecology; these examples are augmented by deeper levels of philosophical interrelatedness discussed in *Laudato Si'*. Please permit me to briefly mention two kinds of philosophical interrelatedness that remain essential to realizing environmental sustainability. First, perhaps the most fundamental of these interrelations is that between human beings and the rest of nature. The English word "ecology" stems from the Greek word *oikos*, which means "home." In speaking of "our common home," Pope Francis observes, "When we speak of the 'environment' what we really mean is a relationship existing between nature and the society which lives in it. Nature cannot be regarded as something separate from ourselves or as a mere setting in which we live. We are part of nature, included in it and thus in constant interaction with it."[10] In the pope's analysis, understanding nature as our common home with which we constantly interact exposes the exaggerations of modern anthropocentrism, which presumes a human supremacy and domination over nature. In Pope Francis's words, "our 'dominion' over the universe should be understood more properly in the sense of responsible stewardship."[11] The understanding that humanity is to act as the steward of nature rather than as its overlord pertains to nonhuman animals and all of material reality. The title of the pope's encyclical, *Laudato Si'*, may be traced back to the *Canticle of Creation* of Saint Francis of Assisi, who perceived such an I-Thou relationship with nature. Pope Francis, like his namesake, also sees himself as a brother in relation to the sundry manifestations of nature. He is not suggesting that all things in nature are of an equal ontological value but that they should be valued in themselves as more than objects to be exploited by humanity.[12]

Second, integral ecology also encompasses the recognition of the interrelatedness of environmental sustainability and broader social issues such as development, the poor, migration, and social justice. As the pope expresses it, "It is essential to seek comprehensive solutions which consider the interaction with natural systems themselves and with social systems. We are faced not with two separate crises, one environmental and the other social, but rather with one complex crisis which is both social and environmental. Strategies for a solution demand an integrated approach to combating poverty, restoring dignity to the excluded, and at the same time protecting nature."[13] Economic decisions driven by short-term gains, the maximization of shareholder profits, private interest, and the culture of consumerism not only are destroying the environment but are also most detrimental to the least privileged and most marginalized human beings. When Pope Francis served as Archbishop of Buenos Aires, he first used the Spanish phrase *la cultura del descarte* (throwaway culture) to signify that the planet's environmental pain is borne most by the disenfranchised, while major corporate actors owned by wealthy shareholders reap the financial rewards. For example, the poor of the global city who suffer most immediately from contaminated water supplies are understandably reticent to challenge the large corporations, which may be responsible for the water pollution but which also represent a source of employment and wages for the financially disadvantaged families of the city.[14] The stark reality is that technocratic governmental and private-sector policies often serve to widen the inequitable distribution of wealth. Integral ecology acknowledges the connection between damage to the environment, bolstering the wealth of the rich, and depriving the poor of essential human goods such as adequate housing, health care, educational opportunities, and the dignity of work at fair compensation. The principle of integral ecology that everything is closely interrelated affirms another principle: that the sum is greater than the whole of its parts. It seems to me that the totality of the global city is certainly more than the mere pluralism of its multitudinous discrete parts. The call of integral ecology to safeguard water, air, biodiversity, and climate is a call to step back, in order to see the beauty of the global city as a whole.

The Goodness of Our Common Humanity

The second major theme that I wish to highlight from *Laudato Si'* is Pope Francis's plea for the recognition of the goodness of our common humanity. In the pope's words, "all of us are linked by unseen bonds and together form a kind of unseen family, a sublime communion which fills us with a sacred, affectionate and humble respect."[15] It is notable that the pope describes our common human qualities in terms of affection, humility, communion, respect, family, sublimity, and sacredness. A less positive anthropological account might focus on human selfishness and its manifold manifestations to characterize human nature. The pope's understanding stems from the biblical doctrine of the creation of man and woman in the image and likeness of God (*imago Dei*), and it expresses a fundamental optimism about the goodness of the human person and all of material creation. The theological idea means that there is an essential human nature shared by all who enjoy a radical equality. At the same time, the idea accepts, appreciates, and honors diversity based on characteristics such as gender, race, creed, nationality, sexuality, age, physical or mental ability, national origin, and economic status. The papal understanding about the goodness of our common humanity is not limited to a strictly theological perspective.

Although the notion of the goodness of our common humanity remains open to dispute, I suggest that the notion can be heuristically posited based on several different kinds of evidence.[16] First, our common humanity finds scientific support in the process of evolution and specifically in the emergence of anatomically modern humans. Bipedalism, encephalization, and ulnar opposition between the thumb and little finger of the same hand are characteristics common to human beings.

The geographic dispersal of modern Homo sapiens from Africa has been mapped by anthropology, molecular biology, and studies of the human genome. However, no two human beings are genetically identical, and each of us has individual characteristics, qualities, emotions, thoughts, and experiences.[17] Given the theories of natural selection and survival of the fittest, it is not difficult to account for selfishness, cruelty, violence, and other forms of human treachery. However, what in evolutionary theory might help to explain human propensities for kindness, generosity, heroism, and even self-sacrificial love? I wish to avoid any biological reductionism of the human person, but perhaps the evolution of morphological structures affords the basis for the development of higher human characteristics. The evolutionary biologist William D. Hamilton, for example, derived a rule for understanding the genetic basis of altruism, and Iain McGlichrist has drawn on neuroscience to describe the evolution of the left and right hemispheres of the human brain and the implications for the higher human faculties.[18] Although the facts of evolution do not constitute empirical verification, it seems to me that they can be interpreted to lend support to Pope Francis's contemplative insight about the goodness of our common humanity.

Second, the consequences of the denial of the goodness of our common humanity might also be considered. The denial may be detected in the United States Supreme Court decision in *Buck v. Bell* when Justice Oliver Wendell Holmes Jr. wrote to uphold the constitutionality of a state eugenics program permitting the compulsory sterilization of people deemed "unfit" and "intellectually disabled," penning the disreputable words, "three generations of imbeciles are enough."[19] On a proportionately graver level, the National Socialist Party of Nazi Germany infamously denied our common humanity by propagating a theory of racial superiority that it nefariously spread to justify haunting crimes against, inter alia, Jews, Slavs, gays, Gypsies, and the mentally challenged. With events such as World War I and II, the Holocaust, and other wars and genocides, the twentieth century might be described as one of breathtaking technological progress coupled with mind-boggling human suffering and loss of life. Lamentably, the denial of our common humanity perdures to the present days in places throughout the globe, and Pope Francis is of course painfully aware of the tragedy of the human existential situation, such as the environmental destruction of our common home. As with the scientific evidence from evolution, horrific examples of negligence, failure, and cruelty hardly meet the rules of a sound philosophical proof of our common humanity, let alone its goodness, but it seems to me that the consequences of the denial of our common humanity ought not to be dismissed lightly.

Third, I suggest that, paradoxically, the diversity of the global city offers a robust affirmation of the goodness of our common humanity. Please permit me to be personal in this regard. Spending most of my life in New York City has fostered my belief in the goodness of our common humanity, and living the past seven years in Abu Dhabi, with its highly diverse population, has only served to intensify this belief. Growing up in Queens, the most ethnically and linguistically diverse place on the face of the Earth, I encountered from an early age a wide variety of human persons of all races, tongues, and creeds.[20] As a boy, I frequented the neighborhood "candy store" (as small convenience stores are known colloquially in New Yorkese), which, among other delectables, served "egg creams" at the cost of fifteen cents (the same as the fare on the subway). Open from early morning until late in the evening, the candy store was operated by a married couple who were European Jews, and during the summer months, when they wore short sleeves, you could see tattooed concentration-camp numbers on their arms. Even as a young boy, I understood what the numbers meant, and I also intuited that no matter what they had suffered, this remarkable couple provided a clean, honest, friendly, and dependable community center in which the patrons had a sense of participation, membership, and solidarity in our neighborhood community.

In New York, the experience of living daily with immigrants from all over the world has left a profound imprint on my soul. No doubt my religious education about the dignity of each person in the *imago Dei* reinforced this natural insight, which the past seven years of my work in Abu Dhabi has only intensified. Essentially, my insight about the goodness of our common humanity derives from the opportunity of living in places like New York and Abu Dhabi, and like so many residents of global cities, I have been enriched by witnessing on a regular basis the countless acts of respect, altruism, compassion, and self-sacrifice of diverse human beings for one another. While I realize that my heuristic methodology, especially my personal experience, is too weak to establish conclusive evidence, I read the affirmation of *Laudato Si'*, of the goodness of our common humanity, as a sermon of hope for environmental sustainability and for humanity as a whole.

How does Pope Francis see the goodness of our common humanity as hope for environmental sustainability? The pope's anthropological focus on human qualities such as respect, compassion, kindness, love, and a willingness to sacrifice suggests that he believes these altruistic characteristics are essential to humanity's contemporary struggle to resolve the global environmental crisis. In *Laudato Si'*, Pope Francis observes that the ability to contemplate the beauty of nature remains something that we all share in common as human beings. The ability of the human person to contemplate beauty is dependent neither on one's social, economic, or educational status nor on one's race, creed, color, sexuality, language, or national origin. Being at the beach close to the ocean under the sun in the salt air; in the mountains, with snow-covered peaks touching the crisp blue sky; in the dark forest, with freshwater running through rocky streams; and in the silence, solitude, and serenity of the pristine desert, with falcons soaring above us, are prototypical experiences of nature that lift our spirits. Such experiences are, of course, not limited to nature but can occur through a variety of encounters, such as with music, art, literature, poetry, and other human beings. However, the pope is describing more than a superficial feeling based on subjective experience. He is depicting a kind of knowledge available to human persons through contemplative reflection on one's experience that leads to self-transcendence. As he puts it, this "sense of deep communion with the rest of nature cannot be real if our hearts lack tenderness, compassion and concern for our fellow human beings."[21] Beauty is, of course, not the only path to self-transcendence. As my modest candy-store example suggests, the suffering of the proprietors had apparently only deepened their compassion for others, and their respect for their own and others' dignity enabled a sense of participation and solidarity that formed a small local community. However, the contemplation of beauty is a striking transformative path to focus on with regard to the damage done to pristine nature by environmental pollution. Pope Francis thus seems to be asking us to foster a contemplative wisdom that transcends merely political and pragmatic considerations in order to safeguard our common home. Such a contemplative wisdom, in the pope's analysis, represents the hope that individuals and communities might put financial self-interests in proper perspective and see the larger picture, which requires certain levels of sacrifice in the short term for the long-term common good. The hope of Pope Francis stems from what Jacque Maritan describes as the priority of the contemplative over the political. Maritan does not mean that the political is unimportant but rather that it should be informed by a contemplative understanding of what constitutes the good and just society.[22]

The pope detects in technology both a benefit and a threat to possibilities for the goodness of our common humanity. Just as human beings share the capacity to contemplate the beauty of nature, they are also able to appreciate the aesthetic appeal of the architectural wonders of global cities, from skyscrapers to systems of public transportation to sports stadiums to centers of art, music, and culture. Technology makes global cities possible and sustains their splendor. Esteeming humanity's technological prowess, Pope Francis comments, "We are the beneficiaries of two centuries of enormous waves of change: steam engines, railways, the telegraph, electricity, automobiles, airplanes, chemical industries, modern medicine, information technology, and more recently, the digital revolution, robotics, biotechnologies and nanotechnologies. It is right to rejoice in these advances and to be excited about the immense possibilities which they continue to open before us."[23] To be sure, science and technology remain essential to achieving sustainability for global cities. Technology itself is not the problem; rather, its misuse reflects a skewed understanding of our humanity, in which "we have come to see ourselves as the lords and masters" of our common home who are "entitled to plunder her at will."[24] While modern technology appears to offer humanity unlimited autonomy, progress, and power, the pope warns that at the same time, it tends to enslave us ever more to the power of the utilitarian calculus based on self-interest and subjective fulfillment.

The deepening of one's humanity through contemplation of nature's beauty stands in contrast to what the pope describes as the "dominant technocratic paradigm" that reduces nature to an instrumentality that can be exploited.[25] The hegemony of the technocratic paradigm means that "the alliance between the economy and technology ends up sidelining anything unrelated to its immediate interests."[26] Regrettably, financial considerations in the market economy often result in a cost-benefit analysis that overlooks the optimal conditions for human flourishing. In Jürgen Habermas's critique of technocratic governance by scientific experts and bureaucrats, he argues that questions about what constitutes "the good life" should not be reduced to problems to be solved by technical specialists. According to Habermas, the legitimate human interest in the control of nature ought not to enable technology to function as an ideology that eliminates the need for public and democratic discourse about human values. Otherwise, technology is prone to be employed as a mask that hides government decisions that advance the interest of the capitalist market economy.[27] When government officials yield to business concerns about the maximization of profits without considering the long-term consequences on all sectors of society and the environment, the common good invariably suffers. Integral ecology appreciates that the global environmental crisis cannot be adequately addressed by a technocratic approach to economic decisions about investment, production, and wealth creation, which excludes altruistic human values from democratic discourse.

According to the pope, technocracy fosters an excessive "anthropocentrism" that propagates "a Promethean vision of mastery over the world," dismissing care for humanity's common home of nature as "something that only the faint hearted care about."[28] Quoting Romano Guardini, the pope warns that "the technological mind sees nature as an insensate order, as a cold body of facts, as a mere 'given,' as an object of utility, as raw material to be hammered into useful shape; it views the cosmos similarly as a mere 'space' into which objects can be thrown with complete indifference."[29] The result of this technocratic approach has been an unprecedented global crisis of contaminated water, polluted air, toxic waste, climate change, deforestation, loss of biodiversity, and a general decline in the quality of human life. Pope Francis believes that technocracy's dominance prevents a more fruitful dialog about environmental sustainability that the goodness of our common humanity would elicit from moral, technological, economic, social, and legal perspectives.[30] The pope nonetheless remains optimistic that "we have the freedom needed to limit and direct technology" and to liberate ourselves from the "tedious monotony" of a technocracy that threatens the very future of nature and humanity.[31] In his words, the human "desire to create and contemplate beauty" has the potential to enable us to transcend technocratic reductionism and excessive anthropocentrism "through a kind of salvation which occurs in beauty and those who behold it."[32] From the pope's perspective, the contemplation of nature's beauty teaches us about the significance of recognizing our common humanity and its intrinsic goodness as essential to global environmental sustainability.

International Law and Critical Legal Studies

In addition to the contemplative insights that *Laudato Si'* offers about the interrelatedness of things and our common humanity, it expresses a practical call for international law aimed at addressing the global environmental crisis. The third major issue I discuss in this chapter is law and the sustainability of global cities in light of certain aspects of critical legal studies. Ever since I first became acquainted with critical legal studies as a law student, primarily through the writings and lectures of Roberto Unger, I have found it to be intellectually appealing.[33] Although *Laudato Si'* takes a nuanced stance toward technology as a solution to the global crisis, Ileana M. Porras has pointed out that the encyclical seems to express a naïve confidence in the ability of law to correct the problem.[34] Awareness drawn from the critical legal studies movement, which emerged in US law schools starting in the 1970s, cautions against an optative assessment of the possibilities for law. Critical legal studies scholars tend to view law as a means to maintain the status quo by the powerful over less advantaged groups in society. The roots of critical legal studies may be traced to US legal realism, an earlier US movement that started with Oliver Wendell Holmes Jr. and that favored an account of law based on what courts and other legal actors actually do rather than on legal doctrine, formal rules, and neutral logic.[35] Critical legal studies expands realism's pragmatic suspicion of legal formalism with an analysis of law as power that renders the meaning of law easily subject to manipulation.[36] The realists remained firmly in the liberal camp, and they were committed to the search for value neutrality. In contrast, critical legal scholars reject the idea of value neutrality completely.[37] For them, there is no distinction between legal reasoning and political debate, and law *is* politics.[38] As a teacher of jurisprudence, I appreciate that "legal realism" and "critical legal studies" are convenient academic titles to describe two distinct schools of thought, each of which encompasses diverse scholars. Nonetheless, at the risk of generalization, it seems to me that legal realism raised legitimate questions about a conception of law limited to formal principles and rules and that critical legal studies astutely extends this suspicion to international law with regard to power disparities between states and law's vagueness.[39]

First, developing realism's critique of law as grounded in formal principles and abstract logic, critical legal studies exposes the power asymmetries between states. For example, when the International Criminal Court (ICC) was established in July 2002, it was hailed as a global legal structure that would ensure that individuals charged with the gravest offenses of concern to the international community, such as genocide, war crimes, and crimes against humanity, would not go unpunished. As a member of the delegation representing the Holy See during the formation of the ICC, I can attest that, in the development of the statute during the two years of preliminary work conducted at the UN headquarters in New York and subsequently at the treaty conference held in Rome, a great deal of time, energy, and debate were focused on the formal structures and rules by which the ICC would operate. However, critical legal theorists would observe that principles, structures, and logic alone cannot possibly hope to address the reality that three of the world's most powerful nation-states, China, Russia, and the United States, are not signatories to the ICC and therefore are not subject to its jurisdiction. Prosecutions thus far undertaken by the ICC have been limited to individuals from less powerful state actors. Admittedly, the ICC's focus on less powerful states is due in part to the fact that its jurisdiction only applies when a nation-state lacks the resources or political will to prosecute a case on its own. Nonetheless, its exclusive prosecution of individuals from less powerful states, such as those in sub-Saharan Africa, has raised the neocolonialism claim brought by African ICC stakeholders.[40] From the perspective of critical theory, the statute of the ICC subjects weak states to the power interests of strong states, even as the more powerful states are out of the reach of the law.

A policy paper issued by the ICC in September 2016 announced that the prosecution of cases would be directed against environmental crimes.[41] Given that the major world powers may be just as likely as other states to commit environmental crimes, critical legal studies raises questions of the effectiveness of the ICC in affording international law for fair and just criminal sanctions.[42] As with the ICC's prosecution of war crimes, crimes against humanity, and genocide, the prosecution of environmental crimes is likely to be limited to officials from weaker states, even as officials from stronger states enjoy exemption from legal sanction by the international court. Just as there are power asymmetries between states, global cities are not of equal power and standing. So-called alpha global cities such as London, New York, and Shanghai exert more financial influence than do other global cities such as Abu Dhabi, Berlin, and Singapore. The power imbalance is exacerbated by the fact that New York and Shanghai fall within the boundaries of nation-states that are not signatories to the ICC. Power inequality between the global cities is further escalated because global cities are in competition with one another for the resources and capital to command increasing shares of global production and finance. If the critical theorists are correct that international law is more about concrete power than abstract justice, the call in *Laudato Si'* for more just legal means to resolve the global environmental crisis may well be naïve. Only damage to the environment caused by private, corporate, and governmental actors from less powerful states will be prosecuted, while the damage caused by the most powerful actors will go unpunished. The critical theorists would argue that the incentive of the law to discourage environmental crimes applies to the weaker, but not to the stronger, states.

Second, critical legal scholars' claim about the indeterminacy of law exposes the vagueness and manipulability of international environmental law.[43] For example, *Laudato Si'* takes a favorable view of the principle of common but differentiated responsibilities and respective capabilities (CBDRRC).[44] Quoting the bishops of Bolivia, it states about differentiated responsibilities, "the countries which have benefited from a high degree of industrialization, at the cost of enormous emissions of greenhouse gases, have a greater responsibility for providing a solution to the problems they have caused."[45] In reference to differentiated capacities, the encyclical explains, "the poorest areas and countries are less capable of adopting new models for reducing environmental impact because they lack the wherewithal to develop necessary processes and to cover their costs."[46] The question is how the CBDRRC principle is applied in specific legal provisions. Critical legal scholars would point out that terms such as "common," "responsibilities," and "capacities" not only are subject to various interpretations but are also easily manipulated by skilled lawyers who take the so-called bad man's view of the law, in which they ask only what negative consequences law might hold for their clients.[47] Critical legal theorists argue that law is radically indeterminate. They seek to demonstrate the indeterminacy of law and show how any given set of legal principles not only can be used to yield contradictory results but also can be manipulated by the powerful, who can afford high-quality legal representation.

Although Pope Francis's encyclical does not incorporate this mode of critical legal analysis, I believe that he would welcome insights about power asymmetries between states and indeterminacy in international law. As he puts it in *Laudato Si'*,

> Inequality affects not only individuals but entire countries; it compels us to consider an ethics of international relations. A true "ecological debt" exists, particularly between the global north and south, connected to commercial imbalances with effects on the environment, and the disproportionate use of natural resources by certain countries over long periods of time. . . . The foreign debt of poor countries has become a way of controlling them, yet this is not the case where ecological debt is concerned. In different ways, developing countries, where the most important reserves of the biosphere are found, continue to fuel the development of richer countries at the cost of their own present and future.[48]

As discussed earlier, the pope believes that global sustainability requires "an integrated approach to combating poverty, restoring dignity to the excluded, and at the same time protecting nature."[49] While the pope may not be a scholar of legal theory, he is no doubt familiar with pragmatic international diplomacy. In penning *Laudato Si'*, Pope Francis is writing from the Vatican, which ranks as the world's smallest independent city-state, with a population of 466 people on 110 acres located within the boundaries of city of Rome.[50] By the size of its population, the Vatican could hardly be compared to powerful global cities such as Delhi, with its 2018 population of approximately 28.5 million, or to London, at approximately 9 million.[51]

However, when I served as a member of the Vatican delegation at the ICC treaty conference, I experienced firsthand how a small entity can influence international policy on two salient issues. First, the question of whether to include the death penalty in the ICC statute was controversial, and at the start of the treaty conference, the position of powerful nations such as China and Russia favored its inclusion. In the weeks prior to the vote on the death penalty, the Vatican quietly used its diplomatic contacts with a large number of nation-states to explain its position. To be honest, I was pleasantly surprised when the majority of nation-states at the convention voted against the death penalty, but the experience taught me a lesson about moral persuasion in international affairs. Second, I also experienced how the small city-state contributed to the design of a balanced international legal structure. The vast majority of nation-states at the treaty conference backed a strong ICC prosecutor. As a lawyer, my sense was that the support for a strong prosecutor was at the cost of the rights of the defense. At a meeting of our petite delegation, I raised the issue, and again, I was surprised when the Vatican cardinal secretary of state endorsed my position. He instructed me to prepare a Vatican statement, which I presented on the floor of the treaty conference.[52] The majority of members voted for a more balanced design that maintained the office of a strong prosecutor but bolstered the rights of the defense at the ICC. Several nongovernmental organizations (NGOs) at the conference were delighted by my support for the rights of defense, but I suspect that, just as with the death penalty, the credit for the shift among the nation-states could be attributed to the moral persuasion of arguments made behind the scenes through the diplomatic efforts of the Vatican secretary of state. Please forgive me if I sound like an apologist for the Vatican and the power of moral persuasion. I am keenly aware that the Vatican, like any other major international entity, has its own self-interests and considerable (indeed at times shocking and scandalous) shortcomings. Nonetheless, I can attest on the basis of my experience as a low-level Vatican representative that moral persuasion is not completely bereft of power in international relations.

Powerful as the seven global cities that we are studying in this volume may be, they are not city-states like the Vatican, let alone major powers such as China and the United States. Global cities remain part of a national jurisdiction and do not enjoy the sovereignty to be principal actors in international law. The classical view of national sovereignty regards cities as administrative subsidiaries of the nation-state. Pointing to the "deep incapacity" of "aging nation states" to address global problems such as environmental sustainability, Benjamin Barber observes that the relationship between city and state is changing. When the higher jurisdiction of the nation-state fails to discharge the responsibilities of sovereignty, Barber calls for cities "to come together with other cities, both within and beyond their national borders," to address transnational issues posed by the ecological crisis.[53] Barber argues on the basis of social contract theory that when states fail in their responsibility to govern, cities have the right to step in and fill the vacuum. I understand Barber's political theory as complementary to Pope Francis's idea about the goodness of our common humanity. Following Jacques Maritan, referenced in the previous section, the logic of the social contract argument could represent an effective political implementation of a deeper contemplative vision for humanity and the environment.[54] On the pragmatic level, it seems to me that global cities at the minimum enjoy the capacity and responsibility to serve as moral agents in the formation of international law that advances ecological sustainability. While I agree with critical legal studies about the asymmetry in power and the indeterminacy of law, I am nonetheless optimistic about the sustainability of global cities. In light of their riches of law, science, art, technology, culture, and religion, do not global cities have the resources to play a leading moral role in nurturing integral ecology?

Religious Sources of Wisdom

My fourth point in this chapter is to consider the possibility of entertaining religious values in the discussion of the sustainability of global cities. In Pope Francis's reflections about the environment in *Laudato Si'*, he offers a theological perspective to the broader conversation, which understandably tends to focus on science, technology, law, and finance. The question of the role of theological values in public discourse has not been without controversy. On the one hand, there are those, like the highly influential John Rawls, who cast doubt on the possibility of bringing religion into the public square on the grounds that religious values are divisive in a pluralistic society.[55] On the other hand, there are people of religious convictions, with Pope Francis as an attractive example, who believe that theological insight has something to contribute to matters of public policy. The extremes of each side of the debate are represented by either voices that would insist that they can impose a certain set of religious values often representing sundry forms of religious fundamentalism or those who reject religious insight so deeply that its mere utterance must be silenced in any and all circumstances. It seems to me that the voice of Pope Francis, although rooted in theology, is not an extreme voice and indeed might even be welcomed by those who adhere to the pristine Rawlsian position. I say this because I do not think that the reflections of Pope Francis in *Laudato Si'* are divisively sectarian. Rather, the pope's voice seems to me one that raises a clarion challenge based on our common humanity to the present state of economic, technological, and legal choices that have resulted in the abuse of, rather than care for, the environment. While this challenge stems from the Pontiff's religious values, it nonetheless contributes principles applicable to the broader conversation about the environment, and in particular, about the sustainability of global cities.

As Pope Francis queries, "Is it reasonable and enlightened to dismiss certain writings simply because they arose in the context of religious beliefs?"[56] It seems clear that neither technology nor economics is sufficient in itself to resolve the environmental crisis. To believe otherwise would represent a poor understanding of the methodology and limits of each discipline. Modern technocracy does not cultivate "the great motivations which make it possible for us to live in harmony, to make sacrifices and to treat others well."[57] Such motivations remain essential to addressing the environmental crisis, and dialog between science, economics, technology, law, and religious values promises to offer an alternative. To explore this alternative, it would be beneficial to abrogate the false dichotomy in which religion and science are diametrically opposed, in favor of the appreciation that each could prove a fruitful conversation partner in the dialog. For example, science flourished in the Islamic world as nowhere else on Earth during Europe's so-called Dark Ages, while in Europe, science and all learning was largely kept alive in religious monasteries.[58] For a more modern example, one has only to think of Catholicism's reconciliation with the facts of evolution.[59] Religious texts should not be interpreted in a fundamentalist way that contradicts scientific fact, and science should not pretend to possess a monopoly on truth in supplying answers to all of life's deepest questions and mysteries.[60]

Pope Francis suggests that "religious classics can prove meaningful in every age; they have an enduring power to open new horizons."[61] As one who teaches comparative religion at a university situated in the Muslim world, with students drawn from all over the globe, I could not agree more. The pope's insights about the goodness of nature and our relationship to it through gentle care and humble stewardship are in large part based on the Genesis creation accounts contained in the Torah of the Hebrew Bible.[62] Likewise, the Qur'an abounds in lyrical descriptions of the natural world, expressing confidence in the goodness of material reality, in which lies one of the sources of Islam's commitment to science.[63] Nor are the meaningful new horizons contributed to the environmental conversation limited to the Abrahamic religions. As a Franciscan priest with a commitment to daily mediation and the contemplation, I have gained insights from my study of Buddhism.[64] Ancient Hindu texts such as the *Upanishads* and *Bhagavad Gita* have affirmed my desire to live a simple life based on the holiness (*Atman*) deep within the self.[65] Reflecting on Taoist beliefs about the purity, humility, and power (*wu wei*) of water has enriched my Franciscan spirituality about the goodness of nature and our relationship to it.[66] I admire Confucianism's pragmatic philosophy about the significance of maintaining a relational rather than individualistic culture that fosters respectful, intelligent, and compassionate human beings (*chun tzu*).[67]

While the distinctiveness of these religious traditions should be respected, they all warn about the excessive tendencies of egocentrism that would dominate nature rather than live in harmony with it. As a student of comparative religion, I am of course aware that the world's major religious traditions each has had harmful historical manifestations, and my critical awareness of this is especially acute in regard to my own Catholic religion. Nonetheless, I believe that the world's great religious traditions express ancient perennial wisdom about the meaning of life, which, when understood in the best possible light, continues to be salvific in our own times. One need not be a believer to appreciate this wisdom, which inspires personal and social "ecological conversion" that nourishes individual flourishing and the common good. The global city, with its religious pluralism, seems to me to offer an optimal forum for the conversation about environmental sustainability. Pope Francis endorses this conversation because he believes that constructive dialog forms part of a process of ecological conversion that is indispensable to global sustainability. In his words, "A commitment this lofty cannot be sustained by doctrine alone, without a spirituality capable of inspiring us, without an 'interior impulse which encourages, motivates, nourishes and gives meaning to our individual and communal activity.'"[68] In the pope's estimation, ecological conversion is not a matter of adopting a particular religious doctrine or even of believing in god(s). Rather, it is a broad social conversion that moves human persons away from self-interest and the desire to dominate nature for financial reasons. It entails individual and social transformation of consciousness motivated by the goodness of our common humanity. At the same time, Pope Francis proposes that the wisdom offered by the world's major religious traditions regarding care for our common home and transcending self could serve as one possible source of the insight necessary to ecological conversion.

Conclusion

As we discuss global sustainable cities, I have raised four points based on Pope Francis's encyclical letter *Laudato Si'*. First, observable phenomena such as the travel of desert dust and maintaining clean city streets exemplify the physical evidence supporting the pope's call for an integral ecology. The evidence is bolstered when one considers the fact that human beings live together with nonhuman animals in our natural common home, the harmony and well-being of which requires respect and stewardship rather than abuse and domination by its human inhabitants. Pope Francis's notion of integral ecology is further verified as a strong thesis by the ways in which environmental problems are interrelated to other social issues such as development, migration, and poverty. Global cities are not only hubs of burgeoning population and robust finance but also centers of science, technology, media, law, medicine, politics, and culture. The mutual reciprocity of these constitutive aspects of the global city serves to remind us that the whole is greater than the sum of its parts. It follows from *Laudato Si'* that the rich resources of global cities situate them to serve as exemplars of integral ecology, which entails the recognition that discrete answers to specific kinds of pollution limited solely to science and technology are insufficient to address the global crisis. This is not to propose that discrete solutions to environmental problems on the basis of the full benefits of science and technology should be abandoned. Of course, one must proceed on a case-by-case basis and do what one can to alleviate specific instances of environmental damage. At the same time, I think that global cities possess the requisite assets to play a leading role in facilitating integral ecology.

Second, global cities are characterized by a rich diversity of individuals and communities, and they depend on the goodness of our common humanity for tolerance and harmony in civic life. Of course, no global city has a perfect record in this regard, but it is remarkable that global cities prosper as epicenters of commerce and culture, given the inherent tensions in their multifarious mix of humankind. In *Laudato Si'*, Pope Francis describes the goodness of our common humanity in terms of qualities such as mutual respect, compassion, participation, solidarity, sacrifice, and courage. He observes that human beings share the capacity to contemplate the beauty of nature and that the contemplation of beauty has the potential to transcend subjective human experience, transforming individual self-interest into higher altruistic qualities. In the pope's view, such individual and social transformation remains essential in order to avoid the impending ecological disaster. No doubt, some people will want to argue contrary to Pope Francis's thesis about the goodness of our common humanity, the transformative power of the contemplation of beauty, and the need for social conversion. Although it may not be quite as robust an argument as the one about integral ecology, I believe that that the thesis remains key to moving away from profit-driven attitudes toward nature based on incomplete economic models that fail to account for the larger human good. Pope Francis faults technocracy for allowing self-interests to dominate decisions about sustainability without consideration of what constitutes the good and just society. Again, our discussion of global cities gives me hope. In light of their diversity of citizens and record of tolerance, cooperation, and harmony, it seems to me that global cities are ripe for fostering a transformation in social consciousness about environmental sustainability.

Third, the rule of law remains indispensable to the public order of global cities, and likewise, their environmental sustainability is linked to international agreements and law. As Pope Francis indicates in *Laudato Si'*, the limits of an approach to the ecological crisis that neglects integral ecology and human goodness are painfully evident in international law. The insights of critical legal scholarship enhance Pope Francis's call for effective international environmental law. Just as with science, technology, and economics, law is a necessary but insufficient part of the integrated approach. Critical legal scholarship cautions that international law too often fails to address power asymmetries and manipulation of law by the powerful, thus weakening international agreements designed to promote environmental sustainability. My own reading of critical legal studies is not that it is simply deconstructionist and negative about the possibilities for law but rather that its critiques about power asymmetries and about the indeterminacy of law constitute a radical call to retrieve the nexus between law and moral value. The sustainability of global cities depends on effective international law rooted in the moral value of integral ecology. Although global cities do not enjoy national sovereignty, the traditional relation between states and cities is changing, and the political power of cities is rising. Not only can they exercise political power to shape international law, but global cities can exercise moral leadership for the transformation of law and legal structures designed to safeguard the environment. The contemplative vision that underpins the formation of effective international law reflects the soul of the global city as not only a center of commerce but also one of learning, culture, and art derived from the rich diversity of its citizens.

Finally, global cities tend to be religiously pluralistic hives in which various sects thrive along with large numbers of nonbelievers. This is the case even for cities such as London and Abu Dhabi, which are situated in national jurisdictions that espouse an established religion. *Laudato Si'* proposes "ecological conversion" as the sine qua non on which rests global environmental sustainability. Pope Francis here is not talking about conversion to a specific religious creed but about a social conversion of attitude toward our common home. The ecological conversion proposed in *Laudato Si'* flows from the altruistic human values that the pope believes are available to us through the goodness of our common humanity. Although ecological conversion is not dependent on religion, religion enjoys a tradition of offering the motivation necessary to individual and social conversion. When interpreted in the best possible light, religious literature fosters altruistic human qualities. Given global cities' religious diversity, their social fabric abounds in this perennial wisdom. With respect for a healthy separation between church and state, it is not a question of imposing religion in a fundamentalist way, nor is it a matter of restricting religious insight so deeply that its utterance in the public square must be silenced in any and all circumstances. Rather, as part of a courteous dialog with persons of different faiths and with non-believers alike, religious values promise to assist in promoting human flourishing in the pluralist society of the global city. Permitting religious wisdom into the larger conversation would count as a way to encourage the transcendence of self-interests and thus to set the optimal conditions for global environmental harmony. Perhaps Pope Francis is speaking in somewhat of an optative voice in *Laudato Si'*, but it may well be a voice that conveys aspirations on which dangles the future flourishing of global cities, of environmental sustainability, and of humanity itself.

NOTES

1. Francis. 2015. *Laudato Si': On Care for Our Common Home*. Encyclical Letter. Huntington, IN: Our Sunday Visitor Publishing (hereinafter cited as "*LS*"). Official version Francis. 2015. *Laudato Si'*, De Communi Domo Colenda. *Acta Apostolicae Sedis* 107: 847–945.
2. *LS* 2015, 15, p. 15.
3. UN Department of Economic and Social Affairs. 2018. *The World's Cities in 2018—Data Booklet*, ST/ESA/ SER.A/417, 15 (for Shanghai), 22 (for Abu Dhabi), www.un.org.
4. For New York City, *see* US Department of Commerce, Bureau of Economic Analysis. n.d. "BEARFACTS." Accessed November 18, 2020, https://apps.bea.gov. For Berlin, *see* Federal Statistical Office of Germany. n.d. "Statistisches Bundesamt." Accessed November 18, 2020, www.destatis.de.
5. *See, e.g.*, Sassen, Saskia. 2001. *The Global City: New York, London, Tokyo*. 2nd ed. Princeton, NJ: Princeton University Press.
6. *LS*, 137, p. 93.
7. *LS*, 139, p. 94.
8. For an overview, *see* the Netflix episode of *Connected, Season 1: Dust*. Nasser, Latif, dir. 2020. *Connected, Season 1: Dust*. www.netflix.com. *See also* Moskowitz, Bruce M., Richard L. Reynolds, Harland L. Goldstein, Thelma S. Berquó, Raymond F. Kokaly, and Charlie S. Bristow. 2016. "Iron Oxide Minerals in Dust-Source Sediments from the Bodélé Depression, Chad: Implications for Radiative Properties and Fe Bioavailability of Dust Plumes from the Sahara." *Aeolian Research* 22 (September): 93–106. https://doi.org/10.1016/j.aeolia.2016.07.001.

9. *See* Koren, Ilan, Yoram J. Kaufman, Richard Washington, Martin C. Todd, Yinon Rudich, J. Vanderlei Martins, and Daniel Rosenfeld. 2006. "The Bodele Depression: A Single Spot in the Sahara That Provides Most of the Mineral Dust to the Amazon Forest." *Environmental Research Letters* 1 (014005) (October): 1–5. https://doi.org/10.1088/1748-9326/1/1/014005.
10. *LS*, 139, p. 94.
11. *LS*, 116, p. 79.
12. *See* Jamieson, Dale. 2015. "The Pope's Encyclical and Climate Change Policy: Theology and Politics in *Laudato Si'*." *American Journal of International Law Unbound* 21 (2015): 122–126. https://doi.org/10.1017/S239877230000129X.
13. *LS*, 139, p. 94.
14. *See* Ivereigh, Austen. 2019. *Wounded Shepherd: Pope Francis and His Struggle to Convert the Catholic Church*. New York: Henry Holt, 200.
15. *LS*, 89, p. 62.
16. *See, e.g.*, Sartre, Jean Paul. 1987. *Existentialism and Human Emotions*, translated by Bernard Frechtman and Hazel E. Barnes. New York: Citadel, 15. *See also* Wilkin, Peter. 1999. "Chomsky and Foucault on Human Nature and Politics: An Essential Difference?" *Social Theory and Practice* 25 (3) (Summer): 177–210.
17. Bruder, Carl E., Arkadiusz Piotrowski, Antoinet A. C. J. Gijsbers, Robin Andersson, Stephen Erickson, Teresita Diaz de Ståhl, Uwe Menzel, Johanna Sandgren, Desiree von Tell, Andrzej Poplawski, Michael Crowley, Chiquito Crasto, E. Christopher Partridge, Hermant Tiwari, David B. Allison, Jan Komorowski, Gert-Jan B. van Ommen, Dorret I. Boomsma, and Jan P. Dumanski. 2008. "Phenotypically Concordant and Discordant Monozygotic Twins Display Different DNA Copy-Number-Variation Profiles." *American Journal of Human Genetics* 82 (3) (March): 763–771. https://doi.org/10.1016/j.ajhg.2007.12.011.
18. Hamilton, W. D. 2006. *Narrow Roads of Gene Land, Volume 1: Evolution of Social Behavior*. Oxford: Oxford University Press, 6–8; McGilchrist, Iain. 2012. *The Master and His Emissary: The Divided Brain and the Making of the Western World*. New Haven, CT: Yale University Press, 16–31.
19. *See* 274 U.S. 200, 207 (1927). *See also* Cohen, Adam. 2016. *Imbeciles: The Supreme Court. American Eugenics and the Sterilization of Carrie Buck*. New York: Penguin Books, 278–280.
20. *See* Solnit, Rebecca, and Joshua Jelly-Schapiro. 2016. *Nonstop Metropolis: A New York City Atlas*. Oakland: University of California Press, 194.
21. *LS*, 90, p. 63.
22. *See* Maritan, Jacques. 1966. *The Person and the Common Good*. Notre Dame, IN: University of Notre Dame Press, 26.
23. *LS*, 102, pp. 69–70.
24. *LS*, 2, p. 7.
25. *LS*, 101, p. 69.
26. *LS*, 54, p. 39.

27. *See* Habermas, Jürgen. 1970. *Toward a Rational Society, Student Protest, Science, and Politics*, translated by Jeremy Shapiro. Boston: Beacon, 114–120; Habermas, Jürgen. 1973. *Theory and Practice*, translated by John Viertel. Boston: Beacon, 3–7.
28. *LS*, 116, p. 79.
29. *LS*, 115, p. 68 (quoting Guardini, Romano. 1956. *The End of the Modern World*. London: Sheed and Ward, 55). Michel Foucault expresses a somewhat different perspective in his work on technology. Although he shares the prevailing skepticism of twentieth-century European intellectuals about technology as alienating, Foucault rejects the humanist arguments in favor of the "re-humanization" of technology. *See* Foucault, Michel. 1984. "What Is Enlightenment?" In *The Foucault Reader*, edited by Paul Rabinow. New York: Pantheon, 31–50.
30. *Cf.* Bodansky, Daniel. 2015. "Should We Care What the Pope Says about Climate Change?" In "Symposium: The Pope's Encyclical and Climate Change Policy." *American Journal of International Law Unbound* 109: 127, 129.
31. *LS*, 112, 113, pp. 76–77.
32. *LS*, 112, p. 76.
33. *See, e.g.*, Unger, Roberto Mangabeira. 1975. *Knowledge and Politics*. New York: Free Press; Unger, Roberto Mangabeira. 1977. *Law in Modern Society: Toward a Criticism of Social Theory*. New York: Free Press; Unger, Roberto Mangabeira. 1984. *Passion: An Essay on Personality*. New York: Free Press. For a liberal response to Unger's critical legal theory, see Altman, Andrew. 1993. *Critical Legal Studies: A Liberal Critique*. Princeton, NJ: Princeton University Press, 153–176.
34. *See* Porras, Ileana M. 2015. "*Laudato Si'*, Pope Francis' Call to Ecological Conversion: Responding to the Cry of the Poor—Towards an Integral Ecology." In "Symposium: The Pope's Encyclical and Climate Change Policy." *American Journal of International Law Unbound* 109: 136–141.
35. *See* Holmes, Oliver Wendell, Jr. 1882. *The Common Law*. London: Macmillan, 5 ("The life of the law has not been its logic: it has been experience. The felt necessities of the time, the prevalent moral and political theories, intuitions of public policy avowed or unconscious, even the prejudices which judges share with their fellow men, have had a good deal more to do than the syllogism in determining the rules by which men should be governed." *See* Holmes, Jr., Oliver Wendell, and Frederick Pollack, 1961. *Holmes-Pollock Letters: The Correspondence of Mr. Justice Holmes and Sir Frederick Pollock, 1874–1932*, vol. 2, edited by Mark De Wolfe Howe. 2nd ed. Cambridge, MA: Harvard University Press, 26 (Holmes to Pollack, 1920), where Holmes commented about law: "I believe that force, mitigated as far as may be by good manners, is the *ultima ratio*." For a discussion of the development of US legal realism including its major intellectual contributors, salient themes, and legal examples, see Fisher, William W., III, Morton J. Horwitz, and Thomas A. Reed. 1993. *American Legal Realism*. New York: Oxford University Press.
36. *See* Tushnet, Mark. 1984. "Critical Legal Studies and Constitutional Law: An Essay in Deconstruction." *Stanford Law Review* 36 (1984): 623–626; Trubeck, David.

1984. "Where the Action Is: Critical Legal Studies and Empiricism." *Stanford Law Review* 36: 574–577.

37. Critical legal scholars maintain that liberalism depends on a "reification" of human nature and the rule of law. Human nature and the rule of law are abstracted concepts that attempt to justify law on the basis of a false objective claim about reality. *See* Gabel, Peter. 1980. "Reification in Legal Reasoning." In *Research in Law and Sociology*, edited by Stephen Spitzer, vol. 3. Greenwich, CT: JAI, 25–46.
38. *See* Dalton, Claire. 1985. "An Essay in the Deconstruction of Contract Doctrine." *Yale Law Journal* 94 (1985): 997–1114, commenting that legal doctrine is "a human effort at world making" (1114).
39. *See* Shaffer, Gregory. Forthcoming. "Legal Realism and International Law." In *International Legal Theory: Foundations and Frontiers*, edited by Jeffrey L. Dunoff and Mark A. Pollack. Cambridge: Cambridge University Press, 4.
40. *See* Schuerch, Res. 2017. *The International Criminal Court at the Mercy of Powerful States: An Assessment of the Neo-Colonialism Claim Made by African States*. Berlin: Asser, 2–11, 287–292.
41. *See* International Criminal Court, Office of the Prosecutor. 2016. *Policy Paper on Case Selection and Prioritization*. www.icc-cpi.int.
42. For a discussion of the environmental crimes committed by powerful actors and "the amorality of energy law," *see* Davies, Lincoln L. 2015. "Consumption and the Amorality of Energy Law." In "Symposium: The Pope's Encyclical and Climate Change Policy Energy." *American Journal of International Law Unbound* 109: 147–152.
43. For a summary and critique of the indeterminacy claim, *see* Wilkins, David B. 1990. "Legal Realism for Lawyers." *Harvard Law Review* 104(2): 469, 472–484. The indeterminacy claim is not limited to US legal realism or to critical legal studies. *See* Endicott, Timothy A. O. 2000. *Vagueness in Law*. Oxford: Oxford University Press, 31–55.
44. *See* Rajamani, Lavanya. 2015. "The Papal Encyclical & the Role of Common but Differentiated Responsibilities in the International Climate Negotiations." In "Symposium: The Pope's Encyclical and Climate Change Policy." *American Journal of International Law Unbound* 109: 142–143.
45. *LS*, 170, p, 113.
46. *LS*, 52, p. 37.
47. Oliver Wendell Holmes Jr. described the "bad man" view of the law in his article "The Path of the Law." Holmes, Oliver Wendell, Jr. 1897. "The Path of the Law." *Harvard Law Review* 10: 457–478. *See* Gordon, Robert W. 1997. "The Path of the Lawyer." *Harvard Law Review* 110: 1014. *See also* Luban, David. 1997. "The Bad Man and the Good Lawyer: A Centennial Essay on Holmes's 'The Path of the Law.'" *NYU Law Review* 72: 1547–1583.
48. *LS*, 51, 52, pp. 36–37.
49. *LS*, 139, p. 94.
50. *See* UN Department of Economic and Social Affairs. 2020. *Population and Vital Statistics Report, Statistical Papers Series A, Vol. LXXII*, 10, https://unstats.un.org.

51. For the respective populations, *see* UN Department of Economic and Social Affairs 2018, 18 (for Delhi), 23 (for London).
52. *See* Coughlin, John J. 1998. "Intervention of the Holy See to the Plenary Session of the Treaty Conference for the International Criminal Court." *L'Osservatore Romano* (English weekly ed.), July 22, 2.
53. Barber, Benjamin. 2017. *Cool Cities: Urban Sovereignty and the Fix for Global Warming*. New Haven, CT: Yale University Press, 16–18.
54. *See* Maritan 1966, 26.
55. Although Rawls's pristine position was highly skeptical of the possibility of admitting religious values into public political discussion, he subsequently modified his position, stating, "Reasonable comprehensive doctrines, religious or non-religious, may be introduced in public political discussion at any time, provided that in due course proper political reasons—and not reasons given solely by comprehensive doctrines—are presented that are sufficient to support whatever the comprehensive doctrines are said to support." Rawls, John. 1997. "The Idea of Public Reason Revisited." *University of Chicago Law Review* 64(3): 783–784. While Jürgen Habermas shares Rawls's view that public justification must be equally accessible to all citizens regardless of any religious affiliation, he cautions against placing an undue burden on the expression of religious believers, observing, "We cannot derive from the secular character of the state a direct obligation for all citizens personally to supplement their public statements of religious convictions by equivalents in a generally accessible language." Habermas, Jürgen. 2006. "Religion in the Public Sphere." *European Journal of Philosophy* 14(1): 9. *See also* Weithman, Paul J. 2002. *Religion and the Obligations of Citizenship*. Cambridge: Cambridge University Press, 180–211.
56. *LS*, 199, p, 130, quoting Apostolic Exhortation *Evangelium Gaudium* (24 November 2013), at 256, *Acta Apostolicae Sedis* 105: 1123. *See* Shelton, Dinah. 2015. "Dominion and Stewardship." In "Symposium: The Pope's Encyclical and Climate Change Policy." *American Journal of International Law Unbound* 109: 133.
57. *LS*, 200, p. 131.
58. Smith, Huston. 1991. *The World Religions*. New York: HarperCollins, 238.
59. *See* Jamieson 2015, 122.
60. *See* Cottingham, John. 2014. *Philosophy of Religion: Towards a More Humane Approach*. Cambridge: Cambridge University Press, 36–47.
61. *LS*, 199, p. 130, quoting Apostolic Exhortation *Evangelium Gaudium* (24 November 2013), at 256, *Acta Apostolicae Sedis* 105: 1123.
62. *See* Genesis 1–2, cited in Levenson, Jon D. 2014. "Genesis." In *The Jewish Study Bible*, edited by Adele Berlin and Marc Zvi Brettlet. 2nd ed. Oxford: Oxford University Press, 12–13 (commentary on Genesis 1:27 and 2:7). For an overview of Jewish ecological theology, see Gerstenfeld, Manfred. 2001. "Jewish Environmental Studies: A New Field." *Jewish Political Studies Review* 13: 3–62. For a feminist critique of the Genesis creation accounts, see Pagels, Elaine. 1989. *Adam, Eve, and*

the Serpent: Sex and Politics in Early Christianity. New York: Vintage Random House, 3–31.

63. *See, e.g.*, Surat Luqman, 31:10–20; Surat Al-An'am, 6:38; Surat Qaf, 50:7–8; Surat Al-Nahl, 16:12. In Nasr, Seyyed Hossein, Caner K. Dagli, Maria Massi Dakake, Joseph E. B. Lumbard, and Mohammed Rustom, eds. 2015. *The Study Quran: A New Translation and Commentary*. New York: Harper One, 1001–1004, 352, 1266, 658. *See* Smith 1991, 238. *See also* Shelton 2015, 133–134.
64. The *Satipatthana Sutta* guides one in the practice of right mindfulness as a way to ending the pain of *dukkha* on the path to the enlightenment of nirvana. *See* Horner, I. B. 2014. "The Five Faculties Separately." In *Buddhist Texts through the Ages*, translated and edited by Edward Conze. London: Oneworld, 56–59. *See also* Conze, Edward. 2003. *Buddhist Meditation*. New York: Dover, 62–109. *See also* Shelton 2015, 134.
65. The *Katha Upanishad*, 1: 20–29, contains the conversation between Nachiketas and Yama, the god of death, in which Atman is described. *See* Olivelle, Patrick, trans. 2008. *Upanishads*. Oxford: Oxford University Press, 232–251. Likewise, in the *Bhagavad Gita*, 2: 20–22, Krishna teaches Arjuna about the universal self, *Atman*, the discovery of which leads to *moksha*. *See* Patton, Laurie L., trans. 2008. *The Bhagavad Gita*. New York: Penguin Classics, 21–22.
66. Attributed to Lao Tzu, the *Tao Te Ching* elucidates humanity's being at one with the universe and calls for self-mastery to foster compassion, moderation, humility, and peace. *See* Mitchel, Stephen, trans. 2006. *Tao Te Ching*. New York: Harper Perennial, 25.
67. Describing Confucius as a noble person, *chun tzu*, the *Analects* list his warmth, humility, earnestness, perseverance, candor, integrity sympathy, and humanism. *See The Golden Mean of Tsesze*, translated by Ku Hungming, in Yutang, Lin, ed. 1942. *The Wisdom of China and India*. New York: Random House, 848–849. Mencius attributes a compassionate core to all human beings. *See* Mencius, Book VI, Part I, translated by James Legge, in Lin 1942, 772–773.
68. *LS*, 216, p. 141, quoting Apostolic Exhortation *Evangelium Gaudium* (24 November 2013), at 261, *Acta Apostolicae Sedis* 71: 1124.

2

Charting the Legal Landscape

Cities' Legal Authority to Develop Environmental Law

JOSEPHINE VAN ZEBEN

This chapter considers cities' legal authority to develop environmental law and the ways in which legal authority might diverge from expressions of "soft" power, as reflected in political and diplomatic action. This divergence matters, as the achievement of most political environmental commitments are conditional on legal action, such as the translation of internationally agreed greenhouse gas emission reduction commitments under the Paris Agreement into legally binding reduction targets for individual firms at the national level.[1] This begs the question as to whether cities' ambitions in the area of environmental policy can be matched with their legal powers.

Discussing this question in a comparative way, without narrowing our scope to one specific jurisdiction, presents several methodological challenges. First, cities vary widely in size and legal status, with some equivalent to federal states or provinces with far-ranging legislative powers (and obligations) and others limited to basic implementation powers without independent lawmaking abilities. Much of a city's legal authority is dictated by its national constitutional order. At the same time, cities may be responsible for implementing international obligations entered into by their national government. These international obligations may thereby affect a city's legal authority and powers with respect to specific issues, even if no general or direct decision was taken with respect to the status or powers of the local authority.

Second, the political ambitions of cities vary. For many cities, if not most, their primary mandate and aim is to provide services to local constituencies through national and local pathways. These cities have limited ability or ambition to incorporate and influence international or regional processes. These "local" cities will still be confronted with (international) environmental policy and law in relation to intrinsically transboundary environmental problems, such as air pollution. But there are also environmental problems that benefit from purely local action, such as soil pollution and waste management. It would therefore be wrong to consider internationally active cities as the sole blueprint for urban environmental action. Moreover, the visibility of international cities does not necessarily align with their domestic legal powers on these issues, and therefore, their ability to implement and/or achieve their environmental ambitions unilaterally may be limited. Finally, this chapter highlights the fact that, willingly or unwillingly, cities are part of a legal "multiverse," composed of the national, regional, and international laws, especially when it comes to environmental issues. This means that both locally and internationally oriented cities are subject to multiple layers of rules that affect their actions.

With these considerations in mind, the question of the role of cities in environmental lawmaking is pressing and timely and will require input from scholars all over the world. This chapter provides a starting point for discussion by highlighting some important shared considerations. The remainder of this chapter is structured as follows: The first section sets out several conceptualizations of "hard" city power, illustrated by international examples. The second section delves into the experiences of European metropoles as case studies regarding the scope of cities' ability to develop policy and considers the extent to which the constraints that European cities face in making environmental law might also constrain cities in other parts of the world. The conclusion discusses differences in hard and soft city power with respect to environmental regulation and discusses their implications for the future.

"Hard" Power: Local Legal Authority

The legal status of local governments is most commonly set out in national constitutions at the hand of clearly defined "local rights."[2] In this bundle of rights, a city's legal *status* (i.e., its place within the constitutional makeup of a nation-state) must be distinguished from legal *competence* (i.e., the power of local governments to act regarding a certain issue). To illustrate the importance of this distinction, one may consider the German constitutional provisions: local autonomy is guaranteed in Article 28 of the Basic Law of the Federal Republic of Germany and corresponding provisions of the *Länder* constitutions.[3] Article 28 guarantees that municipalities have the right to manage their own affairs within the limits set by the law. Typically, "the law" refers to the *Länder* constitutions, which may determine the responsibilities, rights, and duties of municipalities. These *Länder*-specific provisions may not remove or restrict the rights of local authorities so as to infringe with their autonomy.[4] However, this protection provided by Article 28(2) of the Basic Law does not contain any guarantees with respect to the continued existence of individual municipalities, which can be dissolved through an act of parliament.

In most Western countries, local authorities have general competence over issues within their jurisdiction and over their inhabitants.[5] These general powers can be limited by statute and influenced by the level of discretion that local governments have over certain policy areas. In addition, local government actions tend to be subject to judicial review by (administrative) courts and/or by regional or state authorities.[6] Relatedly, the most challenging questions with respect to the "hard," or legal, power of local authorities tend to be linked to their relationship with other levels of government: specifically, the ability of these other levels to overrule, restrain, or even dissolve local authorities. In the United States, these potential conflicts are captured by the possibility of "preemption," which creates a situation where local action is limited by setting uniform standards, either minimum or maximum policies, at the state or federal level.[7] As a result, while in principle, a city like New York has broad legislative competence, in practice, state and federal environmental law impose numerous constraints on the city's authority. The potential for encroachment is arguably even greater in authoritarian countries such as China and the United Arab Emirates, where municipal governments are essentially subdivisions of higher levels of government rather than independent entities.[8]

It would be difficult, if not impossible, to provide an overarching definition or typology of cities that would capture all the different types of local authorities present across the world and their respective legal status and competence. That said, the main responsibilities of local authorities tend to fall into two categories: the provision, or coordination, of public services, and a democratic function, acting as a conduit between individuals, local interest groups, and governmental actors at the local but also higher levels of government. The scope of both of these tasks varies widely, for example, due to differences in the type of governmental system of the relevant country, such as a democratic or authoritarian regime. Legal provisions on local services and a democratic function often interrelate: greater democratic autonomy can, though need not, come with more extensive responsibilities regarding service provision. This makes sense insofar as in order for local governments to credibly fulfill their democratic function, they should be able to act on the democratic input they receive by providing corresponding "output," for example, through the provision of services in line with the priorities set by their electorate.

In line with Article 28(2) of the German Basic Law, municipalities have important service-provision responsibilities, such as the provision of water, electricity, heating, and gas, as well as democratic duties, such as the obligation to ensure the election of local parliaments.[9] By comparison, Chinese cities are largely responsible for the implementation of central policy rather than giving expression to local preferences. While Chinese cities play a crucial role in tailoring national, including environmental, policies to local conditions, this is done in close collaboration with the central government, which also appoints many local-level officials.[10] Though it is common for central governments to play an agenda-setting and legislative role when it comes to environmental policy, the potential for conflict between the local and national level differs between these countries on the basis of their legal status and their democratic role: in China, direct conflict is unlikely due to the close contact between the local and central level with respect to implementation and the fact that local officials are beholden to central government rather than local constituencies. By contrast, in Germany, local authorities have substantial discretion in fulfilling their responsibilities regarding local service provision, which allows for responsiveness to local electorates but can also lead to conflict with the *Land* of which a municipality is part, the German federal government, and/or the European Union. Because all of these higher levels of government play a role in the articulation of environmental policy, this means that there can be conflicts between the local governments and higher levels of government over environmental matters.

TABLE 2.1. Local Government Models

	Feature			
Model	Autonomy	Finance	Legal competence	Judicial review
Relative Autonomy	High	Direct taxation	General competence	(Administrative) courts and/or regional/state authorities
Agency	Limited	Grants or indirect taxation	Only those related to delegated duties	
Interaction	Shared	Grants or indirect taxation	General competence with consultation duty	

Source: Adapted from van Zeben, Josephine. 2017. "Local Governments as Subjects and Objects of EU Law: Legitimate Limits." In *Framing the Subjects and Objects of Contemporary EU Law*, edited by Samo Bardutzky and Elaine Fahey, 123–145. Northampton, MA: Edward Elgar.

One way of mapping the different types of interactions that we can see between national and subnational governments is the models developed by Michael Clarke and John Stewart: the *relative autonomy* model, the *agency* model, and the *interaction* model (see table 2.1).[11] These models can be differentiated based on the degree of autonomy of the local government—high in the relative autonomy model, restricted to implementation in the agency model, and shared in the interaction model—and the local government's ability to raise revenue through taxation, with direct taxation only being envisaged under the relative autonomy model, while under the other models, the local government relies mostly on grants or indirect taxation.[12] These two parameters, autonomy and finance, directly feed back into the democratic and service-provision roles played by local governments. It is worth noting that these models speak to a general division of powers; with respect to environmental policy specifically, we see that it is relatively common for central governments to take the lead, for example, in order to prevent NIMBY-ism (not in my backyard).[13] At the same time, land-use decisions, including zoning, often remain local.[14]

Most countries tend to apply one type of model to their local authorities, often linked to their governance model. For example, federal states more frequently adopt a relative autonomy model than unitary states do.[15] However, significant variation continues to exist between federal/unitary states, with variations at times also existing *within* countries as a result of the historical position of certain local governments:

for example, in the United Kingdom, as a historically highly centralized city, London holds a special position with a relatively large measure of autonomy.[16]

Case Study: Environmental Action by European Cities

City-level action features heavily in European Union (EU) environmental policy, and cities themselves are keen to support EU-level environmental initiatives. From the EU's perspective, the rapid and ongoing urbanization of the EU makes the need for city involvement self-evident. From a political perspective, the support for both the EU and environmental action tends to be high in urban areas, which are also the primary places for EU citizens to relocate to outside of their own member states.[17] In addition, the role of cities in mitigating the causes of environmental and climate problems, as well as their front-line position in needing to adapt to the consequences of environmental and climatic change, is an international phenomenon.[18] As a result, urban policies directly affect, both positively and negatively, the drivers of environmental and climate change and, by extension, heavily impact the surrounding rural areas and ecosystems. European cities nevertheless often lack legal powers to match their environmental ambitions and political commitments.[19] Notwithstanding the cities' role in the provision of social and political services, the implementation of EU law, and the growing number of EU local-government policies, such as the EU Urban Agenda, this section discusses how "European" cities remain, legally speaking, almost entirely "national." This section discusses the position of cities within the European legal order, the creation and implementation of EU environmental and climate law generally, and the role of cities in those areas.[20] It also considers the extent to which the experience of European cities translates to the experiences of the non-European case-study cities examined in this volume (namely, Abu Dhabi, Beijing, Delhi, New York, and Shanghai).

The Position of Cities within the European Legal Order

The European Union is a multilevel system of governance made up of constituent nation-states—called member states (MS).[21] The EU grew out of an agreement between a group of sovereign nation-states to combine some of their powers, with the goal of better achieving shared goals, such as peace and prosperity. Although the EU is composed of nation-states, it is not a nation-state itself; rather, it is an intergovernmental and supranational organization with limited sovereignty. The ability of the EU to legislate or to make legally binding decisions on a specific topic is framed in terms of competence. Competences are powers that must be conferred, in whole or in part, from the member states to the EU (the principle of conferral).[22] This marks a key difference between the EU and nation-states: the EU's sovereignty—and therefore its competence to act in a certain policy area—is not intrinsic or complete; rather, it must be expressly conferred from the member states and is limited to those expressly conferred powers.

The European Union's multilevel governance system allows for the involvement of a broad range of actors, including nonstate actors and individuals, which have been excluded from traditional forms of international law.[23] However, local governments remain firmly within the exclusive sphere of competence of the member states, as underlined in the European treaties, and therefore depend on national law for their place within the EU's legal system.[24] The legal fact, or fiction, that constitutional arrangements pertaining to the status and competences of local government are, and should be, outside the scope of EU law has gone largely unquestioned, also by the European courts.[25]

In practice, local governments are routinely affected by, and affect, EU law in their roles as public-service providers and conduits for political participation. Some EU involvement pertaining to local governments' provision of services has been overwhelmingly positive, for example, through investment in local infrastructure and development.[26] The net effect on the democratic role of local governments is less clear.[27] While some localities have been hugely successful at increasing their influence by bypassing the member state and forging their own relationship with Europe—such as the small towns of Oulu, Oss, Montpellier, and Sandwich, which have been incredibly successful in attracting start-up activities, more so than EU capitals—others have seen their domestic positions eroded through EU membership and centralization of democratic processes to the national level.[28] The effect of this situation is that local governments do not have any type of direct legal relationship with the EU *as a local government*. They are subject to EU law in some of the same ways that individuals are, but, as organs of the state, their link to the EU is moderated through their member state.

Importantly, the situation of the European cities is not unique. Cities around the world generally do not have constitutional status; the comparative constitutional law scholar Ran Hirschl emphasizes that "most constitutional orders currently in existence . . . treat cities . . . as 'creatures of the state.'"[29] This is also true of the non-European case-study cities examined in this volume. Looking first at US cities, as in the EU, the federal government in the United States often does not distinguish between state and local governments (indeed, local governments are not mentioned in the federal Constitution), and states define the extent of power that their cities possess. Moreover, like in the EU, both the US federal government and the individual states can preempt local environmental policies.[30] This has created conflicts over environmental policies because, as in Europe, urban electorates often prefer a higher degree of environmental protection than does their state or the nation as a whole.[31]

New York City has encountered many of these constraints in its attempts to develop a progressive environmental agenda. Like most states in the United States, New York State maintains broad authority to preempt local environmental policies and has used this authority. For instance, in 2017, the state preempted a city ordinance that required consumers to pay a fee for carryout plastic bags.[32] The state constitution also prohibits New York City (and indeed all cities within the state) from imposing new taxes without state authorization; thus, New York City cannot impose "green taxes" on electricity or other polluting products, which could be a critical tool for curtailing pollution.[33] The federal government further restricts the policy tools available to New York City via preemption. For example, federal law prohibits state and local governments from setting energy efficiency standards for many types of appliances such as those used in heating, cooling, and ventilation.[34] This is a substantial constraint on New York City because energy use in buildings accounts for the lion's share of the city's greenhouse gas emissions.[35]

The non-Western case-study cities examined in this volume are arguably even more constrained. In the UAE, the city of Abu Dhabi is administered by emirate-level governmental bodies, and it is the central and emirate-level government that act as the main legislative and policy-making bodies, including for environmental action, not the city.[36] As for the Chinese cities, the Chinese Communist Party selects much of the leadership of Beijing and Shanghai, and some scholars have argued that the central government has exerted increasingly tight control over local governments in recent years.[37] Thus, while both Beijing and Shanghai have exhibited a degree of independence from the central government by developing their own innovative policies, it seems doubtful that they would be able to do so if the central government disapproved.[38] Delhi presents yet another distinctive case because it is both a Union Territory, which means it is a subdivision of the central government of India, and a state with its own independent legislature.[39] Even so, the state and central government bodies share administrative responsibility for the territory, leaving the state legislature constrained in the policies it can pursue.

EU Environmental and Climate Law

As mentioned, the EU can only legislate in areas where it has competence. There are three types of competence: exclusive, shared, or supportive.[40] In areas of exclusive competence, the EU is the only actor empowered to legislate; in areas of shared competence, both the EU and the member states may legislate. Supportive competences allow the EU to facilitate and coordinate action between the member states but do not empower the EU to legislate or take any independent actions.[41] The relationship between the EU and its member states represents a fine balance between the need for effective cooperation and the safeguarding of national sovereignty in a supranational system. The principles of conferral, subsidiarity, and proportionality—and their political and legal protection mechanisms—are the key to maintaining this balance.[42] In the field of environmental law, the balance often tips in favor of the EU legislature due to the transboundary effects of most environmental issues.

Environmental policy is an area of shared competence in which the EU is very active, and the preemption of member states' action on environmental issues is common; once the EU has exercised its competence in a shared area, member states are preempted from acting. The main exceptions to this are situations in which the member states want to take more ambitious action regarding environmental protection than the EU has taken. In most cases, except where there has been full harmonization, this remains possible even when the EU has already acted, provided that these protection measures do not create any problems for the EU internal market or are based on a demonstrable environmental need supported by scientific evidence.[43]

The EU has legislated extensively on environmental and climate issues (based on the authority of Articles 191–193 of the Treaty on the Functioning of the European Union [TFEU]), which has led to the almost complete preemption of member states' legislation in this area. However, member states are still responsible for the implementation of EU law, which—in the case of the EU law being adopted as a directive rather than a regulation—provides significant discretion for the member state.[44] In addition, the member state can decide where the responsibility for implementation and enforcement of EU environmental law lies within the member state: it can delegate this authority to a local government or maintain the power at the central level. If there turns out to be a breach of EU law due to lack of implementation, the member state as a whole will be held responsible by the EU.[45] Any resulting fines can then be claimed back from local authorities by the central government, but this, again, also depends entirely on the arrangements within the member state.[46] From the perspective of the EU, the member state is a unitary body, and it does not distinguish between the central and local levels.

Within the EU, a city's abilities to regulate issues such as air pollution (including greenhouse gas emissions), water quality, and climate change adaptation, for example, by regulating land use, is therefore a direct result of both the member states' own ability to regulate in these areas on the basis of possible EU preemption and the legal authority granted to a city within its national constitutional order.

European Cities in EU Environmental Law and Policy

On December 9, 2020, Frans Timmermans, the executive vice president for the European Green Deal of the European Commission, presented its "Sustainable and Smart Mobility Strategy."[47] The focus of the strategy, part of the EU's flagship climate and environmental program, "the

European Green Deal," focuses on the transformation of the EU's transport system, with a view to cut 90 percent of transport emissions by 2050.[48] Within the strategy, specific reference is made to aim to have one hundred climate-neutral European cities by 2030.[49]

The inclusion of cities in this strategy, or in the EU's climate and environmental ambitions more generally, is not surprising. The EU's Seventh Environmental Action Programme, adopted in 2013, already underlined the importance of urban issues, as by 2020, 80 percent of EU citizens were projected to live in urban areas.[50] Apart from sectoral initiatives such as the Smart Mobility Strategy, there are numerous other EU environmental and climate policy instruments that directly or indirectly reference city-level action. Notably, European mayors launched the "Green City Accord" initiative in support of the Green Deal on October 22, 2020, as well as the Mayors Alliance, which draws together forty mayors from across Europe.[51]

The Green City Accord calls on mayors to make a political commitment to action on air quality, quality and efficiency of water, protection of biodiversity, noise reduction, and advances with respect to circular economy. Specifically, cities have to submit an initial report that includes baseline information on these issues, targets for improvement by 2030, and an overview of next steps.[52] Cities then have to report on progress every three years. Though there are no penalties for noncompliance, cities that do particularly well can be rewarded with the Green Capital and Green Leaf Awards, which are connected to a financial prize between €200,000 and €600,000.[53] Past winners of the Green Capital Awards include well-known international cities such as Hamburg, Stockholm, and Lisbon but also smaller, lesser-known cities such as Essen (Germany), Vitoria-Gasteiz (Spain), and Nijmegen (the Netherlands).[54]

From a more bottom-up perspective, London—which has been forced to leave the EU, together with the United Kingdom—provides an interesting case study.[55] London unquestionably holds a unique position among UK cities. Greater London is home to 8.5 million people, and its economy, together with the financial industry of the "City," is responsible for roughly 22 percent of the UK's GDP.[56] Culturally, London—home to landmarks such as the Big Ben, the Tower of London, Buckingham Palace, the Houses of Parliament, and Westminster Cathedral—is one of the most diverse parts of the country, with 36.7 percent of its population born outside the UK, boasting the lowest rates of racism in the UK.[57] London was similarly unique in the European Union, which, compared to other parts of the worlds, is home to relatively few metropoles; within the EU, only Paris is comparable to London with regard to inhabitants and share of national GDP, and both are the only European metropolitan areas with more than 10 million inhabitants.[58]

In governance terms, "London" is the 1,580-square-kilometer area governed by the Greater London Authority (GLA), established through referendum in 2000.[59] The GLA represents the fourth system of metropolitan government for London within thirty-five years.[60] Local self-government has not been a given for London; in the fourteen years preceding the establishment of the GLA, the abolishment of the Greater London Council left London without a citywide system of government.[61] City government in London has always been two-tiered. There is the citywide council and the boroughs, and the division of power between them is complex and changes over time. The GLA is headed by the mayor of London, who is directly elected every four years, as is the London Assembly that supports the mayor. At the time of the adoption of the GLA Act in 1999, London was the first city in the United Kingdom to have a directly elected mayor.[62] The powers of the GLA are set out in the Greater London Authority Acts (GLAA of 1999 and 2007) and include control over the Metropolitan Police.[63] The GLA's powers are shared with London's thirty-two boroughs, all of which predate the GLA and form the first level of local government in London.[64] In addition, there is the City of London, which has a special position within this already-complicated system on the basis of its historical importance.

With respect to environmental regulation, Part 8 of the GLAA (2007) sets out the main obligations and powers of the GLA. It only contains two chapters: one on waste regulation and one on climate change and energy. Both waste regulation and climate change are heavily, and almost exclusively, regulated by the EU, with the responsibility placed on the member states to implement the policies. The GLAA makes clear that, within the UK, the GLA is responsible for developing and implementing policies related to waste and climate change, all of which had to be in line with European regulation until January 1, 2021.[65] If not, EU citizens or the European Commission could bring an action against the

UK (rather than the GLA, which under EU law is not a separate entity from the UK) for violating EU law.[66]

Notably, the GLA is tasked with taking into account policies with respect to both mitigation of and adaptation to climate change. Air pollution is not specifically mentioned as a separate duty of the GLA within the GLAA. This is not to say that this duty does not exist but simply that it is not distinct from general duties imposed on local authorities within the UK. There are several challenges that London and the GLA face in regulating these issues: (1) preemption at the EU and national level; (2) limited taxing powers and thus a limited budget to implement these policies; and (3) the complex structure of Greater London itself, with specific powers divided between a number of smaller boroughs and authorities.

A draft agreement was signed between the EU and the UK to regulate their relationship after the transition period triggered by Brexit, starting January 1, 2021.[67] The draft agreement, signed on December 24, 2020, counts 1,259 pages. Nevertheless, it is difficult to predict the exact consequences of Brexit for London and its ability to create environmental law or set environmental policies. Prior to leaving the EU, the UK expressed that it wanted to pursue both an ambitious environmental agenda *and* deregulation.[68] It is hard to see exactly how these two will coexist. The UK has adopted more stringent environmental regulation than the EU has at times and has championed certain environmental initiatives within the EU.[69] It has, however, also delayed and obstructed certain policies.[70] Importantly, the history of environmental regulation in the UK before joining the EU was one of deregulation—to the detriment of the environment.

Though London is exceptional in many ways, its experiences in environmental regulation are shared by many other European cities insofar as they underline the tensions between political, economic, and social importance, on the one hand, and legal dependence on central government, on the other. Recent work on the impact of the Covenant of Mayors—a voluntary network of local governments with over seven thousand participants aimed at promoting, coordinating, and documenting climate initiatives, particularly related to the 2015 Paris Agreement—shows that national climate policy and domestic territorial organization remain crucial for local climate policy.[71] This work, undertaken in Germany and France, shows that even in member states with relatively high measures of autonomy for local authorities—such as Germany—the policy coordination that the Covenant of Mayors accommodates between participants cannot act as a replacement for "hard" legal powers; we must continue to differentiate between political (local) "mobilization" and "effective climate action."[72] Specifically, while the Covenant of Mayors has been perceived as very successful in establishing a multilevel governance approach to climate change action, which includes the local level, the regulatory impact of the Covenant has been limited.[73] In both Germany and France, participation in the Covenant was seen as "an extra" for municipalities that already had the resources and autonomy to be well positioned for a strong local climate-policy performance: not a replacement for domestic provisions accommodating such action.[74]

Conclusion

This chapter has sought to highlight the importance of the distinction between hard (legal) and soft (political, cultural, economic) power of cities and how it affects their ability to take action on environmental matters. The point is not to suggest that either hard or soft power is more important for environmental regulation—often we see cities taking effective environmental action in the absence of explicit legal authority to do so—but to highlight the different roles that these kinds of powers play and what tensions their divergence may create. The most important impact that we see is that policy ambitions often remain solely as ambitions in the absence of corresponding legal competence to act on them. As many cities across the world are either faced with the absence of environmental legislative competence or the threat of preemption by higher levels of governance with respect to environmental law, this is a common reality.

This also continues to be true for European cities, despite their instrumental role in the implementation and enforcement of European laws and policies. It has been argued that, in light of these roles, European local governments are developing into entities that may "actively seek to change and succeed in changing . . . dynamics in ways which facilitate European policy mobilization," rather than actors that are "essentially inconsequential and passive players until either an incidental mobilization, or a central government decision is taken which passes decision-making powers down [to them]."[75] While this is undoubtedly true for some cities, especially powerful regional hubs, most smaller local governments continue to be restricted by the institutional arrangements of their respective member state.

It is difficult to generalize about the legal position of cities around the world, but in countries outside the EU, cities also appear to be constrained in realizing environmental objectives. In the United States, the position of a city in large part depends, positively or negatively, on the views of its state on local autonomy and to a degree on whether federal law preempts local authority. In other parts of the world, the relationship between cities and their governments may be different still—for example, in China, where local autonomy is restricted to ensure the successful local implementation of central policies.

One of the main consequences of the status quo for the future development of environmental law relates to the potential benefits that one expects of local environmental action. For some time, the focus has increasingly been on a "polycentric" approach to environmental problems, for instance, climate change.[76] The failure to overcome collective-action problems at the international level for many environmental challenges and the ambitions of local groups to address environmental issues directly have combined into a strong surge in initiatives at the small and medium scale.[77] At the same time, is seems unlikely that these initiatives would "crowd out" international agreements on transboundary environmental issues, making their proliferation a win-win situation.

However, the continued precariousness of hard power with respect to environmental issues sets the scene for increased conflict between local and central/regional levels of government. Moreover, national and regional policies with respect to local environmental action tend to focus on local experimentation and/or tailored implementation. This is no doubt beneficial to local conditions, but this focus on "scaling down" can come at the expense of focusing on how to "scale up" successful local experiments. At times, local soft power can help with this, as many global cities fulfill an exemplary function to others. However, the potential of other cities to follow suit depends, at least in part, on their hard power.

Cities play an important and distinct role when it comes to environmental management. The question as to whether this role should be expanded should not only be answered in reference to the failure of national and international processes, however tempting this may be.

NOTES

1. For the EU's implementation of its Paris Agreement commitments, see European Commission. n.d. "2030 Climate and Energy Framework." Accessed January 6, 2021, https://ec.europa.eu.
2. Norton, Alan. 1991. "Western European Local Government in Comparative Perspective." In *Local Government in Europe*, edited by Richard Batley and Gerry Stoker. London: Palgrave, 21–40.
3. *See in detail* Haschke, Dieter. 1998. "Local Government Administration in Germany." *German Law Archives*, https://germanlawarchive.iuscomp.org.
4. Haschke 1998.
5. Haschke 1998.
6. Haschke 1998 (citing Mouritzen, Poul Erik, and Kurt Houlberg Nielsen. 1988. *Handbook of Comparative Urban Fiscal Data*. Odense Universitet Dansk Data Arkiv).
7. *See, e.g.*, Wyman, Katrina M., and Danielle Spiegel-Feld. 2020. "The Urban Environmental Renaissance." *California Law Review* 108(2): 305–377. https://doi.org/10.15779/Z38DZ0325P.
8. Kostka, Genia, and Jonas Nahm. 2017. "Central-Local Relations: Recantralization and Environmental Governance in China." *China Quarterly* 231: 567–582; Savarani, chapter 4 in this volume.
9. Kostka and Nahm 2017.
10. Lin, Alvin. 2020. "The Authority of Cities in China to Develop and Implement Environmental Regulations and Policies, with the City of Beijing as an Example." Memo for NYU Abu Dhabi course (on file with author), 5.
11. Models adapted from Clarke, Michael, and John Stewart. 1990. "The Future of Local Government: Issues for Discussion." *Public Administration* 68: 249–258. https://doi.org/10.1111/j.1467-9299.1990.tb00758.x. Cited in Stoker, Gerry. 1991. "Introduction: Trends in Western European Local Government." In *Local Government in Europe: Trends and Developments*, edited by Richard Batley and Gerry Stoker, 1–20. London: Macmillan Education.
12. Clarke and Stewart 1990.
13. *See, e.g.*, McGurty, Eileen Maura. 1997. "From NIMBY to Civil Rights: The Origins of the Environmental Justice Movement." *Environmental History* 2(3): 301–323. https://doi.org/10.2307/3985352.
14. On Germany, *see, e.g.*, Haschke 1998. *See also generally* Treaty on the Functioning of the European Union (TFEU), art. 192(2).
15. *See* Stoker 1991.

16. *See in detail* Travers, Tony. 2015. *London's Boroughs at 50*. London: Biteback. *See also* Seymour, Mark, Maartje Abbenhuis, Simone Gigliotti, and Giacomo Lichtner. 2014. "Fault Lines: Cohesion and Division in Modern Europe." *Australian Journal of Politics and History* 60(3): 329. https://doi.org/10.1111/ajph.12064.
17. On support for the EU and environmental action, *see* Wike, Richard, Jacob Poushter, Laura Silver, Kat Devlin, Janell Fetterolf, Alexandra Castillo, and Christine Huang. 2019. "European Public Opinion Three Decades after the Fall of Communism." Pew Research Center, October 14, www.pewresearch.org; De Moor, Joost, Katrin Uba, Mattias Wahlström, Magnus Wennerhag, and Michiel De Vydt, eds. 2020. *Protest for a Future II: Composition, Mobilization and Motives of the Participants in Fridays for Future Climate Protests on 20–27 September, 2019, in 19 Cities around the World*. Gothenburg University Library. https://doi.org/10.17605/osf.io/asruw. On relocation, *see* Castro-Martín, Teresa, and Clara Cortina. 2015. "Demographic Issues of Intra-European Migration: Destinations, Family and Settlement." *European Journal of Population* 31: 109–125. https://doi.org/10.1007/s10680-015-9348-y.
18. Intergovernmental Panel on Climate Change. 2018. Special Report on Global Warming of 1.5°C. *See also* Wyman, chapter 6; and Spiegel-Feld, chapter 14 in this volume.
19. This chapter builds on earlier work on the topic of the status of local authorities in the EU more generally. *See* van Zeben, Josephine. 2017. "Local Governments as Subjects and Objects of EU Law: Legitimate Limits." In *Framing the Subjects and Objects of Contemporary EU Law*, edited by Samo Bardutzky and Elaine Fahey, 123–145. Northampton, MA: Edward Elgar.
20. A large part of the EU's environmental competence, regulated in Articles 191 to 193 of the Treaty on the Functioning of the European Union, refers to climate policy. To exclude climate law and policy would therefore provide an incomplete view of this area.
21. The EU as we know it now has only existed since the Lisbon Treaty (adopted in 2007, in force since 2009). The EU replaces the European Economic Community, the European Atomic Energy Community, and the more recent European Community, which were in place during certain periods since 1958. For a detailed discussion on the topic of the position of cities within the European legal order, *see* Finck, Michele. 2017. *Subnational Authorities in EU Law*. Oxford: Oxford University Press; van Zeben 2017.
22. Treaty on European Union (TEU), art. 5.
23. Van Gend en Loos v. Nederlandse Administratie der Belastingen, Case 26/62, ECR 1 (1963).
24. Regarding the European treaties, *see* TEU, art. 4(2) ("The Union shall respect . . . [the member states'] national identities, inherent in their fundamental structures, political and constitutional, inclusive of regional and local self-government."). This position is comparable to that of the United States. *See* Frug, Gerald E. 1987. "Empowering Cities in a Federal System." *Urban Law* 19(3): 553, 554 (referencing *Hunter v. Pittsburgh*,

207 U.S. 161 (1907)). For case law of the EU Court of Justice on the position of local governments, see Fernanda, Nicola. 2012. "Invisible Cities in Europe." *Fordham International Law Journal* 35(5): 1282. This chapter highlights the ways in which local government has been marginalized in *legal* analyses of the European Union. Local governments do form part of more inclusive frameworks such as that provided by multilevel governance scholarship. *See, e.g.*, Marks, Gary, Liesbet Hooghe, and Kermit Blank. 1996. "European Integration from the 1980s: State-centric v. Multi-level Governance." *Journal of Common Market Studies* 34(3): 341; Hooghe, Liesbet, and Gary Marks. 2003. "Unraveling the Central State, but How? Types of Multi-level Governance." *American Political Science Review* 97(2): 233.

25. *See* TEU, art. 4(2) ("The Union shall respect the equality of Member States before the Treaties as well as their national identities, inherent in their fundamental structures, political and constitutional, inclusive of regional and local self-government."). For scholars arguing in favor of a more inclusive approach, see Weatherill, Stephen, and Ulf Bernitz. 2005. *The Role of Regions and Sub-national Actors in Europe*. Oxford, UK: Hart, 3–6. (It should be noted that their focus is on regions, rather than local governments, which differ with respect to their legislative capacities. Nevertheless, parallels may be drawn as to the normative arguments in favor of their inclusion into the EU's legal order.) *See also* Scott, Joanne. 2002. "Member States and Regions in Community Law: Convergence and Divergence." In *Convergence and Divergence in European Public Law*, edited by Paul Beaumont, Carole Lyons, and Neil Walker. Oxford, UK: Hart, 131–150. For the purpose of liability of breaches of EU law, for instance, the Court of Justice of the European Union has consistently held that the member state is liable regardless of which body was "responsible" for the breach according to the internal division of powers of the member states. *See* Davis, Roy W. 2006. "Liability in Damages for a Breach of Community Law: Some Reflections on the Question of Who to Sue and the Concept of 'the State.'" *European Law Review* 31(1): 69.
26. The budget for regional development for 2014–2020 is €454 billion. For a complete overview, see European Commission. n.d. "European Structural and Investment Funds." Accessed January 6, 2021, http://ec.europa.eu.
27. *See also* Loughlin, John. 2001. *Subnational Democracy in the European Union: Challenges and Opportunities*. Oxford: Oxford University Press.
28. *See, for example*, the work of the Conference of European Regions with legislative powers (REGLEG), *available at* www.regleg.eu. *See, e.g.*, Stothard, Michael. 2019. "Some of Europe's Most Successful Startup Hubs Are in Tiny Towns, Not Capital Cities." *Sifted*, June 17, https://sifted.eu. *See* Jeffery, Charlie. 2000. "Sub-national Mobilization and European Integration: Does It Make Any Difference?" *Journal of Common Market Studies* 38(1): 1–23. *See also* Martinico, Giuseppe. 2013. "The Impact of the Treaty on Stability, Coordination and Governance on the National Constitutional Structure: The Regional Example." *Michigan Journal of International Law Emerging Scholarship Project* 1: 101–113.

29. Hirschl, Ran. 2020.*City, State: Constitutionalism and the Megacity*. New York: Oxford University Press, 9 (referring to "cities' constitutional non-status"), 10.
30. Frug, Gerald E., and David J. Barron. 2008. *City Bound: How the States Stifle Urban Innovation*. Ithaca, NY: Cornell University Press.
31. Briffault, Richard, Nestor M. Davidson, Paul A. Diller, Olatunde Johnson, and Richard C. Schragger. 2017. "The Troubling Turn in State Preemption: The Assault on Progressive Cities and How Cities Can Respond." *Journal of the ACS Issue Briefs* 11: 3–22.
32. McKinley, Jesse. 2017. "Cuomo Blocks New York City Plastic Bag Law." *New York Times*, February 14, www.nytimes.com. New York State subsequently passed its own plastic-bag law. Environmental Conservation Law (ECL), art. 27, tit. 28.
33. Scharff, Erin A. 2018. "Green Fees: The Challenge of Pricing Externalities under State Law." *Nebraska Law Review* 97: 168–224.
34. *See* Klass, Alexandra B. 2010. "State Standards for Nationwide Products Revisited: Federalism, Green Building Codes, and Appliance Efficiency Standards." *Harvard Environmental Law Review* 34(2): 335–368.
35. Spiegel-Feld, chapter 14 in this volume.
36. *See* Savarani, chapter 4 in this volume.
37. *See* Lin, chapter 9 in this volume.
38. *See* Lin, chapter 9; Mao et al., chapter 12; and den Hartog, chapter 16 in this volume.
39. Ganguly, chapter 7 in this volume.
40. *See* TFEU, arts. 2(1)(2) and (5).
41. This is an important role for the EU in areas that are otherwise considered important for national identity and sovereignty and therefore *not* part of EU competences, such as educational policy. *See* TFEU, art. 6.
42. TEU, art. 5.
43. TFEU, art. 114.
44. *Regulations* are arguably the "strongest" legal act that the European Union can adopt: they have general application, are legally binding in their entirety, and are directly applicable. The latter is particularly important, as it means that even without any additional implementing measures taken by the member states, the provisions of a regulation can be relied on by, and applied to, actors within the EU. *Directives* are by far the most common type of legal act within EU environmental law and are strongly preferred by member states. They are also legally binding but only to the member states to which they are addressed and only "as to the result to be achieved." This means that directives set out an environmental aim and leave member states free to choose the method through which that aim is achieved. *See also* TFEU, art. 288.
45. *See* Weatherill and Bernitz 2005.
46. *See in detail* Panara, Carlo. 2015. *The Sub-national Dimension of the EU: A Legal Study of Multilevel Governance*. Liverpool, UK: Springer.

47. European Commission. 2020. "Sustainable and Smart Mobility Strategy: Putting European Transport on Track for the Future." COM (2020) 789 final. Accessed December 9, 2020, https://ec.europa.eu.
48. European Commission. 2019. "The European Green Deal." COM (2019) 640 final, https://eur-lex.europa.eu.
49. European Commission 2020, 2.
50. Decision No 1386/2013/EU of the European Parliament and of the Council of 20 November 2013 on a General Union Environment Action Programme to 2020 "Living well, within the limits of our planet" Text with EEA relevance, OJ L 354, 28.12.2013, 171–200.
51. *See* European Commission. n.d. "Green City Accord." Accessed December 10, 2020, https://ec.europa.eu; Mayor's Alliance for the European Green Deal. n.d. *Mayor's Alliance for the European Green Deal Manifesto*. Accessed January 6, 2021, https://eurocities.eu.
52. European Commission. 2020. "Green City Accord: Clean and Healthy Cities for Europe." https://doi.org/10.2779/476324.
53. On the European Green Capital, *see* European Union. n.d. "European Green Capital." Accessed December 10, 2020, https://ec.europa.eu; on the European Green Leaf Award, *see* European Union. n.d. "European Green Leaf Award." Accessed December 10, 2020, https://ec.europa.eu.
54. *See* European Commission. n.d. "Winning Cities." Accessed December 22, 2020, https://ec.europa.eu.
55. In the Brexit referendum, 59.9 percent of London's voters voted to "Remain," while 40 percent voted "Leave." Kirk, Ashley, Malcolm Coles, and Charlotte Krol. 2017. "EU Referendum Results and Maps: Full Breakdown and Find Out How Your Area Voted." *The Telegraph*, February 24, www.telegraph.co.uk.
56. Office for National Statistics. 2014. "Regional Gross Value Added (Income Approach) NUTS3 Tables." July, www.ons.gov.uk.
57. Krausova, Anna and Carlos Vargas-Silva. 2013. "London: Census Profile." Migratory Observatory at the University of Oxford, May 20, www.migrationobservatory.ox.ac.uk. Taylor, Matthew, and Hugh Muir. 2014. "Racism on the Rise in Britain." *The Guardian*, May 27, www.theguardian.com.
58. *See* United Nations, Department of Economic and Social Affairs, Population Division. 2014. "World Urbanization Prospects." https://population.un.org; Eurostat. 2016. *Urban Europe—Statistics on Cities, Towns and Suburbs—the Dominance of Capital Cities*, https://ec.europa.eu.
59. The referendum result was overwhelmingly (72 percent) in favor of the establishment of an elected mayor and the introduction of an elected assembly.
60. Travers, Tony. 2003. *The Politics of London: Governing an Ungovernable City*. New York: Macmillan International Higher Education.
61. *See generally* Whitehouse, Wesley. 2000. *GLC: The Inside Story*. Sunbury-on-Thames, UK: James Lester.

62. Since then, London's example has been followed by several other cities. A key example is Manchester, which received its first directly elected mayor in an agreement with the government that also saw the decentralization of powers over the National Health Service.
63. Greater London Authority Act 1999 and Greater London Authority Act 2007.
64. For a detailed overview of the boroughs' history and powers, *see* Travers 2015.
65. Since January 1, 2021, the relationship between the UK and the EU is regulated by the EU-UK Trade and Cooperation Agreement. At time of writing (December 30, 2020), this agreement is still in draft form, available here: EUR-Lex. n.d. Document 22020A1231(01). Accessed December 22, 2020, https://ec.europa.eu. It is not yet clear what implications this agreement will have for environmental policy in the UK, though it looks to be the case that the UK can mostly set its own environmental policies. This may impact its ability to export goods to the EU.
66. The CJEU will no longer have jurisdiction over such conflicts after January 1, 2021, as detailed in the EU-UK Trade and Cooperation Agreement. EUR-Lex, n.d.
67. EUR-Lex, n.d.
68. Carrington, Damian. 2019. "MPs Warn Post-Brexit Environment Plans Fall 'Woefully Short.'" *The Guardian*, April 25, www.theguardian.com.
69. *See in detail* Hilson, Chris. 2018. "The Impact of Brexit on the Environment: Exploring the Dynamics of a Complex Relationship." *Transnational Environmental Law* 7(1): 89–113.
70. Hilson 2018.
71. Bendlin, Lena. 2020. *Orchestrating Local Climate Policy in the European Union: Inter-Municipal Coordination and the Covenant of Mayors in Germany and France*. Berlin: Springer.
72. Bendlin 2020, 253.
73. Steinberg, Paul F. 2015. "Can We Generalize from Case Studies?" *Global Environmental Politics* 15(3): 152–175. https://doi.org/10.1162/GLEP_a_00316.
74. Bendlin 2020, 256.
75. Jeffery 2000, 8.
76. *See, e.g.*, Ostrom, Elinor. 2009. "A Polycentric Approach for Coping with Climate Change." Research Working Paper No. 5095. World Bank, Washington, DC.
77. Ostrom 2009.

PART II

Protecting Water Quality and Providing Safe Drinking Water

3

No City Is an Island

Water Management in Berlin

PATRYCJA DŁUGOSZ-STROETGES AND BARBARA ANTON

Berlin flows with water. Today, almost eight thousand kilometers of water mains, located beneath Berlin's roads, deliver an average of 580,000 cubic meters of clean drinking water per day to households, industry, and businesses. Despite such a large amount of water moving beneath the city's surface daily, Berlin is located in a relatively dry area of Germany, with an average annual precipitation and evaporation of only six hundred millimeters.[1] While Berlin's five rivers—Spree, Havel, Dahme, Panke, and Wuhle—flow through the city, they do not play a major role as a source of water supply.[2] Instead, the city relies on the groundwater aquifer, which provides high-quality groundwater for nearly all of Berlin's potable use.[3] A grand total of 25 percent of the city's area is in a water-protection zone that has immense ecological and recreational value.[4]

Water Supply and Clean Water Provision

Supplying clean water, wastewater treatment, and drainage-system (sewerage and stormwater) management in and for Berlin is the responsibility of the Berlin Water Works, or Berlin Wasserbetriebe (BWB), a utility company that is more than 150 years old and has a turbulent history. BWB has around eight hundred deep wells, some reaching depths of 170 meters (just under half of the height of the Berlin's Fernsehturm [TV Tower/East Berlin], at 368 meters, and deeper than the Berlin's Funkturm [Radio Tower/West Berlin] raises from the ground, at 147 meters). All but one of the city's nine waterworks, where the water is treated before it is distributed to consumers, are located inside the city's administrative area.[5] Water in Berlin is not treated chemically; rather, in order to ensure clean water, the utility provides spring-quality groundwater to the city via a natural treatment process. This includes the removal of ligand (iron) and mangal, which, when combined with oxygen, can cause pipe blockage.[6] In this process, the water passes through sand filters and is stored in large storage tanks, from which it is pumped directly to households. The company also monitors water quality daily, testing around forty-five thousand samples per year for impurities.[7]

History of Berlin's Water: BWB

In order to understand the wider context in which Berlin is placed, one must accept that its specific geography and its history of political upheavals make the case of Berlin a distinctive one.[8] The world wars, the isolation of West Berlin, and subsequent reunification all had an enormous impact on the city's infrastructure. For over a century and a half, the history of water management in Berlin has been coupled with the history of BWB.

This history dates back to the 1850s, when Berlin made its first attempts to construct a sewage system, and the "Berlin Waterworks Company" was commissioned by the municipality, with royal support.[9] There were a number of reasons for the municipality to take this step at the time, not least due to the ravaging cholera epidemics. Over the following decades, a vast sewer network was established, encompassing the city—not yet united at the time—and its immediate surroundings.[10]

During each of the world wars, all the utilities suffered greatly, both with regard to the physical infrastructure and financially. Damage to the Waterworks Company was substantial, and compensation by the Main Office for War Damage to the providers was inadequate.[11] But, just like the rest of the city, they slowly rebuilt themselves—in the case of West Berlin, heavily subsidized by the federal government based in Bonn.

Impact of Administrative Division

At the height of the Cold War, essential amenities for an isolated West Berlin were supplied by or via East Germany. This presented challenges, at times with grave consequences and security implications.

When, in 1952, for example, negotiations on the price of water supply broke down, the authorities in West Berlin, assuming that they had the capacity to provide the required water quantities from within the city's border, decided to shut off the supplies coming in from the East. What transpired in the few weeks following this decision revealed West Berlin's vulnerability to water shortages: some districts were left without access to water, forcing the city to put emergency measures in place, in the form of temporary pipelines and tanks. Three weeks later, the negotiations reopened, and a new price was agreed.[12] This experience also resulted in increased efforts by the West Berlin administration to become more self-sufficient and reduce its water dependency on the East. Extensive plans were created, funded, and completed in the following years. Innovative measures and techniques were implemented, increasing drinking-water reserves, reducing leakages in pipelines, and decreasing domestic water consumption. To give an example, by 1978, West Berlin Waterworks managed to increase its water-supply capacity by a staggering 227 percent compared to 1949.[13] Even today, Berliners use substantially less water per capita—110 liters per day—than do residents of most capital cities in the Organisation for Economic Co-operation and Development (OECD).[14]

Water after Reunification

The fall of the Berlin Wall in 1989 and the reunification of Berlin dramatically changed the city's landscape, both politically and physically, but did little to change Berlin's cautious approach to its water management. There was plenty for the new city administration to do, as Berlin, especially its eastern part, was left with infrastructure in urgent need of repair and modernization. In 1990, no less than forty-two thousand West Berliners and sixty-three thousand East Berliners were still not connected to the public sewers.[15] Enthusiastic and optimistic expectations of rapid population growth and economic development put additional stress on the city's administration and infrastructure. While ultimately proven incorrect, it was estimated that by 2010 the city's population would increase from 3.4 to 4.0 million. As a consequence of this projection, in 1992, BWB launched an infrastructure program of 12.8 billion Deutschemark (DM), with most of the funds allocated to modernizing and expanding the city's wastewater system by 2000. By 1993, the sum was increased to 20 billion DM, to be invested over the following decade, only to be reduced again to 12.8 billion DM after concerns emerged regarding the affordability of the project.[16] Nevertheless, between 1990 and 1996, BWB spent a billion DM a year.

Even then, increasing the water-supply capacity was not an easy undertaking. Limitations posed by relying solely on local ground- and surface-water sources proved challenging, as did the expectation that the water consumption of West Berlin would eventually be matched by East Berliners as well, meaning there would be a need for an additional 90 million cubic meters of clean water per year by 2010 (from 340 million cubic meters to 430).[17] These concerns, paired with a fear of water shortages (without questioning the projections), resulted in the production of the 1993 Water Supply and Wastewater Disposal Concept, which "warned that shortage of water could even jeopardize Berlin's future: 'Providing the population with water is getting increasingly uncertain, resulting in water supply becoming an early limiting factor on population and urban development.'"[18] In the aftermath of these events, and as the cost of infrastructure investments could no longer be softened by subsidies from the federal government in Bonn, the financial burden fell on the city's consumers, who experienced hefty tariff hikes. Between 1990 and 1996, for example, West Berliners' price of water nearly tripled, from 1.30 to 3.45 DM per cubic meter, thus making Berlin's water more expensive than that of most other German cities, and this even before the privatization of the utilities (see the following section).[19]

Toward the end of the millennium, the city had the opportunity to source water from Brandenburg, a state that encompasses Berlin. The city administration ultimately decided, however, that all necessary water should be sourced from within the city's boundaries, thus ensuring that Berlin is self-sufficient by operating a closed-water-cycle approach. In addition, a charge of €0.31 per cubic meter on the service provider BWB was levied by the city's senate to counter over-extraction.[20] As a consequence of these decisions by the city administration, a check-and-balance system that ensured the long-term sustainability of Berlin's water resources began to take shape.[21]

Privatization of Water and Its Remunicipalization

There was one major disruption in the recent Berlin water-management history, which affected water governance in the city for a number of years: the partial privatization of BWB and the creation of Berlinwasser Holding S.C. From 1999, private investors, namely, the French multiutility enterprise Vivendi (renamed Veolia in 2003), the German energy company RWE, and, with a smaller part, the insurance company Allianz, together owned 49.9 percent of the company's shares. Despite the State of Berlin retaining the remaining 50.1 percent, this move proved far more controversial than the privatization of other utilities in the city.[22] Even though the ultimate sale price of 3.1 billion DM was considered high, and a number of concessions were made—including no compulsory redundancies, fixed water and wastewater tariffs (until 2003), and a commitment to invest 5 billion DM in BWB over the following decade—there was a public sense of defeat.[23] The sale contract guaranteed high revenues, meaning that "the sale price was effectively a loan to the city to be paid off with future revenues from BWB."[24] But BWB was already a highly profitable enterprise, even with the steep decline in water use that followed reunification.[25] This, along with the lack of transparency in negotiations, raised concerns as to the viability of the move. It culminated with a general public discontent with privatization after the 2003 water-tariff hikes.

To address some of the issues raised, the Berliner Wassertisch (Berlin Water Roundtable) was founded in 2006. It was created as a network of organizations and citizens committed to reversing the privatization deal. Eventually, after a referendum in 2011, both the Social Democratic Party (SPD) and Christian Democratic Union (CDU) admitted that the sale of BWB was a mistake, and on November 7, 2013, the Berlin parliament approved a proposal to buy back the shares from the two remaining private shareholders: RWE and Veolia.[26]

In retrospect, the privatization made little sense, both financially and politically.[27] But it did have a positive knock-on effect on the public's influence on the city's decision-making. A number of other initiatives were created or supported by BWB, such as an open forum for discussing urban water issues (Stadtgespräch Wasser) or the Berlin Centre of

Competence for Water (KWB), whose mandate is the promotion of science, research, and development in the water sector. Both are still active today, and over the past decade, the former has proven successful in both sparking a public debate on Berlin's water issues and "creating a cross-party consensus on the importance of water to the city's development, and more specifically, in getting decentral rainwater management embedded in the coalition agreement of 2016."[28] The Berlin Water Roundtable in 2013 created a Wasserrat (Water Council), working toward democratizing BWB, under the slogan, "after remunicipalization: democratization."[29] Its Water Charter listed a set of goals for the use and management of water to be transparent, socially just, and ecologically sound. The planned Zukunftsrat (Futures Council) is to be a part of this process, in order to safeguard the long-term security and sustainability of water provision in the city.[30]

Sustainable Water Management in Berlin

Since the city's reunification, it has engaged in and advocated for the promotion of sustainable water use. Starting in 1989, the short-lived red-green coalition in West Berlin introduced a number of innovative policies for the protection of environment and climate. After the reunification, these policies affected the entire city, and some of them ended up being supported and pursued by the following SPD and CDU governing coalition formed in 1991 and consecutive governments. Berlin was thus to become a model for a sustainable city, "demonstrating how it was possible to steer urban development along a path that did not equate growth with ever-increasing resource use and environmental pollution."[31] This approach laid the foundations for the years to come and for the continued endeavor to become a center of water excellence.

The persistent objective pursued by the Berlin Senate is to make sure that the city can rely on its own water resources. For over a century, water management in Berlin has been designed in an at least partially closed water cycle: water supply is replenished by surface water through bank filtration and aquifer recharge via infiltration ponds. As the natural discharge in Berlin's rivers is low, the treated wastewater is a significant contributor to the overall flow.[32]

With regard to wastewater treatment, particular attention is paid to the removal of nutrients to avoid eutrophication, and the city is making investments to upgrade the wastewater-treatment plants with tertiary filtration and ozonation.[33]

At the same time, climate change, rapid urbanization, and pollution are increasingly interfering with the city's water supplies. Various projects are under way to reduce pollution of surface waters and to release the pressure on the water cycle. A number of research-and-development projects, funded by the German Ministry of Education and Research (BMBF), are focused on developing "new multi-barrier treatment processes for planned potable water reuse schemes," for example, augmenting existing drinking-water resources by using better recycled wastewater.[34] Multiple benefits are foreseen through the usage of environmental buffers for the periodical storage of water, as well as from the recycled water, including the creation of water reserves and the conservation of wetland ecosystems and wildlife habitats.

Another critical aspect of water management in Berlin concerns the handling of stormwater. Stormwater management has become more and more important as the extreme weather events of the past few years have demonstrated that the city's infrastructure is not yet fully prepared for what is expected to come in the future.[35]

Decentralized Water Management: A Way Forward

The traditional way to deal with rainwater is to simply let it run off via a sewage system. This creates an array of problems. For instance, the system may not be able to cope technically with uncontrollable overflows of combined sewers, which leads to direct discharges of untreated stormwater. Such an aggravation of runoff during heavy precipitation has a disruptive effect on the infrastructure as well as the local ecosystems: the grit and grime of urban living can be swept into water bodies together with other materials, like pollen, and overwhelm aquatic life.[36]

A more innovative approach that is increasingly being applied is to use a decentralized rainwater-management system. This requires more than a single procedure and a rather complex system of a variety of selected and, if necessary, combined measures, all in accordance with local needs. A number of technical options are widely available and more frequently used for the on-site management of stormwater runoff: green roofs, porous paving, or rainwater harvesting systems. The unavoidable runoff can be either returned to subsoil or, if subsoil is not sufficiently capable of infiltration, caught by retention systems. A local water balance can thus be maintained.[37]

In July 2017, the Berlin Senate decided to further promote "decentralized rainwater management as an effective part of climate change adaptation."[38] Among others, the resolution requires an annual reduction ("decoupling") of 1 percent of the built areas from which stormwater is directly discharged into the combined sewer system. Additionally, all new residential areas are to be planned based on the principle of decentralized rainwater management. The recently founded Regenwasseragentur (Rainwater Agency) will support these projects.[39]

There are already numerous examples of decentralized water management applied in Berlin. Since more than two decades ago, green roofs and infiltration systems have been successfully implemented in some large development areas like the Rummelsburger Bucht and Adlershof.[40] In Rummelsberg, for example, all buildings are covered by six- to eight-centimeter-deep green roofs, and under all new pavements, an eighty-centimeter-deep layer of soil makes dispensing with stormwater sewers possible.[41] It can be safely said that Berlin is past pilot projects, and a green infrastructure approach is being copied across the city, with the hope that the positive effects will be replicated.[42] In recent years, building owners have been encouraged by means of financial incentives to regreen their courtyards and plant grasses and mosses on rooftops to absorb rainwater.[43] *Stadtschwamm*, or a sponge city, is a concept increasingly being used to describe these practices in Berlin.[44]

A staggering fourteen hundred projects with more than fifteen units of housing are currently being pursued, adhering to the principles of a sponge city.[45] Furthermore, while striving to create a more sustainable and livable city, Berlin has the goal of constructing a citywide network of landscaped streets and linear parks. Those would link the city's neighborhoods both to one another and to the major parks, while also helping to ventilate built-up areas and absorb extreme weather events. The city also wants to create more ponds, ditches, urban wetlands, and green strips along roads.[46]

Figure 3.1. Gleisdreieck Park in Berlin in March 2021: a park designed on the principle of the sponge city. (Photo: P. Długosz-Stroetges)

Communication with stakeholders remains an important topic for water management in Berlin. The measures currently being planned and implemented under the sponge city approach are discussed and explained via the "City Talks" organized by the Rainwater Agency. It gives city dwellers an opportunity to deliberate the impacts of, for example, green roofs or other nature-based solutions, while at the same time having a chance to ask questions that, if left unaddressed, could create undesirable tensions.[47]

Dealing with (Global) Challenges: Utilizing Opportunities

All across the globe, the impacts of climate change are being felt—and Berlin is no exception. Water shortages due to heat waves were not considered a major issue in Berlin in the past but are becoming daily concerns nowadays. In 2020, two hundred out of the city's five hundred lakes ran dry.[48] Moreover, forecasts suggest that in twenty years, Berlin will become 2 to 3 degrees Celsius warmer than it is now.[49] As coping with heat waves as well as extreme rainfalls becomes the new normal, the city becomes increasingly concerned with their impact on the partially closed water cycle that underpins water provision in Berlin. The application of the sponge city approach, like using rainwater to cool the city, helps to alleviate the impacts of climate change, but it must also be supported by investments in the city's aging infrastructure. While the latter poses a great challenge to any city, there is a great opportunity to innovate, apply new technologies, and "decouple."[50]

Extreme weather events also provide an additional opportunity: they at times give the much-needed momentum for a paradigm shift. When, in the summer of 2017, a bill (mentioned earlier) on water management was proposed to the Berlin Senate, there was little enthusiasm for it. But a few days before the vote, a major rainfall caused flooding across the city. The bill, initially considered unimportant, was voted for by all parties. The floods illustrated tangibly and vividly the importance of sound water management for the city and its citizens.[51] However, changing legislative frameworks is not enough; among others, there is a need to lessen administrative burdens, as well as the burdens on the application of technical solutions, which at times hamper innovation.[52]

Both politically and with regard to water management, West Berlin in particular has enjoyed island-like qualities. Its relative autonomy, both physically and as a political unit in a federal state, does not, however, prevent it from being buffeted by the winds of global change. Its physical water infrastructures had to adapt to the consequences of this change, and this will continue to be the case. Meanwhile, the example of water-utility privatization shows that Berlin was also not immune to the ideological influence of neoliberal globalization.

In spite of Berlin's unique features, it remains an interesting case study for local governments globally—as no locality can escape the challenges that global change presents for their water management. But Berlin also showcases that even great challenges can be turned into opportunities, given political support, adaptable administration practices, and public engagement.

NOTES

1. Over the past few years, due to climate change, precipitation has decreased, while evaporation has increased; in 2020, precipitation was around 500 millimeters, while in 2018, it was only 380 millimeters. Sieker, Heiko. 2021. Interview by the author. February.

2. This makes up for a total of 360 kilometers of waterfront (including both rivers and canals). Künzel, Michael. 2004. "Berlin—Stadt am Wasser." Berliner Denkmaltag, Berlin. www.stadtentwicklung.berlin.de.
3. Salian, Prit, and Barbara Anton. 2011. "Making Urban Water Management More Sustainable: Achievements in Berlin." SWITCH—Managing Water for the City of the Future, http://switchurbanwater.lboro.ac.uk.
4. IWA. 2015. "City Water Stories: Berlin." February 21, https://iwa-network.org.
5. The one is Stolpe, located on the outskirts of Berlin.
6. Also, as well as causing white laundry to turn gray, it makes water taste like blood.
7. Brears, Robert C. 2017. "Berlin Transitioning towards Urban Water Security." In *Urban Water Security*, 151–164. Chichester, UK: Wiley.
8. Salian and Anton 2011. For a discussion of the institutional and governmental structure in Germany and Berlin, see Ohlhorst and Schreurs, chapter 13 in this volume.
9. Gachea, Frédéric, Michael Städler, and Alexandre Brun. 2017. "Paris, London, Berlin: 3 Metropolises Facing the Challenges of Urban Waters." *UPLanD* 2(3): 231–240. https://doi.org/10.6092/2531-9906/5417.
10. Boscheck, Ralf. 2002. "European Water Infrastructures: Regulatory Flux Void of Reference? The Cases of Germany, France, England and Wales." *Intereconomics* 37(3): 138–149.
11. For example, "of the 22.7 million Reichsmark claimed by the Berlin water utility for damage to property and water losses during the war, only 9 million Reichsmark were paid in compensation by the Main Office for War Damage." Moss, Timothy. 2020. *Remaking Berlin: A History of the City through Infrastructure, 1920–2020*. Cambridge, MA: MIT Press, 315.
12. Merritt, Richard L. 1968. "Political Division and Municipal Services in Postwar Berlin." In *Public Policy*, edited by John D. Montgomery and Albert O. Hirschman. Cambridge, MA: Harvard University Press, 165–198, cited in Moss, Timothy. 2004. "Geopolitical Upheaval and Embedded Infrastructure: Securing Energy and Water Services in a Divided Berlin." Working paper, Erkner, Leibniz-Institute für Regionalentwicklung und Struktureplanung. www.irs-net.de.
13. Bärthel, Hilmar. 1997. *Wasser für Berlin: Die Geschichte der Wasserversorgung*. Berlin: Verlag für Bauwesen, cited in Moss 2004.
14. Berliner Wasserbetriebe. n.d. "From the Well to the Tap." Accessed February 2021, www.bwb.de.
15. Pawlowski, Ludwig. 1999. "Abwasserentsorgung in Berlin zehn Jahre nach dem Fall der Mauer." *Korrespondenz Abwasser* 46(11): 1694; Bärthel, Hilmar. 2003. *Geklärt! 125 Jahre Berliner Stadtentwässerung*. Berlin: Verlag für Bauwesen, 251.
16. Moss 2020, 555.
17. Tessendorff, Heinz. 1990. "So Soll es Bleiben: Berliner Wasser—Alles Klar!" In *Umwelt- und Naturschutz für Berliner Gewässer. Kongreß "Berlin auf dem Trockenen?" Zukunft der Wasserversorgung*, 40–55. Berlin: Senatsverwaltung für Stadtentwicklung und Umweltschutz, 52; Moss 2020, 556.

18. Moss 2020, 557.

19. Moss 2020, 554–556.

20. Lanz, Klaus, and Kerstin Eitner. 2005. *WaterTime Case Study—Berlin, Germany.* International Water Affairs.

21. Salian and Anton 2011.

22. *Cf.* Moss 2020.

23. Beveridge, Ross, and Matthias Naumann. 2013 "The Berlin Water Company: From 'Inevitable' Privatization to 'Impossible' Remunicipalization." In *The Berlin Reader: A Compendium on Urban Change and Activism*, edited by Matthias Bernt, Britta Grell, and Andrej Holm, 189–203. Bielefeld, Germany: Transcript-Verlag, 195.

24. Moss 2020, 586.

25. In Berlin, between 1989 and 2007, the demand for drinking water dropped by 45 percent. SENGUV and Berliner Wasserbetriebe. 2008. *Wasserversorgungskonzept Berlin 2040*. Pressekonferenz. www.bwb.de.

26. SPD is the Social Democrat Party, and CDU is the Christian Democrat Party in Germany. RWE and Veolia indicated that a sale would be desirable since BWB started to become a corporate image problem for both. Moss 2020, 598.

27. "The windfalls from privatizing the city's utilities did not significantly alleviate the city's debt problem. Yet by selling off such substantial shares the city forfeited annual income from the two profitable utilities. . . . BWB's private investors received profits from the utility totalling about 670 million euros between 1999 and 2006: that is, one-third of the sale price they had paid after only eight years." Moss 2020, 587–588.

28. Moss 2020, 599.

29. On the BWB democratization efforts, see Moss 2020, 599.

30. Indeed, one contentious point is whether the Futures Council should only engage water experts or should be composed of selected or elected water consumers. Moss 2020, 600.

31. Moss 2020, 559.

32. IWA 2015.

33. Even during the extreme rainfall of June 2017, the systems in Adlershof, for example, coped well. Sieker, Heiko. 2018. "Regenwassermanagement in Berlin." www.sieker.de.

34. IWA 2015, 1.

35. KWB. 2019. "Annual Report 2019." www.kompetenz-wasser.de.

36. Dead fish, for example, can be spotted in the river Spree after heavy rainfalls. *Cf.* Sieker 2018.

37. Sieker 2018.

38. Sieker 2018.

39. Sieker 2018.

40. Sieker 2018.

41. Bloomberg Quicktake. 2017. "Berlin Is Becoming a Sponge City." YouTube, August 23, www.youtube.com/watch?v=uWjGGvY65jk.

42. Lui, Li, and Marina Bergen Jenson. 2018. "Green Infrastructure for Sustainable Urban Water Management: Practices of Five Forerunner Cities." *Cities* 74: 126–133. https://doi.org/10.1016/j.cities.2017.11.013.
43. For example, "Since 2000, the charges for domestic wastewater and precipitation water are being billed separately in Berlin. . . . For instance, for greened roof areas only 50% of the respective area is considered in calculating the precipitation water fee." SenStadtUm (Senate Department for Urban Development and the Environment). 2016. "Anpassung an die Folgen des Klimawandels in Berlin—AFOK." April, www.berlin.de.
44. Broadly speaking, the concept underscores the importance of avoiding sealing up of the ground in favor of applying permeable surfaces, allowing for absorption, retention, and evaporation of water. *Cf.* Li, Hui, Liuqian Ding, Minglei Ren, Changzhi Li, and Hong Wang. 2017. "Sponge City Construction in China: A Survey of the Challenges and Opportunities." *Water* 9(9): 594; World Future Council. 2016. "Sponge Cities: What Is It All About?" January 20, www.worldfuturecouncil.org. Concretely, the Sponge City strategy, proposed for Berlin by a group of German architects led by Carlo Becker, suggests utilizing nature to combat the heat-island effect or floods. Becker insists that rainwater is too useful a resource to let it to simply flow out of the city; instead, it should be allowed to absorb into green infrastructure. As it then gradually evaporates, a natural cooling effect is achieved. *Cf.* Bloomberg Quicktake 2017.
45. Sieker 2021.
46. Eighteen percent of the city's area is covered by parks. Some, like the Gleisdreieck Park, are already designed on the sponge city principle. *Cf.* IWA 2015.
47. IWA 2015.
48. Sieker 2021.
49. Sieker 2021.
50. Sieker 2021; Siemens. 2020. "Toward Water 4.0 with the Digital Twin." June, https://new.siemens.com.
51. Sieker 2021.
52. This is underlined by one academic who is also a practitioner, Heiko Sieker: "One must dive deep into the darkness of the city's administration" in order to understand the administrative limitations on the implementation of innovative solutions (green infrastructure, digitalization, etc.), some of which seem or are impossible to overcome. Sieker 2021.

4

Restoring Freshwater Resources in Abu Dhabi

Challenges and Solutions

SARA SAVARANI

Water is essential to life, and one of the basic requirements for a habitable city is a safe and consistent source of freshwater. The water situation in the Emirate and City of Abu Dhabi, however, presents an urgent concern. Freshwater is essential for various purposes in the day-to-day operations of the emirate, including industrial processes, forest and agriculture irrigation, and municipal and household use.[1] And yet access to a high-quality freshwater supply has become increasingly difficult to maintain—due primarily to a depletion of available resources and a simultaneous rise in demand. An arid climate, limited precipitation, and lack of surface-water sources all limit the amount of freshwater that is naturally available in the region. And while the emirate has, until recently, been able to rely on groundwater aquifers as a consistent source of freshwater, these sources are being depleted faster than they can recharge.

While environmental conditions contribute to this situation, the rise in demand on the emirate's water resources has further exacerbated the problem. Abu Dhabi, both the capital city and the largest emirate of the United Arab Emirates (UAE), has seen tremendous population and economic growth in recent years, placing additional strain on an already stressed water supply. Furthering the strain on the water system is the emirate's lavish water use, with households using around three times more water daily than the global average. Given the historical and projected growth, and the resulting imbalance in supply and demand, the national, emirate, and local governments have begun exploring various strategies, including integrated water-management and water-reduction measures, to provide a sustainable water supply. These strategies, however, often present their own environmental concerns—desalination, for example, is extremely energy- and carbon-intensive—and other potential solutions, such as recycling wastewater or artificially recharging existing aquifers, have their own barriers to overcome.[2] Furthermore, in developing a comprehensive and sustainable strategy for water management in Abu Dhabi, the emirate seeks to balance its sustainable-water-management efforts with continuing social and economic development.

This chapter notes both the successes and challenges that the emirate has faced in addressing its water concerns, including through resource management and pursuing alternative water procurement strategies. It stresses that, while the emirate has taken affirmative actions to address its water concerns, these actions are insufficient and present their own environmental impacts. Given that the Emirate of Abu Dhabi and its departments hold much of the authority in water management, this chapter focuses primarily on the efforts of the emirate, rather than the city.

Water Sources and Use in Abu Dhabi

The availability of freshwater has long posed a problem in the UAE. This is largely due to challenging meteorological conditions: the climate is hot and arid, with average temperatures between 23 and 35 degrees Celsius, highs of 50 degrees in the summer, and an average total rainfall amount of just above eighty millimeters per year.[3] This harsh climate has an impact on both the availability and demand of freshwater resources in the region, which fluctuates depending on the season. While summers in Abu Dhabi are hot and arid and there is both less freshwater available and a higher demand, the opposite is true in winter, when the emirate is faced with an excess in production and a need for water storage may arise.[4]

Abu Dhabi, like much of the UAE, has little to no surface water available. Rather, the primary source of water for many years has been groundwater aquifers. These aquifers currently provide about 65 percent of the total water used in the UAE.[5] However, groundwater is largely a nonrenewable resource; while deep aquifers are nonrenewable, shallow aquifers are renewable, provided there is an external source of water, such as rainfall, to recharge them.[6] As such, it is vulnerable to overexploitation. In recent years, this supply has begun to deplete at a rate more than twenty times the natural recharge rate of aquifers due to overpumping of aquifers, done to satisfy the emirate's excessive demand.[7] Recent estimates have concluded that the emirate's groundwater supply will only last roughly fifty more years—and even less time in areas with heavy irrigation practices.[8] Further threatening the existing groundwater supply is the increasing salinity of groundwater, which makes it unsuitable for both potable use and agricultural irrigation.[9] This particular problem is expected to worsen in the future, as climate change has the potential to increase saline intrusion into the existing groundwater supply and reduce rainfall, further unbalancing aquifer salinity levels.[10] Even without the impacts of climate change, however, the emirate is expected to experience increased stress on its water supply.

While water concerns are inextricably linked with the country's natural geography and topography, much of the current shortage of water that the city is experiencing can also be traced back to the recent growth and development in the region. In recent years, the UAE has become a hub of the Middle Eastern economy. Commercial exploitation of oil in Abu Dhabi fueled dramatic economic growth, and the subsequent rapid urbanization and population growth have increased the demand on the city's already-scarce water resources.[11] This population is expected to continue to increase; a recent UN forecast estimates that the population will grow from over 9.8 million in 2020 to over 11 million in 2030.[12] With this increase in population comes an increasing demand for water across all sectors.

The agricultural sector consumes the largest amount of water in the emirate—nearly two-thirds of total consumption—with over twenty-four thousand farms in operation.[13] And while demand for water in the agricultural sector is increasing due to the growing population and the resulting surge in demand for food, the rising salinity levels threaten the ability of the emirate to grow crops and meet this growing demand.[14] This poses a challenge for food security in the emirate, particularly as the government attempts to reduce imports and grow more food locally.[15] The agricultural sector has begun to make efforts to conserve its limited water resources, including by moving away from water-intensive crops and introducing drip irrigation. However, more must be done to ensure a sustainable water and food supply and to reduce or maintain salinity levels in existing water reserves.

The second largest sector responsible for water consumption in the emirate is municipal water use, which includes domestic and household use. The rate of water consumption per capita in the UAE is one of the highest in the world, at 550 liters per day—nearly three times higher than the global average.[16] Due to the hot climate, water use in this sector is attributed to, inter alia, watering landscaping and other greenery, running air-conditioning units during hot summer months, and supplying drinking water year-round.[17] While this sector faces many of the same concerns as water used for the agricultural sector, it faces the additional challenge of requiring public opinion to support the shift to new sources of drinking water. For example, while recycled wastewater may be useful for agricultural and industrial processes and prove to be a viable alternative to groundwater wells, it is unpalatable among the broader public as a source of drinking water. Rather, desalinated water has become the primary source of household and drinking water in the emirate and makes up about 21 percent of the total water use for all purposes in the emirate.[18] Desalinated water, however, is expensive to produce, and other options have been considered.[19] These options will be discussed in more detail later in the chapter.

Governance Structure in Abu Dhabi

In order to understand the extent to which the UAE, the Emirate of Abu Dhabi, and the City of Abu Dhabi can address their water concerns, it is important to understand the authority that each holds in relation to the others. Within the UAE, there are three main levels of governance. At the top is the federal government, spanning the entire Federation of the UAE. The UAE is then divided into seven emirates, including Abu Dhabi, which each has its own rulers and bodies of government. Notably, while emirates are not the smallest level of government, they are often referred to as local governments and play a significant role in local affairs.[20] The emirates further divide their territories into distinct regions and municipalities, though they vary in how they do so. The Emirate of Abu Dhabi divides its territory into three distinct regions, including the Abu Dhabi City Municipality. Each region is administered by an emirate-level department. This section explains some of the authority that each level of government holds, both generally and in the context of water regulation and management.

The UAE Federation

The UAE is a federation of seven semiautonomous emirates governed by its constitution, which establishes a civil law system and distributes powers between three branches of government—executive, legislative, and judicial—and across five institutions.[21] Under the UAE Constitution, federal laws are deemed supreme over the laws of the individual emirates, and the emirates are prevented from passing any laws on any matters that have been designated as exclusive to the federal government. Article 121 of the UAE Constitution includes a list of matters for which only the Federation may enact laws—though notably, there is no reference to the management of groundwater or other domestic water resources.[22]

The Federation also has a series of ministries that are responsible for passing laws in specific areas. The Ministry of Energy and Infrastructure (MOEI), for example, is the federal body responsible for laws and policies related to water-resource management.[23] The MOEI has passed laws at the federal level that address these matters, but the local governments of the emirates play a significant role in implementing these laws.

The federal government has endeavored to promote a sustainable water supply in the country. For example, the MOEI in 2017 released its "UAE Water Security Strategy 2036," with a goal to ensure that the country has a safe and sustainable water supply, including during emergencies. This strategy covers various means of ensuring such a supply exists, including by reducing overall demand for water resources, increasing the amount of water that is produced, and increasing the amount of water that is stored. As such, the strategy implements three key programs to achieve these goals: focusing on water demand management, water supply management, and the production and distribution of water during emergencies.[24]

Abu Dhabi, the Emirate

Emirates may pass their own laws for any matters that have not been explicitly identified as reserved for the Federation in the constitution or have been identified as exclusive to the Federation but for which the Federation has not yet passed any laws.[25]

While the UAE Constitution does limit the emirates' authority to a certain degree, each emirate retains the authority to organize its own government, as well as the authority to pass and implement their own laws governing all matters that have not already been assigned to the exclusive authority of the federal government.[26] The constitution also calls on the emirates to implement the laws that the Federation passes by passing their own "local laws, regulations, decisions, and decrees necessary for such implementation."[27] Emirates also retain sovereignty over their natural resources, including water.[28]

Each emirate is led by a ruling family, which can make royal decrees and has its own Executive Council, which is chaired by the Crown Prince and passes laws and decrees for the emirate, sets its general policy and development plans, and approves and monitors government projects. There are also several emirate-level departments that implement the initiatives of the Executive Council, similar to the ministries at the federal level. The main regions within the Emirate of Abu Dhabi are administered by municipalities, including the capital city of Abu Dhabi.

While there are federal laws related to environmental protection, the emirates have retained much authority under these laws to pass their own environmental laws and policies. The Emirate of Abu Dhabi has in response passed numerous laws, resolutions, and decrees that address various environmental matters, including ones regulating water.[29] For example, Law No. 2 of 1998 established the governmental bodies that regulate the water and electricity sector.[30] Furthermore, both Law No. 6 of 2006 and Law No. 5 of 2016 were enacted by the Emirate of Abu Dhabi to designate groundwater as property of the local government, facilitating the government's ability to pass measures to manage and protect it.[31]

Laws of the emirates are implemented by the emirates' departments. The Environment Agency in Abu Dhabi (EAD), for example, plays a significant role in environmental protection and conservation, ensuring that other government bodies are considering the environment in their actions, and is responsible for developing environmental policies and regulating actions.[32] As such, one of EAD's priorities was to preserve groundwater resources, for both the present and future generations.[33] EAD has also been designated as the "competent agency" for environmental protection laws and regulations for the Emirate of Abu Dhabi—effectively tasking it with ensuring that the federal and local laws are implemented in the emirate and allowing the agency to pass its own policies for these matters. EAD is responsible for much of the emirate's regulation and management of groundwater, including setting water quality standards, regulating groundwater abstraction, managing drilling permissions, and inspecting wells.[34] EAD also promotes various efforts to educate the public on groundwater matters.[35]

Another department actively involved in managing water in Abu Dhabi is the Department of Energy (DOE).[36] DOE is responsible for providing the public with water and electricity. The Abu Dhabi Water and Electricity Company (ADWEC), in turn, is responsible for forecasting the water and electricity demand—including by gathering and analyzing demand regarding seasonal variations in both domestic and "bulk water demand"—which is then used by the departments for water planning and management. While not a perfect estimate, this allows the government to approximate the demand on the resource and take steps to adequately meet the emirate's needs.

Abu Dhabi, the City

In comparison to the federal and emirate governments, the Abu Dhabi Municipality has a fairly narrow scope of responsibilities, as it is merely a subdivision of the emirate's government. The municipality is primarily tasked with implementing planning and infrastructure projects in the city, such as bridges, drainage systems, roads, and transportation, and providing municipal services to the public.[37] Today, the municipality is overseen and managed by the emirate's Department of Municipalities and Transport (DMT), which is responsible for guiding, managing, regulating, and monitoring urban planning and development in the municipalities.[38] As such, the municipality's role in managing the water crisis in Abu Dhabi is minimal. Thus, the remainder of this chapter focuses on the efforts that the emirate, rather than the city, has taken in addressing Abu Dhabi's water concerns.

Government Response to Water Issues

The federal and emirate governments have already taken steps to address the growing water crisis in Abu Dhabi. At a national level, the UAE launched its "Water Security Strategy 2036" in 2017, which aimed to reduce demand on water resources by 21 percent, increase the water supply and storage capacity in the emirate, and increase the reuse of treated water up to 95 percent.[39] However, much of the regulatory and planning efforts regarding water procurement, use, and storage, including for groundwater, desalinated water, and recycled water, have taken place at the emirate level.[40]

At the emirate level, EAD has focused on both conserving groundwater resources and implementing integrated water-management strategies.[41] As part of Abu Dhabi's most recent five-year strategic plan, it set specific water-related goals to meet during the period between 2016 and 2020. Overall, the strategy aimed to reduce the total volume of groundwater that was extracted annually from 2.198 million cubic meters to 1.82 million cubic meters. Other key objectives include educating stakeholders on groundwater-related matters, strengthening the legal and regulatory framework surrounding groundwater, exploring options to conserve groundwater, and pursuing a strategy of integrated water management, which promotes a water-management system that holistically coordinates the development and management of water resources.[42] With these steps, the emirate hopes to extend the period that groundwater reserves remain an effective and usable water source, as well as to reduce the amount of water that is used for agriculture, to reduce or maintain the salinity levels of existing reserves, and to increase the use of recycled water, where appropriate and effective. Abu Dhabi's strategic planning will continue for its electricity and water sectors for the next five years; a new strategic plan, covering the period from 2021 to 2025, was approved by EAD in late November 2020.[43] This plan includes various strategies that focus on the sustainable management of the emirate's groundwater resources.

The emirate has undertaken various efforts, both in furtherance of and independent of this strategic plan, to address its water crisis. While not an exhaustive list, this section describes some of the efforts that the Emirate of Abu Dhabi and its departments have taken.

Mapping and Planning for Water Management

One of the first significant efforts that EAD undertook was to conduct a comprehensive assessment of the current status of groundwater resources in the emirate. EAD launched a Groundwater Wells Inventory and Soil Salinity Mapping Project in 2015—the first effort of its kind to inventory all the existing groundwater wells in the emirate, used across all sectors. It also marked the first phase of the emirate's initiative to conserve groundwater reserves, promote the sustainability of water resources, and develop a comprehensive plan to address farmland affected by the increasing salinity of water reserves.[44] EAD also completed a groundwater-monitoring program in 2016, conducted in partnership with the United Stated Geological Survey, and in 2017 launched the Groundwater Quality Baseline Survey (GWQBS) project to implement the program. The monitoring program was geared at both the level and quality of the groundwater and aimed to develop a baseline for groundwater quality, as well as to assess the level of contamination of the water supply and to gauge the natural recharge rate of groundwater reserves. In concluding the survey, the groundwater quality baseline was set in 2018.[45] Also in 2018, EAD published a Groundwater Atlas documenting the current state of groundwater reserves in Abu Dhabi, which was intended to be used for planning and regulation. Much of this data came from the Mapping Project. The Atlas aimed to enhance water plants with the latest technology to increase capacity and efficiency. All data from the Atlas, which EAD has stored on a central database, has been made available to researchers, decision-makers, and other stakeholders, allowing them to make decisions regarding sustainable groundwater management that are grounded in data.[46]

Desalination

As water scarcity has become a major concern in Abu Dhabi, the local government has begun to rely more on alternative sources of water and to consider a strategy of integrated water management. One of the favored solutions of the emirate is desalination, a process by which salt is extracted from saltwater in order to produce potable freshwater.

There are some obstacles to turning desalination into a sustainable solution. Financially, desalination is expensive. Advances in technology have reduced the cost significantly, and certain technologies have seen production prices drop from roughly $1.00 per cubic meter in 2011 to between $0.50 and $0.80 per cubic meter in 2016.[47] However, costs remain high; in total, desalination processes cost the UAE around $18 million every day.[48] Energy consumption is another major, related problem; desalination is an energy-intensive procedure, primarily due to the demanding power generation needed to operate the plants.[49] While the government is working to establish more power generation, this imposes another cost on an already-costly process. The operation of desalinated water plants also introduces concerns about noise pollution and the discharge effects on marine life, as well as concerns about the impact of coastal management on the desalination infrastructure itself; algal bloom, for example, can clog intake filters, causing plants to cease operating and close.[50]

Yet despite these concerns, Abu Dhabi has embraced desalination as a potential solution to its water crisis. The emirate is a major procurer of desalinated water; approximately 9 percent of the total desalinated water use in the world is used in Abu Dhabi. It also has a significant number of desalinated water plants already in operation; approximately a third of Abu Dhabi's total water capacity comes from nine major plants in the emirate.[51] Taken together, these plants produce approximately 4.13 million cubic meters of water per day. The emirate is also expanding its desalination infrastructure and is currently developing the largest desalination plant in the world—the Al Taweelah Power and Water Complex—which will produce over 900,000 cubic meters of water per day.

Desalination is not, on its own, a perfect replacement for groundwater reserves. Effective strategies will have to be implemented to store water after it is desalinated in order to maintain a stable water supply, particularly in the event of an emergency. In such a situation, the emirate can store surplus water produced by desalination plants in aquifers by using artificial storage techniques.[52] The government currently stores surplus water for up to five years in designated aquifers and recently launched a project to establish the largest underground reservoir in the world, using 26 million cubic meters of desalinated water, reserved for emergencies.[53]

If Abu Dhabi has at all been successful in dealing with its problem of water scarcity, it is largely due to its embrace of desalination. But while desalination may effectively solve the local water problem, it has serious consequences for the global climate and could also harm marine biodiversity—making this a fairly limited victory.

Integrated Water Management

Abu Dhabi has also increasingly turned to expanding its wastewater capacity and artificially recharging existing aquifers with aquifer storage and recovery (ASR), both for regular use and for emergencies. When used effectively, ASR can increase or maintain groundwater levels—a strategy that has been contemplated and explored at the national level.[54] DOE has also begun to contemplate using wastewater processing plants as an alternative source of freshwater for agricultural water use, which may be cheaper to run than desalination plants. Relatedly, DOE announced its intent to set up a wastewater monitoring lab to consider the safety of the wastewater supply and to address its impacts on the environment.[55]

Encouraging Water Savings in the Emirate

Abu Dhabi has also employed several strategies to encourage the adoption of efficient and water-saving practices among the most water-intensive sectors. For example, the emirate introduced subsidies for farmers who reduce water usage by producing less water-intensive crops.[56] The agricultural industry at large is also taking steps to introduce water-saving techniques, such as drip irrigation, to further reduce water use.

Other programs have been introduced that specifically target buildings to reduce their water use. For example, the Buildings Retrofit Programme, introduced by DOE, aims to save on both water and electricity use, with an expected savings of 2.7 terawatt-hours of electricity and 9 cubic megameters of water by 2030. Another initiative, implemented by Abu Dhabi distribution companies and others, sought to install roughly

21,900 efficient water-ablution taps in 1,577 mosques, which would reduce their water consumption by over 30 percent.[57] Finally, DOE has also introduced the Energy and Water Efficiency Policy for Government Buildings, an implementation framework that specifically targets government buildings and aids them in adopting water-efficiency measures.

The emirate has also historically used financial incentives to encourage water savings. When the current water system was set up in the emirate in 1960, while Abu Dhabi was under British control, water was provided to residents for free. However, since then, various pricing mechanisms have been introduced. At first, a flat rate of AED 50 for the water use of expatriates was introduced. This was replaced in 1997, when a new price mechanism was introduced after the installation of water meters in buildings, and non-Emirati residents began paying AED 2.2 per one thousand liters. In 2015, a tiered price system was introduced, prices were raised, and the fees were extended to include Emirati residents as well.[58]

Finally, the emirate has launched several educational campaigns that are intended to educate and nudge behaviors among public and domestic water users. The former Abu Dhabi Water and Electricity Agency (ADWEA), for example, launched a water and electricity campaign, "Rethink Your Lifestyle," which used the public dissemination of information, including through television and newspaper advertisements, brochures, and displays in public places, to encourage residents to save water. DOE also launched an Energy Efficiency Advisor service in 2021, which is aimed at providing support to residents in taking steps to reduce water consumption at home. Finally, DOE also launched a "Use It Wisely" campaign in July 2020, to encourage sustainable water and electricity practices among individuals by promoting water- and electricity-saving tips and competitions across social media platforms.[59]

Going Forward: Challenges and Solutions

As Abu Dhabi continues to face increasing demand on its water supply, the emirate will have to take more steps to fully address the problem. Climate change, for example, is likely to aggravate the water challenges facing the emirate. The temperature is projected to increase by approximately 2 to 3 degrees Celsius in the UAE over the next forty to sixty years, further increasing demand on the scarce resource.[60] According to United Nations Development Programme estimates, climate change could result in even more severe water shortages, as well as increased aquifer salinity—negatively impacting the ability to irrigate agricultural land and grow food.[61] However, some scholars do note that many of the issues that Abu Dhabi faces with regard to its water resources are not necessarily due to the impacts of climate change; many of them would continue to worsen even in its absence, and one study notes that climate change will have only a small impact on both water demand and groundwater storage.[62] The key aggravators, rather, are the increasing population and the high domestic water use—factors that will continue to rise in the coming years unless a very substantial change of course takes place.

Regardless of the causes, the key to addressing Abu Dhabi's water-security concerns going forward will continue to be resource management and technology innovations. While Abu Dhabi has begun to implement some strategies to address its water challenges, these strategies have their own impacts, environmental and otherwise, that must be considered. The emirate's favored strategy, desalination, is not only costly but also carbon-intensive, worsening the conditions that threaten the water supply to begin with. Furthermore, many of the measures that Abu Dhabi had taken so far have not materially reduced demand and do not adequately address one of the key factors at play: the lavish water use in the emirate. Government action will need to be targeted, and both water-conservation and integrated-water-management efforts will need to be expanded. Without such comprehensive and sustainable strategies, Abu Dhabi will face a dire water crisis in the years to come.

NOTES

1. Al-Katheeri, E. S. 2008. "Towards the Establishment of Water Management in Abu Dhabi Emirate." *Water Resource Management* 22: 205–215. https://doi.org/10.1007/s11269-006-9151-y.
2. Environment Agency, Abu Dhabi. 2017. *Abu Dhabi State of Environment Report 2017*, 13, www.ead.ae.
3. Environment Agency, Abu Dhabi. 2019. *Groundwater Quality Baseline Survey*. www.ead.ae.
4. Al-Katheeri 2008.

5. Environment Agency, Abu Dhabi. 2018. *Groundwater Atlas of Abu Dhabi Emirate*. www.ead.ae. Note that this percentage includes all total water use, not just providing drinking water, which is primarily sourced from desalinated water. Further discussion on the implementation of desalinated water plants appears later in this chapter.
6. Al-Qaran, S., and M. Mohamed. n.d. "Improving Strategic Groundwater Reserves in UAE via Managed Aquifer Recharge with Recycled Water." United Arab Emirates University.
7. Al-Katheeri 2008.
8. Environment Agency, Abu Dhabi. 2016. *Strategic Plan 2016–2020*, www.ead.ae.
9. Al-Katheeri 2008.
10. Some of the impacts of climate change are as of yet uncertain and may vary depending on the region. UAE Ministry of Climate Change and Environment. 2021. *The UAE State of Climate Report: A Review of the Arabian Gulf Region's Changing Climate & Its Impacts*. www.moccae.gov.ae.
11. Szabo, Sylvia. 2011. "The Water Challenge in the UAE." *Dubai School of Government Policy Brief* 29 (December): 1–8; Al-Katheeri 2008.
12. United Nations. 2017. *World Population Prospects: Data Booklet*, 12, www.un.org; World Bank. n.d. "Population, Total—United Arab Emirates." World Bank: Data. http://data.worldbank.org.
13. Environment Agency 2019. While the emirate imports much of its food, it has also sought to produce certain crops, such as dates, both to supplement its food supply and to export for its economy. Shahin, S., and M. Salem. 2015. "The Challenges of Water Scarcity and the Future of Food Security in the United Arab Emirates (UAE)." *Natural Resources and Conservation* 3(1): 1–6. https://doi.org/10.13189/nrc.2015.030101.
14. Al-Katheeri 2008.
15. Environment Agency 2019.
16. Günel, Gökçe. 2016. "The Infinity of Water: Climate Change Adaptation in the Arabian Peninsula." *Public Culture* 28(2): 295.
17. While drinking water is a significant portion of private-sector water use, many people in the emirate prefer bottled water over tap water, due to concerns about the water quality and potential for pipe contamination—though bottled water has its own environmental impacts, in the form of pollution and waste.
18. Kizhiser, Mohamed, Mohamed M. Mohamed, Walid El-Shorbagy, Rezaul Chowdhury, and Adrian McDonald. 2021. "Development of a Dynamic Water Budget Model for Abu Dhabi Emirate, UAE." *PLoS ONE* 16(1): e0245140. https://doi.org/10.1371/journal.pone.0245140.
19. Al-Katheeri 2008.
20. UAE Constitution, arts. 123, 125; UAE. 2021a. "The Local Governments of the Seven Emirates." December 6, https://u.ae.
21. UAE 2021a; UAE. n.d. "About the Constitution." Accessed August 30, 2021, https://u.ae. These five institutions are the Federal Supreme Council (FSC), the

President of the Union, the Council of Ministers of the Union, the Federal National Council (FNC), and the Federal Judiciary. UAE Constitution, art. 45. Note that the authorities have revised their names since the constitution was established. The legislative powers of the UAE are vested in the FSC, which acts as the main legislative body and "draws up general policies and approves various federal legislations," and the FNC, which acts as the consultative council and discusses and approves, amends, or rejects drafted laws or constitutional amendments. UAE. n.d. "The Federal Supreme Council." Accessed August 30, 2021, https://u.ae; UAE Constitution, arts. 68–77; UAE. n.d. "The Federal National Council." Accessed August 30, 2021, https://u.ae.

22. UAE Constitution, art. 121.
23. UAE Ministry of Climate Change and Environment. n.d. "About the Ministry." Accessed August 30, 2021, www.moccae.gov.ae; UAE Ministry of Climate Change and Environment. n.d. "Duties and Responsibilities." Accessed August 30, 2021, www.moccae.gov.ae.
24. UAE. 2021b. "The UAE Water Security Strategy 2036." June 6, https://u.ae.
25. Article 116 of the UAE Constitution states that "the Emirates shall exercise all powers not assigned to the Federation by this Constitution." UAE Constitution, art. 116. Article 122 further emphasizes that "the Emirates shall have jurisdiction in all matters not assigned to the exclusive jurisdiction of the Federation, in accordance with the provision of the two preceding Articles." UAE Constitution, art. 122. *See also* UAE 2021a. Yusuf. 2015. "Legal and Judicial System in the United Arab Emirates." *Shoeb Saher*. www.shoebsaher.com ("Individual Emirates are permitted to enact their own legislation in matters that are not exclusive to the federation, as well as in those matters in relation to which—albeit exclusive to the federation—the federation has not yet exercised its legislative powers.").
26. UAE Constitution, arts. 116–119; UAE 2021a.
27. UAE Constitution, art. 125. *See also* UAE. 2021c. "Where to Find Local Laws of the UAE's Emirates." August 16, https://u.ae.
28. Luomi, Mari. 2014. *The Gulf Monarchies and Climate Change: Abu Dhabi and Qatar in an Era of Natural Unsustainability*. Oxford: Oxford University Press.
29. Law No. (5) of 2016.
30. Law No. 2 of 1998 Concerning the Regulation of Water and Electricity Sector, http://leap.unep.org.
31. Environment Agency 2018.
32. Environment Agency, Abu Dhabi. n.d.-a. "About Us." Accessed August 30, 2021, www.ead.gov.ae; Environment Agency, Abu Dhabi. n.d.-b. "What We Do." Accessed August 30, 2021, www.ead.gov.ae.
33. Environment Agency 2018.
34. Environment Agency 2018; Environment Agency 2016.
35. Environment Agency 2018.
36. Reuters. 2018. "Abu Dhabi's Utility Folded into New Energy Department—Spokesman." March 25, www.reuters.com. The utility was formerly known as the

Abu Dhabi Water and Electricity Agency (ADWEA) but was folded into DOE in 2018.

37. UAE Department of Municipalities and Transport, Abu Dhabi City Municipality. n.d. "Who We Are." Accessed August 30, 2021, www.dmt.gov.ae. The Abu Dhabi City Municipality was first established in 1962 as the Department of Abu Dhabi Municipality and Town Planning, and a royal decree was issued several years later, in 1969, to appoint the first municipal board "with the task of providing comprehensive services to the public and ensure proper planning of the developing city, with regularized road networks, maintenance services, sewerage, lighting works, launching the Agriculture Development Plan in the Emirate and establishing public markets in various areas." UAE Department of Municipalities and Transport, n.d.
38. UAE Department of Municipalities and Transport, n.d. The current Department of Municipalities and Transport was established by Law No. 30 of 2019, after several years of restructuring and creating different departments to manage municipal affairs.
39. UAE. 2021d. "Water." August 16, https://u.ae.
40. Environment Agency 2016.
41. Environment Agency 2016.
42. "Integrated Water Resources Management (IWRM) is a process which promotes the coordinated development and management of water, land and related resources in order to maximise economic and social welfare in an equitable manner without compromising the sustainability of vital ecosystems and the environment." Global Water Partnership. 2011. "What Is IWRM?" December 7, www.gwp.org.
43. Dadlani, Disha. 2020. "Five-Year Plan Approved for Abu Dhabi Water and Electricity Sector." *Construction Week*, November 7, www.constructionweekonline.com.
44. Environment Agency 2018.
45. Environment Agency 2019.
46. Environment Agency 2018.
47. World Bank. 2019. "The Role of Desalination in an Increasingly Water-Scarce World." Washington, DC.
48. Günel 2016.
49. Giwa, A., and A. Dindi. 2017. "An Investigation of the Feasibility of Proposed Solutions for Water Sustainability and Security in Water-Stressed Environment." *Journal of Cleaner Production* 165: 721–733. https://doi.org/10.1016/j.jclepro.2017.07.120.
50. Giwa and Dindi 2017, citing Dawoud, M. A., and M. M. Al Mulla. 2012. "Environmental Impacts of Seawater Desalination: Arabian Gulf Case Study." *International Journal of Environment and Sustainability* 1(3): 22–27.
51. *Utilities*. 2020. "Nine Major Water Desalination Plants Enhance Abu Dhabi's Water Security." July 23, www.utilities-me.com.

52. Al-Katheeri 2008.
53. Al-Katheeri 2008; Szabo 2011.
54. Günel 2016.
55. Trade Arabia. 2021. "G42 to Set Up New Wastewater Monitoring Lab in Abu Dhabi." June 20. tradearabia.com; *Utilities*. 2021. "Abu Dhabi Department of Energy Launches Wastewater Monitoring Lab." June 20, www.utilities-me.com; Godinho, Varun. 2021. "New Abu Dhabi Lab to Use AI Tech to Analyse Wastewater for Infectious Diseases." *Gulf Business*, June 21, https://gulfbusiness.com.
56. Szabo 2011.
57. *Gulf Today*. 2021. "Abu Dhabi DoE Hosts Workshop on Energy, Water Efficiency Policy." June 6, www.gulftoday.ae.
58. Günel 2016.
59. Zawya. 2021. "Abu Dhabi Department of Energy Launches Free Live Energy Efficiency Advisor Services to Support Households in Reducing Consumption." April 28, www.zawya.com.
60. UAE Ministry of Climate Change and Environment 2021.
61. UNDP. 2010. *Mapping of Climate Change Threats and Human Development Impacts in the Arab Region*. www.arabstates.undp.org.
62. Günel 2016.

5

Safeguarding Delhi's Water

VIDYA VIJAYARAGHAVAN

The National Capital Territory of Delhi, "one of the largest urban agglomerations in the world, has faced significant water problems" for more than a decade.[1] Compared to other major urban cities, Delhi's unique administrative setup and geographical difficulties present complex challenges that compound existing stressors on its water supply. The city's confined land mass, huge influx of population, haphazard spatial development, growing industrialization, and deteriorating and outdated infrastructure are some of the chief stressors that have notoriously led to an insurmountable water situation. All these factors, coupled with the increasing impacts of climate change, have overwhelmed the ability of the city to cope with the challenge of water scarcity over the years.[2]

Historically, Delhi has witnessed a series of failures in building its water and sewerage systems.[3] The city has limited drinking-water sources within its boundaries; these sources cater to only 10 percent of its population. The remaining 90 percent is supplied with water from surface-water sources—namely, the Yamuna, Ganga, and Bhakra Beas systems, which are snow-fed northern rivers originating outside the city.[4] These external water resources are made available to Delhi through numerous interstate agreements with neighboring states.[5] However, the city is currently facing several management challenges regarding its water and wastewater systems.

With respect to Delhi's wastewater system, the amount of sewage generated by the city exceeds the capacity of its wastewater-treatment plants on a daily basis. More than half of the sewage generated in the city is dumped into its main lifeline for water supply: the Yamuna River, which is now one of the most polluted rivers in the world.[6] Delhi alone is responsible for an alarming 78 percent of Yamuna River's pollution.[7] Furthermore, the existing sewage-treatment plants in the city are not fully functional or utilized. Only 78 percent of the houses in the city are connected to a sewer system, and the rest discharge their waste directly into open drains and the Yamuna River. For many decades, the city has lacked a systematic and holistic approach to plan, formulate, and implement a robust water and sewerage system.

This chapter begins with an overview of the management of the water supply and wastewater systems in Delhi. It then discusses key challenges the city faces in managing both of the systems and lays out the institutional framework that currently exists. Finally, it discusses the policy interventions that the government has pursued to tackle these challenges and concludes with some suggestions for future management of the systems.

Delhi's Water-Supply System

Delhi's water-management system dates back to medieval times. The ancient city "had about 800 waterbodies, both natural and man-made" and an extensive range of water resources: stepwells, ponds, streams, natural drains, and a network of check dams.[8] All of these sources ultimately drained into the Yamuna River and provided residents of the city with drinking water throughout the year.[9] However, these traditional systems were replaced with a new system under British governance in the early twentieth century, leaving the old system unattended.[10]

Over the years, the city, which was once self-sufficient, started looking for outside sources of water. Several memoranda of understanding (MoUs) were signed with other states to obtain water supply. One such prominent MoU, signed in 1994, shared Yamuna River's water with four other states: Haryana, Himachal Pradesh, Rajasthan, and Uttar Pradesh.[11] Another MoU obtained water supply from the Bhakra reservoir, upper Ganga canal, and Munak canal in an agreement with Himachal Pradesh, Uttar Pradesh, and Haryana, respectively.[12] The 1994 MoU also created the Upper Yamuna River Board, which regulates the allocation of available flows among the beneficiary states and monitors the return flows in accordance with the agreements.

For the purpose of governance, Delhi has been divided into five local bodies: the Municipal Corporation of Delhi, the New Delhi Municipal Council (NDMC), the South Delhi Municipal Corporation, the East

Delhi Municipal Corporation, and the Delhi Cantonment Board (DCB). The majority of Delhi, about 95 percent of its land, is under the jurisdiction of the city's municipal corporations (MCs).[13]

The governance of Delhi's water infrastructure was not centralized in the city until the early 1990s, when the management of water infrastructure shifted from municipal to state control. Until 1958, the water and sewage management of the city was managed at the local level by an agency named the Delhi Joint Water and Sewage Board. Then, under the Delhi Municipal Corporation Act, the Delhi Water Supply and Sewage Disposal Undertaking was constituted, which managed the water and sewerage system until 1996 under the aegis of the MCs. In 1998, the power shifted from the MCs to the state government of Delhi, which enacted the Delhi Jal Board Act of 1998 (DJB Act), establishing the Delhi Jal Board (DJB) and granting it the powers to procure, supply, and distribute water and to treat wastewater.[14]

Around 95 percent of Delhi receives its water supply from DJB; for the remaining 5 percent of Delhi, water supply and distribution are managed by other local bodies under the NDMC and DCB.[15] DJB procures nearly 91 percent of raw water from surface-water sources and gets the rest from underground sources. The water-supply network is mainly composed of pipelines and underground reservoirs. To ensure safe consumption, DJB first treats the raw water at water treatment and recycling plants, before then supplying the water to households through pipelines and other means.[16]

While the DJB Act lays out the responsibilities and infrastructural arrangement for water and wastewater management, Delhi has not kept pace with the city's massive population growth over the years. The city's population increased by an average of 50 percent until 2001, and by the time of the 2011 census, the population had grown to around 16.78 million.[17] Currently, the population is about 26 million and will continue to increase, putting immense pressure on the availability of adequate water supply for the city.[18]

Given this massive population, a significant number of households in the city—around 17 percent—lack direct access to a piped water supply.[19] DJB claims that these areas are supplied instead via private water tankers, which stop on the roads and require that people line up to fill their containers with water.[20] However, these tankers are only available for a limited number of hours, do not cater to a sufficient number of households in their areas, and lack a proper monitoring system. Moreover, the tankers are poorly maintained, leading to large amounts of water contamination and water wastage—another unattended problem in the city, with around 40 percent of the supply lost due to "leaky pipes, illegal tapping of pipelines, poor maintenance of tankers and . . . wastage by people during consumption, such as using [a] reverse osmosis system for water purification."[21]

Depletion and unbridled extraction of groundwater sources by private companies, industries, and unauthorized construction activities further exacerbate the problem.[22] Many of the waterbodies are not able to be revived after they are depleted, and of the groundwater sources that remain, many cannot provide safe drinking water, due to high salinity and levels of fluoride, nitrate, and arsenic that exceed the limits prescribed by the Bureau of Indian Standards.[23]

Delhi also faces unpredictable rainfall patterns due to effects of global warming and climate change, which makes it important to capture and store as much precipitation as possible. This is a huge challenge, given the congested urban space of the city.[24]

Though stormwater drains have been built, covering nearly 75 percent of the city's road length, due to the contamination of their water with solid waste, they have proven ineffective.[25] Furthermore, rainwater harvesting, which could be an essential technique to recharge groundwater supply, has been poorly implemented in the city. While Delhi has introduced a mandate for all private buildings and housing societies over five hundred square meters to install rainwater harvesting systems, the vast majority have not done so.[26]

All of these factors have drastically reduced the amount of water that is available for consumption and have greatly widened the gap between supply and demand in Delhi. In 2019, the city's needs were about seventy-two gallons per capita daily of water, which included supply needed for both domestic and nondomestic consumers, such as industries, hotels, and fire stations.[27] DJB could fulfill only 74 percent of this total demand, leaving a reported supply-demand gap of about 323 million gallons per day.[28] Also, nearly one-third of the population of the city lives in unplanned settlements, which pose particular challenges when supplying water.[29] A piped water-supply system is not legally available for this population, and the "dense constructions and haphazard expansion of households" in the area make it difficult to design and implement water infrastructure.[30] As a result, many of these settlements rely on water tankers for their water supply, leading to an "inequitable distribution" of water supply in Delhi, "where poor households are acutely affected."[31]

Aggravating the situation further, Delhi has a long history of interstate water disputes with its neighboring states. By far the most contentious dispute is with the state of Haryana. In 1996, the Supreme Court of India had to interfere when Haryana withheld water from Delhi and ordered Haryana to release 450 cusecs of water per day. Later, in 2018, the Delhi state government accused Haryana of noncompliance with the 1996 order, alleging that the state had taken advantage of its position as an upper riparian and supplied only 330 cusecs of water per day to Delhi.[32] The Supreme Court ordered both states to amicably resolve the issue with the involvement of the Upper Yamuna River Board. Ultimately, Haryana agreed to supply water to Delhi as per the 1996 agreement.[33] Such disputes illustrate the challenges that Delhi has encountered in sharing water resources with other states. These interstate battles over water may have diverted the state government's attention away from resolving some of the fundamental water-management issues within the city boundaries.

Apart from these challenges, another major issue that Delhi faces is its poor water quality. The expansive population growth in the city, coupled with developmental activities such as industrial growth, has choked the natural drains and wells with large amounts of waste, sewage, industrial chemicals, agricultural runoff, and pollutants, turning them toxic.[34] In addition, untreated effluents from industrial units in Haryana are released into the Yamuna River, further polluting the water supply.[35] And, as the Bureau of Indian Standards reports, the administrative district of the city, New Delhi, has the most unsafe tap water among the twenty-one state capitals in India.[36]

Issues with the sewerage system plague the city's water quality as well. Households situated along the river will regularly dump their garbage directly into the water, converting the rivers and seasonal streams into sewage drains.[37] The water supply in northern Delhi became so heavily polluted, in fact, that it was no longer safe for consumption, after sewage from the Najafgarh drain to the north flowed down the Sahibi River and into the Yamuna River, ultimately reaching Delhi.[38] Moreover, the water quality in the city is threatened by sewage contamination at multiple points. As sewer and water pipelines often run parallel to each other, treated water is at constant risk of being contaminated by sewage pipeline leaks.[39] Untreated sewage may also contaminate stormwater drains, further worsening the water quality across Delhi.[40] Excessive concretization and encroachments have also led to frequent flooding and waterlogging. The scarcity of safe and good-quality drinking water has caused high exposure to water-related diseases such as cholera, typhoid, diarrhea, and gastrointestinal diseases. In fact, nearly 15 percent of the registered deaths in Delhi are attributed to infectious and parasitic diseases.[41]

In all, the exceedingly poor water management in the city has forced its population to look beyond the government to private water-supply sources to fulfill its drinking-water needs. Such sources include privately owned borewells and tubewells, private water tankers, and packaged water suppliers.[42] These sources dominate the water supply in areas where DJB falls short.[43] Of the private suppliers, the most notorious are the illegal gangs known as "water mafias." These water mafias dig boreholes without permission, source millions of gallons of water from illicit borewells—with the connivance of police and local politicians—and extract thousands of liters of groundwater, depleting the sources beyond repair and selling the extracted water for profit in water-deficit areas.[44] Some newspaper reports indicate that these mafias are connected to DJB officials and that they divert the DJB-owned water tankers from their mandated routes to sell water commercially.[45] Even worse, these mafias do not take any precautions while digging borewells to prevent dangerous contaminants, such as iron and arsenic, from mixing with water, causing permanent damage to water sources and making them unsafe for consumption.[46]

Delhi's Wastewater-Management System

As with the water infrastructure, Delhi's sewage infrastructure has not been able to keep pace with the rapid development of the city. DJB is the responsible authority to collect, treat, and dispose of sewage generated in the majority of the city, excluding the areas of the NDMC and DCB.[47]

However, a very high proportion—nearly 80 percent—of consumers' water supply ends up as wastewater.[48] As a result, managing the sewerage system is a huge task for DJB and the other planning authorities. Moreover, Delhi's current sewerage network only covers 78 percent of its population, and its water-treatment plants only treat 66 percent of the generated wastewater, on top of facing "various issues such as blockages, disconnected trunk and peripheral lines, old and deteriorated sewers, among others."[49] While the city does have some green spaces, which could mitigate some of these problems, they are "disconnected with the natural drainage network."[50] As a result, huge amounts of untreated wastewater are discharged into the Yamuna River.

Although there are separate sewage and stormwater drains in Delhi, due to a lack of sewage drains, the stormwater drains often become contaminated and waterlogged by excess sewage.[51] The city's stormwater infrastructure also lacks the capacity to deal with the runoff generated during high-intensity rainfalls. In total, there are 201 natural drains and 22 major drains that flow into Delhi—a major portion of which are not lined or have damaged lining.[52] There is no system to prevent pollution, sewage, construction waste, or garbage from being dumped into these drains, and so during rainy seasons, they overflow, leading to flooding. And because most of these drains finally discharge into Yamuna River, the polluted water from storm drains worsens the river's pollution.[53]

Eleven different agencies share responsibility for maintaining or managing different parts of the stormwater drainage system, making addressing sewage concerns particularly challenging to coordinate.[54] For example, the state's Irrigation and Flood Control Department constructs and maintains large drains, but the Public Works Department takes care of specific drains in particular areas; meanwhile, the MCs manage and dispose of silt removed from drains in their areas, but DJB is responsible for drains that discharge more than 1,000 cusecs.[55] And none of these agencies seem to have taken substantial measures to fulfill their mandates.

Financing for Water and Sewerage

The water and sewerage systems in Delhi are financed mainly by three sources: funds allocated by the central government; funds from the state government of Delhi; and water-tariff and sewerage-maintenance charges levied by DJB.

At the state level, the DJB Act requires that DJB frame an annual budget, and the finances for this budget come mainly from two sources. The first source of funding is the capital budget, comprising both loans and grants-in-aid from the state government of Delhi and funding from central government schemes, as stated earlier. For 2020 and 2021, the budget increased by 70 percent, and a huge sum of 3,724 crore rupees was designated for DJB—in particular, for supplying water to unauthorized colonies and installing four decentralized wastewater treatment plants, with a capacity of four million gallons per day, along the Yamuna riverbank.[56]

The second source of funding is the revenue budget, which includes income from revenue receipts of the city's internal sources: water tariffs, fees from sewerage service, infrastructure charges, and other administrative and financial charges. For instance, DJB has a fixed charge for water use based on the user's monthly consumption of water, calculated in kiloliters.[57] Domestic users can use up to twenty kiloliters per month without charge, provided the water use is monitored by functional water meters. Where there is no meter, a flat rate is imposed, based on an "assumed average consumption" value for consumers in unauthorized colonies or villages. There is also a monthly sewerage-maintenance charge for collection and treatment, set at 60 percent of the "charges of volumetric water consumption."[58]

Recently, DJB issued a mandatory annual sewage-pollution charge for all households, including both domestic and nondomestic consumers in Delhi. To date, sewage charges have been levied only from households with a functional sewer connection and an active water connection. This new charge divides the residential colonies in the city into eight categories, which determines the amount of the charge. All households will be required to pay these sewerage charges whether or not they have sewer or water connections, including unauthorized colonies. The charges will be collected by the power-distribution companies.[59]

Despite these various funding sources for the city's water and sewerage systems, DJB is often plagued with financial difficulties, which seem to have grown worse with time.[60] While in 2010 DJB had a surplus budget of more than 60 crores, it has since developed a deficit of more than 600 crores. In fact, in September 2020, DJB announced a severe financial crunch and sent notices to several central government departments in Delhi, directing them to pay their outstanding dues to the tune of 6,811 crore rupees.[61] Furthermore, according to a report by the comptroller and auditor general of India, DJB owes around 20,000 crores in loan repayments.[62] All of this demonstrates how handicapped DJB is in improving the water and sanitation infrastructure of Delhi. These financial difficulties also, to an extent, explain why DJB has struggled through the years to reduce the supply-demand gap in water infrastructure and maintenance of the sewerage system in the city.

Policy and Planning Interventions

The inability of the government to provide safe and adequate drinking water and sanitation services, along with the enactment of the National Water Policy in 2002, has sparked debates regarding the privatization of the water system in Delhi.[63] Since 2005, the state government of Delhi has repeatedly attempted to fully privatize DJB's water and sanitation services, viewing it as a panacea for water-mafia problems and public discontent with DJB's services.[64] However, these efforts were thwarted by NGOs and welfare associations, which feared that privatization could lead to increased water tariffs, making water less accessible to the poor, and reduced accountability to the public at large.[65] Many groups advocated for a fundamental right to water access. The government succumbed to public opinion, and since then, there have been no further efforts to privatize the water system, though many projects are handled by private companies and agencies. For instance, in 2019, DJB issued a €145 million contract to SUEZ, funded by the central government, to build and operate a wastewater-treatment plant, replacing the existing Okhla wastewater plant, in order to restore the water quality of Yamuna River.[66]

Recently, the Delhi state government has proposed exploring new techniques and state-of-the-art technologies to tackle the failures in the city's water management.[67] One significant move introduced by DJB was the draft Delhi State Water Policy in 2016 (Draft Delhi Policy), which attempted to develop a comprehensive and strategic approach to the water sector in Delhi. The Draft Delhi Policy surveys the existing national water policies and samples the other states' water policies to draw useful inferences for policy formulation in Delhi.[68] The policy calls for the development of a comprehensive information system for water-related data, the use of science and technology to aid in crisis management, the installation of rainwater-harvesting structures for all new constructions, the reuse of urban water effluents, and the development of the least water-intensive sanitization and sewerage system, among other effective and innovative ideas. However, the state government of Delhi has not yet officially enacted this policy.[69]

DJB also released the Delhi Water and Sewer (Tariff and Metering) Regulations, 2012, which provide for the establishment of well-designed rainwater-harvesting structures and impose a penalty for noncompliance with the rainwater-harvesting mandate in buildings.[70]

With respect to drainage management, the last drainage master plan designed for Delhi was adopted in 1976, and since then, the city has been growing exponentially, with political apathy toward the city's flooding, waterlogging, and mindless encroachments of stormwater drains. The new Drainage Master Plan for National Capital Territory (NCT) of Delhi, which was designed in 2018 and included recommendations to stop encroachments and sewage flow into stormwater drains and to establish a single agency for drain management, has seen no further progress, and there have been no signs of the state government implementing this plan.[71]

Other recent developments regarding the city's water and sewerage systems include the Master Plan for Delhi, 2021, the National Capital Region (NCR) Regional Plan, 2021, the Manual of Stormwater Drainage System, and three draft bills of the Central Water Resources Ministry: the National Water Framework Bill, the River Basin Management Bill, and the Model Bill to Regulate and Control the Development and Management of Ground Water.[72] These policies and bills broadly cover the current challenges in the water and drainage management in the city and provide methods to plan, formulate, and implement strategies to improve the systems in the future. As this illustrates, there is no dearth of policy interventions and action plans for water and sewerage management in Delhi. Rather, the main challenge facing Delhi seems to be the effective implementation of these policies and the corresponding augmentation of financial and human resources that is necessary to tackle this humungous and complex system.

It is paramount for a big metropolis like Delhi, which is overflowing with people and activities, to be brought under the control of iron hands to swiftly push forward reforms. Therefore, what could be a possible solution is to call for a central government intervention to integrate and centralize the water and sewerage systems. Water is a state subject as per the Indian Constitution, and it is therefore managed by the state. However, this requires a more integrated, centralized approach, focusing on revamping the existing water- and sewerage-management system, reviving natural stormwater management, and introducing approaches to integrate the infrastructure systems. The city also urgently needs a sustainable urban drainage system.

Some of the existing grey infrastructure for water management could also be replaced and integrated with a green and blue infrastructure water-management system. Some of the major cities in the world and especially those in the United States, such as New York, Portland, and Seattle, have implemented an integrated blue-green-grey approach that provides urban services using natural systems.[73] This could be mimicked in Delhi, provided that enough attention is paid to its unique geographical and administrative complexity.

With respect to stormwater harvesting in Delhi, a recent report by the Centre for Science rightly points out that the stormwater-management system should be viewed as more than capable of dealing with excess rainwater and should be considered an important resource to conserve water. The report proposes that a strategic "Water-Sensitive Urban Designing and Planning" (WSUDP) approach—which would integrate and optimize the available water resources, conserve the local water bodies, install effective wastewater plants, build robust stormwater-harvesting systems, provide water-recycling units, and create awareness of water usage in all households—is the dire need of the hour.[74] The report states that this WSUDP approach has been successful in many parts of the world and is optimistic about its application in Delhi, since the city is one of the greenest metropolises in the Global South, with more than sixteen thousand parks and gardens.[75]

Thus, with an effective and integrated management plan and the required political impetus, it is possible to put the water and sewerage systems of the city back on track. However, only time will tell us if the city is heading toward crisis or will be able to save itself from one.

NOTES

1. The National Capital Territory (NCT) of Delhi is both a city and a Union territory, which functions similar to other states in India. It is jointly administered by the central government, the NCT elected state government, and five municipal corporations. New Delhi is one of nine districts of NCT of Delhi and also the capital of India. Any reference to "Delhi" in this chapter indicates the whole of the city/state, in other words, the NCT of Delhi. Rohilla, Suresh K., Shivali Jainer, Dhruv Pasricha, and Shivani. 2020. "Roadmap for Implementation of Water-Sensitive Urban Design and Planning in Delhi: Stormwater Harvesting in Parks and Open Spaces." Centre for Science and Environment, New Delhi, November 18, www.cseindia.org.
2. Climate change impacts are already visible in Delhi, including erratic rainfall along with a decline in river flow, declining groundwater output, and reduced water for environmental flows. *See generally* Indian Natural Trust for Art and Cultural Heritage. 2016. "Water Policy for Delhi 2016." http://naturalheritage.intach.org.
3. For instance, "the first master plan of Delhi made in 1981 grossly underestimated the targets for water and sewerage capacity as 160 and 142 million gallons per day respectively as opposed to the real need for 496 and 397 million gallons per day." Sheikh, Shahana, Sonal Sharma, and Subhadra Banda. 2015. *The Delhi Jal Board: Seeing beyond the Planned*. New Delhi: Centre for Policy Research, 2, www.cprindia.org.
4. Aijaz, Rumi. 2020. "Water Supply in Delhi: Five Key Issues." ORF Occasional Paper No. 252, Observer Research Foundation, June, www.orfonline.org.
5. Maria, Augustine. 2008. *Urban Water Crisis in Delhi: Stakeholder Reponses and Potential Scenarios of Evolution*. Paris: Institut du Developpement Durable et des Relations Internationales. www.iddri.org.
6. Gautam, Rajneesh Kumar, Islamuddin Islamuddin, Nandkishor More, Saumya Verma, Spriha Pandey, Neha Mumtaz, Rajesh Kumar, and Md. Usama. 2017. "Sewage Generation and Treatment Status for the City of Delhi, Its Past, Present and Future Scenario—A Statistical Analysis." *International Journal for Research in Applied Science & Engineering Technology* 5(v) (May): 926–933. https://doi.org/10.15171/ajehe.2018.02; Rinkesh. 2020. "Top 19 Most Polluted Rivers in the World in 2020." Conserve Energy Future. www.conserve-energy-future.com.
7. The statistics are found in a report of the Central Pollution Control Board, cited in Rohilla, Suresh K., Shivali Jainer, and Mahreen Matto. 2017. *Green Infrastructure—A Practitioner's Guide*. New Delhi: Centre for Science and Environment, http://cdn.cseindia.org.
8. Jacob, Nitya, 2020. "Traditional Water Systems of Delhi." Sahapedia, July 13, www.sahapedia.org; Rohilla et al. 2020.
9. *See* Nitya 2020; Ritter, Kayla. 2019. "Groundwater Plummets in Delhi, City of 29 Million." Circle of Blue. www.circleofblue.org.

10. *See* Tarini, Manchanda. 2013. "Water Movements in Delhi." Ritimo, June 12, www.ritimo.org; *see also* Reidenbach, Matt, and Hana Thurman. 2018. "Past, Present, and Future Water Resources in a Megacity: Delhi, India and the Yamuna River Project." *Uva Darden—Global Water Blog*, November 16, https://blogs.darden.virginia.edu.
11. Government of Rajasthan, Water Resources. 2021. "Inter State Treaties/Agreements." March 29, http://water.rajasthan.gov.in.
12. Rohilla et al. 2020.
13. Sheikh, Shahana, and Subhadra Banda. 2015. "The Intersection of Governments in Delhi." Cities of Delhi, Centre for Policy Research, April 1, www.cprindia.org.
14. Section 9(1)(A) mandates that DJB "treat, supply and distribute water for household consumption or other purposes to those parts of Delhi where there are houses, whether through pipes or by other means." For more information, see the official website of Delhi Jal Board. Delhi Jal Board. n.d. "Welcome to the Delhi Jai Board." Accessed March 1, 2022, http://delhijalboard.nic.in.
15. Sheikh and Banda 2015, 1–2.
16. Rumi 2020.
17. Office of the Registrar General and Census Commissioner, India. n.d. "2011 Census." Accessed December 14, 2020, https://censusindia.gov.in.
18. World Population Review. 2021. "Delhi Population 2021." April 2, https://worldpopulationreview.com.
19. Rumi 2020.
20. Sheikh and Banda 2015, 4.
21. USAID, Safe Water Network. 2016. "Drinking Water Supply for Urban Poor: City of New Delhi." October, www.safewaternetwork.org. An average reverse-osmosis purifier system wastes nearly three liters of water for every one liter of purified water. Rumi 2020.
22. Rumi 2020.
23. Rumi 2020. According to one study, "232 waterbodies out of 629 waterbodies in Delhi have no scope for revival due to excessive encroachments." Rohilla et al. 2017, 13.
24. Rumi 2020.
25. Rohilla et al. 2020.
26. Of the 15,706 units registered, only 1,000 have installed rainwater-harvesting systems. Singh, Paras. 2019. "Losing Ground: Just 1,200 Rainwater Harvesting Units in a City of 2 Crore." *Times of India*, March 23, https://timesofindia.indiatimes.com.
27. Rumi 2020.
28. Rumi 2020.
29. Sheikh and Banda 2015, 1, 6.
30. Rumi 2020.
31. Rumi 2020; Tarini 2013.
32. *TimesNowNews.com*. 2019. "Delhi HC Lambasts Haryana for Blocking Flow of Clean Water into Yamuna: A Look at Rows over Sharing of Water." January 31, www.timesnownews.com.

33. Sureshwar D. Sinha v. Union of India, I.A. No. 43621 of 2018, Writ Petition 537/1992, https://indiankanoon.org; *New Indian Express*. 2018. "Supreme Court Directs Delhi, Haryana to Sort Out Yamuna Water Sharing Dispute." April 23, www.newindianexpress.com; Shubangi. 2021. "War of Waters: Water-Sharing Disputes in Haryana." Bhajan Global Impact Foundation, March 30, http://bhajan foundation.org.
34. Ritter 2019.
35. Rumi 2020.
36. Patel, Shivam. 2020. "Explained: Understanding BIS' Draft Standard for Drinking Water Supply." *Indian Express*, August 23, https://indianexpress.com; *The Hindu*. 2019. "Delhi Has the Most Unsafe Tap Water." November 16, www.thehindu.com.
37. Nitya 2020; Rumi 2020.
38. Re-Centering Delhi Team. 2016. "Mapping Sewage Infrastructure." Yamuna River Project. www.yamunariverproject.org.
39. Rumi 2020.
40. Babu, Nikhil M. 2020. "Is Delhi Losing Battle against Urban Flooding?" *The Hindu*, July 26, www.thehindu.com.
41. Bidhuri, Swati, Mohd Taqi, and Mohd Mazhar Ali Khan. 2018. "Water-Borne Disease: Link between Human Health and Water Use in the Mithepur and Jaitpur Area of the NCT of Delhi." *Journal of Public Health* 26: 119–126. https://doi.org /10.1007/s10389-017-0835-y.
42. There are approximately 250 private water suppliers supplying water through tankers in Delhi. "People buy packaged water not only because they are health conscious but also to stock drinking water in their houses when DJB supply falls short. Industries also purchase packaged water in bulk in order to meet drinking water requirements." Daga, Shivani. 2003. *Private Supply of Water in Delhi*. New Delhi: Centre for Civil Society—Markets and Regulations.
43. Daga 2003.
44. Sethi, Aman, 2015. "At the Mercy of the Water Mafia." *Foreign Policy*, July 17, www. foreignpolicy.com; Majumder, Sanjoy. 2015. "Water Mafia: Why Delhi Is Buying Water on the Black Market." *BBC News*, July 28, www.bbc.com; Nathoo, Leila. 2016. "Breaking Delhi's Water Mafia: How Access to Clean Water Got Political in India." *Independent*, March 21, www.independent.co.uk.
45. Sheikh and Banda 2015, 7.
46. Sethi 2015.
47. Sheikh and Banda 2015, 2.
48. Rumi 2020.
49. Rumi 2020.
50. Rohilla et al. 2017, 20.
51. Ghosh, Shaunak, and Proma Chakraborty. 2019. "Delhi's Ready for Rain, but Its Drainage System Might Not Be." *News Laundry*, July 3, www.newslaundry.com.
52. Rohilla et al. 2020, 9.

53. CPD Delhi. n.d. *Chapter 10: Storm Water Drainage*, 10-1, Accessed December 14, 2020, https://ccs.in; Mishra, Siddhanta. 2020. "Breathing New Life into Yamuna." *New Indian Express*, December 7, www.newindianexpress.com.
54. Babu 2020.
55. CPD Delhi, n.d., 10-2.
56. NDTV. 2020. "Delhi Government Hikes Water, Sanitation Budget by 70%." March 24, www.ndtv.com.
57. DJB, Government of NCT of Delhi. 2018. "Water." March 19, http://delhijalboard.nic.in.
58. Rumi 2020.
59. Saini, Gaurav. 2020. "Now All Households in Delhi Will Have to Pay Sewage Charges." *Mint*, September 27, www.livemint.com.
60. DJB has reported deficits in its annual budgets every year in the past six years and has reported its inability to pay for its loans. For example, in 2019–2020, there was a net deficit of 187 crores. *See* DJB Budget 2019–2020, http://delhijalboard.nic.in.
61. *The Hindu*. 2019. "Increasing Expenses, Stagnant Revenue Leave DJB with Higher Deficit." October 21, www.thehindu.com; *Hindustan Times*. 2020. "7 State Agencies Owe DJB Rs 6,811 Crore in Bills: Raghav Chadha." September 30, www.hindustantimes.com.
62. *Business Standard*. 2019. "Delhi Jal Board Yet to Repay Debt of Over Rs 20,000 Crore: CAG Report." December 2, www.business-standard.com.
63. *The Hindu*. 2016. "Privatisation of Water Supply in Delhi Opposed." October 18, www.thehindu.com.
64. Bhaduri, Amita. 2012. "Privatisation of Water Services in New Delhi: Myth and Reality—Report by Water Privatisation—Commercialization Resistance Committee." India Water Portal, July 9, 4, www.indiawaterportal.org.
65. Singh, Sanghita. 2005. "Water Privatization Protest in Delhi Just Got a Facelift." *DNA India*, October 17, www.dnaindia.com.
66. WaterWorld. 2019. "Suez Wins €145M Contract to Build, Operate India's Largest Wastewater Treatment Plant." July 26, www.waterworld.com.
67. *The Hindu*. 2020. "'Delhi Water Supply Will Soon Be at Par with Capital Cities across the World,' Says Arvind Kejriwal." September 26, www.thehindu.com.
68. *See generally* Indian Natural Trust for Art and Cultural Heritage 2016.
69. The Water Policy for Delhi is still available in draft form on the official website of the DJB.
70. Regulation 50 of the Delhi Water and Sewer (Tariff and Metering) Regulations. 2012. http://delhijalboard.nic.in.
71. Babu 2020.
72. Delhi Development Authority. 2010. *Master Plan for Delhi—2021*. New Delhi: VIBA. www.dda.org.in; Ministry of Housing and Urban Affairs, New Delhi. 2021. *Draft Manual on Stormwater Drainage Systems*. New Delhi: Government of India. http://mohua.gov.in.

73. Anand, Ashwathy, and Sahana Goswami. 2020. "Living with Water: Integrating Blue, Green and Grey Infrastructure to Manage Urban Floods." WRI India, September 3, https://wri-india.org.
74. Rohilla et al. 2020, 7.
75. Rohilla et al. 2020, 32.

6

New York City's Water

KATRINA MIRIAM WYMAN

New York City's drinking water comes from surface-water sources located outside the city. Although the city is surrounded by water, that water is not suitable for drinking because it is salty ocean water and brackish river water. The city built its system for importing drinking water, with legal assistance from New York State, starting in the 1830s and continuing into the twentieth century. The city continues to maintain the system, but it has been subject to considerable state and federal regulatory oversight since the latter twentieth century.

The city also started building its system for managing wastewater in the mid-nineteenth century. As in other US cities, the introduction of a water-supply system in the nineteenth century facilitated the development of indoor water closets, which increased the need for a systematic approach to removing wastewater.[1] Like many larger older US cities, New York City built a combined wastewater system in the nineteenth century that collects domestic wastewater and excess rainwater into a single system.[2] The city built portions of the oldest of its fourteen wastewater-treatment plants in 1903; the system was significantly enlarged in the 1930s through the 1950s, but the city continued to dump untreated sewage into the Hudson River until 1986.[3] The federal Environmental Protection Agency (EPA), acting under the federal Clean Water Act passed in the 1970s, finally forced the city to treat sewage before discharging it into the Hudson River.[4] However, even today, when it rains heavily, the capacity of the city's sewage system is exceeded, and the city dumps combined overflows of untreated sewage and rainwater into the waters around the city.[5] Partly to comply with the requirements of the federal Clean Water Act, the city has been working for over a decade to install "green infrastructure" on public and private property to absorb rainwater and thus to reduce combined sewage overflow events.[6]

The city's interest in green infrastructure as a means of dealing with stormwater—including the cloudbursts likely to become more frequent as the climate continues to warm—resembles that of other cities discussed in this volume, such as Berlin.[7]

This chapter begins with the management of New York City's water supply and then discusses the management of wastewater and the funding of the water system. It concludes with some observations about the potential of cities as environmental actors, drawing on the lessons from New York City's management of its water supply and wastewater systems.

Water-Supply System

Until the early part of the nineteenth century, drinking water in New York City (which was then Manhattan) often came from local wells, streams, and ponds, including a famous pond known as the Collect in lower Manhattan.[8] As these sources became increasingly polluted, wealthier people began importing water from outside the city and deeper wells within the city.[9] Over time, pressure increased to find new sources of supply. Water was needed to fight fires, and epidemics were attributed to polluted water.[10] A major issue was whether private actors or the government would build the drinking-water supply system that people increasingly recognized would be necessary. After the private sector proved incapable, the city turned to government.[11] In many US cities in the nineteenth century, developing the water-supply system often was "the first major undertaking of [the] city government" and the first infrastructure project requiring significant public financing.[12]

New York City's drinking water now comes from surface waters in three parts of New York State outside the city's borders: the Croton watershed, located in counties just north of the city; and the Catskills and Delaware areas located farther from the city.[13] The Croton River was dammed, and the Croton reservoir and aqueducts were built to supply the city between 1837 and 1842.[14] The opening of the Croton reservoir in 1842 was a major event in the city.[15] To build the Croton system, the city obtained legislation from New York State creating a Board of Water Commissioners to oversee construction and the financing of that construction, and legislation authorizing the board to expropriate land outside the city for the city's benefit.[16] However, the city paid for the construction of the water-supply system.[17] The original Croton system was expanded and replaced in the ensuing decades.[18] In 1905, the city was authorized by the state to build a system to bring water from the Catskills region.[19] Eventually, the city was also authorized to bring water from the Delaware River.[20] The city's desire to source water from the Delaware led to litigation in the Supreme Court that limited how much water New York City could take from the river, as the lower riparian states that bordered the Delaware sought to protect their access to the water.[21] In 1965, roughly a decade before the city experienced a major fiscal crisis, it finished construction of the newest reservoir in the system (Cannonsville, in Delaware County).[22] Notably, the city provides water to roughly one million people in counties from which the city draws water; it also provides sewage services in some upstate areas from which it draws water.[23]

Most of the land surrounding the source waters for the Croton, Catskills, and Delaware systems is privately owned; only 6.4 percent of the land is city owned.[24] Although New York State law authorized the city to regulate land uses in the watershed area to protect its water supply, the city did little throughout most of the twentieth century to actually regulate these uses, establishing only some anemic regulations.[25] New York State also failed to protect the areas bordering the source waters, and there was considerable development in parts of the watershed, especially in the Croton area, which has suburbanized; the development removed the "forests and wetlands that [had] served as natural filters."[26] According to a 1991 report, approximately one-third of the city's water supply was of "borderline" quality.[27]

Until the federal Safe Drinking Water Act was passed in 1974, the federal government never had a role in directly regulating the quality of New York City's drinking water.[28] The act required EPA to establish maximum contaminant levels that public water utilities such as New York's must satisfy. In 1986, it was amended to require the filtration of surface-water systems in certain situations.[29] As a result of these amendments, EPA forced New York City to build a filtration plant, which opened in 2015, to filter the water coming from the Croton part of the system, which is about 10 percent of the city's water.[30]

Ninety percent of the city's water remains unfiltered thanks to an agreement that the city, New York State, EPA, communities in the Catskills and the Delaware regions, and environmental NGOs negotiated to protect these parts of the watershed area in 1997.[31] Under this watershed agreement, the city agreed to pay for land acquisitions and upgrade sewage treatment in the Catskills/Delaware region; it has spent over $1.7 billion since then to protect the "unfiltered water supply" and receive federal and state waivers of "filtration requirements."[32] If not for this agreement, the city would have been required by EPA to build a filtration plant that would have cost approximately $6 to $8 billion dollars in order to filter the water that it receives from the Catskills/Delaware regions.[33] The desire to avoid building the plant incentivized the city to negotiate with the communities in the Delaware and Catskills regions to institute a plan that was "protective enough" of the interests of those communities, "acceptable to EPA," and "affordable" for the city.[34] Historical "tensions" remain between the city and the communities in the watershed areas, especially in the rural Delaware and the Catskills regions, which are economically depressed and where development is regulated to protect the source waters for the city's water supply.[35]

In addition to requiring the city to build a filtration plant for water coming from the Croton area and driving the city to better protect the source waters in the Delaware and the Catskills regions, EPA has also used its authority under the Safe Drinking Water Act to force the city to take other steps to protect drinking-water quality. The New York State Department of Health now oversees New York City's compliance with the federal Safe Drinking Water Act.[36] Nonetheless, EPA remains involved in overseeing the quality of the city's drinking water. For example, in 2019, after EPA sued the city under the Safe Drinking Water Act, EPA and the city agreed to a consent decree that will require the city to invest almost $3 billion to cover the Hillsview Reservoir, which holds water from the Catskills and Delaware region before it enters the city, and make other improvements.[37]

Wastewater

After New York City began building a system to bring water into its borders in the 1830s, it started building sewers to remove wastewater. As in other cities, the introduction of a water-supply system was accompanied by an increase in demand for water, due in part to the introduction of indoor water closets.[38] Starting in the mid-1800s, New York City built sewers. Initially, in Manhattan, they "were built piecemeal under ward leaders who retained control," but by 1855, sewers covered a good portion of Manhattan.[39] By contrast, sewers seem to have been built in a more coordinated manner in Brooklyn; in 1857, the state legislature created the Brooklyn Board of Sewer Commissioners for the purpose of building a sewer system in Brooklyn.[40]

Many parts of New York City have a combined sewage system that collects both used domestic water and rainwater into a single pipe system.[41] For decades, the city dumped raw, untreated sewage into the waters that surround it.[42] Although the first sewage-treatment plant dates to 1903, it was not until the 1930s that the city built four more, when the federal government provided financial assistance.[43] Seven more plants were opened between 1944 and 1967, and the city's thirteenth and fourteenth were built in 1986 and 1987.[44]

The city continued to dump untreated sewage into the Hudson River until the opening in 1986 of the North River Treatment Plant, which was mostly paid for by federal funding.[45] EPA's implementation of the Clean Water Act in the late 1970s and early 1980s finally forced the city to build the North River plant to treat sewage deposited into the Hudson.[46] In 1979, EPA obtained a court order that required the city to build a plant to treat sewage going into the Hudson by 1986 and to update that plant's secondary treatment by 1991.[47] As mentioned earlier, however, the city still dumps untreated sewage into nearby waterways, including the Hudson River, when the sewage system overflows due to heavy rains.[48] The need for EPA to step in to force the city to build a treatment plant and stop dumping untreated sewage into the Hudson suggests that local governments can be particularly parochial and shortsighted and that a higher-level check can be useful because it reduces the potential for cities, concerned mainly about their residents, to externalize their pollution to others.[49]

The siting of that sewage-treatment plant also suggests another limitation of local governments as environmental actors: the potential for local governments to burden communities within their borders who lack political power with environmental disamenities such as sewage-treatment plants. The city built the Hudson River sewage-treatment plant required by EPA near an African American and Latino neighborhood in Harlem.[50] The location, which had been years in the making, was opposed by Black community leaders, but to no avail.[51] Although the decision to site the plant was made at the local level, EPA facilitated the siting of the plant. In 1979, it "issued a Finding Of No Significant Impact (F.O.N.S.I.) for the North River facility" under the National Environmental Policy Act, and that finding meant that the city never had to undertake an environmental-impact statement analyzing the plant's expected impacts on air quality in the surrounding areas.[52] These impacts proved significant. Opposition to the plant contributed to the formation in 1988 of a well-known environmental justice group in New York City, West Harlem Environmental Action (WE ACT), which, in a momentous environmental justice lawsuit, sued the city for nuisance for the odors coming from the plant.[53] The city eventually settled the suit on December 30, 1997, just before Mayor David Dinkins left office and was replaced by Rudolph Giuliani.[54] Mayor Dinkins, the city's first Black mayor, had opposed the plant before he became mayor; indeed, as Manhattan Borough president, he had commissioned a study that demonstrated that air pollution was coming from the plant.[55] The siting of the North River Treatment Plant provides evidence that local governments may be more responsive to some local interests than others and that this selective responsiveness can harm communities of color and low-income communities. EPA's role suggests that higher levels of governments may also fail to step in to redress inequities at the local level. In the same period when the North River Treatment Plant was built, other major infrastructure was built in New York in response to federal mandates; the siting of some of that infrastructure also contributed to environmental injustice.[56]

While the first generation of local responses to Clean Water Act requirements was the construction of wastewater-treatment plants such as the North River plant, often with federal funding, the Clean Water Act has more recently contributed to local investments in green infrastructure. Congressional amendments to the Clean Water Act in 1987 required EPA to begin regulating stormwater discharges under permitting provisions under the act. However, federal funding was lacking to assist local governments in complying with the new standards.[57] The need to meet federal standards for stormwater management without sufficient federal funding has helped to spur New York City (and other cities) to try to introduce "green infrastructure" as a substitute for, and complement to, "grey infrastructure." Grey infrastructure is the backbone of traditional sewer systems: "pipes and treatment facilities."[58] Green infrastructure keeps stormwater from going into the sewage system by absorbing the water; examples "include green roofs, trees and tree boxes, rain gardens, vegetated swales, [and] pocket wetlands."[59] Like other cities in the United States and elsewhere, green infrastructure has appealed to New York partly on the basis that it may be a cheaper way to deal with stormwater than grey infrastructure.[60] Green infrastructure also has other benefits, including creating more pleasing environments, providing habitat for nonhumans, and addressing the urban heat-island effect.[61]

In 2012, New York City's Department of Environmental Protection and the New York State Department of Environmental Conservation, which administers the federal Clean Water Act requirements in the state, agreed that the city would satisfy stormwater-management obligations through a "combination of grey and green infrastructure."[62] Unfortunately, as of 2016, the city was not meeting the targets for installing green infrastructure included in the agreement.[63] This lack of compliance may signal that green infrastructure is harder to implement than local actors thought in the early 2010s and that additional innovation is needed to scale up the use of such infrastructure. One challenge is installing green infrastructure on private property; while the city has been able to implement green infrastructure on public property, doing so is more complicated on private property, which covers over 50 percent of the land where the city wants to install green infrastructure.[64] The city needs to find ways of incentivizing private landowners to install green infrastructure for the social benefits—most notably, but not exclusively, stormwater management—that it will yield, especially in an era of climate change when extreme rainfalls will increasingly put pressure on the city's drainage system.[65] As advocates have observed, the quest to build green infrastructure to assist the city in complying with the Clean Water Act parallels the negotiation of the watershed agreement to avoid building a filtration plant for much of the city's water: both involve using cheaper, greener approaches than grey infrastructure to bring the city into compliance with federal environmental standards.[66]

Governance and Funding

New York City owns its water-supply and sewage systems.[67] The city finances its water and sewage systems through water rates paid by users. The New York Water Board, whose members are appointed by the mayor, determines the level of those water rates; by statute, they must be set to ensure that the water system is "self-sustaining."[68] The rates for "most small properties" are volumetric and based on the amount of water consumed, although some property owners pay a flat charge in exchange for installing water-conservation measures.[69] The city also has programs to credit the water bills of customers who might have difficulty paying water costs.[70] The charge for sewage is based on the charge for water supplied—159 percent of the charge for water supplied.[71] There is no separate charge for the stormwater that the city must manage due to impervious surfaces; however, advocates have proposed that the city should levy a distinct stormwater charge based on square footage, similar to the stormwater fees and charges imposed by many other local governments in the United States.[72]

The New York City Municipal Water Finance Authority issues bonds to fund capital improvements to the water and sewerage systems, and the city also receives some funding from the state and federal governments for capital projects.[73] The city's Department of Environmental Protection, a mayoral agency, is the bureaucratic entity that operates the water-supply and sewage system.[74]

Implications

The history and management of New York City's water-supply and wastewater systems illustrate some of the promise and perils of local government action to protect the environment.

The local initiative to develop the water-supply system starting in the nineteenth century and the local funding for that system illustrate a local willingness to act when the city stands to benefit from environmental protection. Without a safe and reliable source of drinking water, New York City would not have been able to develop into the large metropolis that it is today. City leaders had the incentive and the resources to secure that water. In light of COVID-19, it is notable that actual experience with epidemics, such as yellow fever and cholera, and fears of such epidemics contributed to motivating US cities such as Philadelphia and New York to build water-supply systems in the nineteenth century.

However, aspects of the history also illustrate some of the limitations of local governments as environmental protectors. Recall the city's need for state authorization to build the water-supply system, the city's decades-long failure to adequately protect the watershed areas from which it draws its water, and the city's dumping of sewage into neighboring waters. Because cities lack the legal authority to regulate outside their borders, they cannot take steps to obtain water or protect resources outside their borders without authorization from higher levels of government with broader geographic authority or the agreement of the affected local areas. Moreover, local governments may lack the funding and resources to adequately regulate, even when they have the legal authority. They may take the easy way out and export pollution problems to other local areas, as New York City did by dumping untreated sewage into surrounding waters.

The interventions of higher-level authorities, such as EPA and New York State authorities in New York City's case, also may benefit the city. Forcing New York City to treat sewage before dumping wastewater into its surrounding waters, such as the Hudson River and New York Harbor, eventually helped to clean up these waters. As a result of the federal Clean Water Act, waters such as the Hudson River and New York Harbor are much cleaner than they have been for a long time.[75] That has increased the attractiveness of the waterfront and facilitated the redevelopment of waterfront areas, including the creation of major new waterfront parks in New York City since the late 1990s.[76] The New York City experience suggests the advantages of multiple levels of government taking an interest in environmental protection and not relying on local governments alone. Federal and state intervention will probably continue to be necessary to help New York City manage the new water-related challenges of climate-change-related flooding and sea-level rise, as well as the perennial problems of supplying safe drinking water and managing wastewater.[77] The city's aging water-supply infrastructure will exacerbate these challenges and probably require the city to increase its investment in repairing and replacing portions of the system.[78]

NOTES

1. Tarr, Joel A. 1996. *The Search for the Ultimate Sink: Urban Pollution in Historical Perspective*. Akron, OH: University of Akron Press, 133–134.
2. *See* Goldman, Joanne A. 1997. *Building New York's Sewers: Developing Mechanisms of Urban Management*. West Lafayette, IN: Purdue University Press, 90; Tarr 1996, 136. Goldman also emphasizes that sewers were initially constructed in a "piece-meal" manner in New York City. Goldman 1997, 51.
3. Cromwell, Bob. n.d. "New York City's Wastewater Treatment System." *Toilets of the World*. Accessed September 3, 2021, https://toilet-guru.com. During the 1930s, the federal government provided financial support for building sewage-treatment plants. Andreen, William L. 2003. "The Evolution of Water Pollution Control in the United States—State, Local, and Federal Efforts, 1789–1972, Part II." *Stanford Environmental Law Journal* 22: 226, https://ssrn.com/abstract=554122. Miller, Vernice D. 1994. "Planning, Power and Politics: A Case Study of the Land Use and Siting History of the North River Water Pollution Control Plant." *Fordham Urban Law Journal* 21: 712n35.
4. Miller 1994, 712.
5. NYC Department of Environmental Protection. n.d. "Combined Sewer Overflows." City of New York. Accessed September 3, 2021, www1.nyc.gov.
6. NYC Department of Environmental Protection. 2021. "NYC Green Infrastructure: 2020 Annual Report." City of New York, April 30, www1.nyc.gov.
7. Crownhart, Casey. 2021. "How Ida Dodged NYC's Flood Defenses." *MIT Technology Review*, September 3, www.technologyreview.com.
8. Goldman 1997, 10. On the history of New York City, see Lankevitch, George. 2020. "New York City." *Encyclopedia Britannica*, October 23, www.britannica.com. *See also* Taylor, Dorceta E. 2009. *The Environment and the People in American Cities, 1600s–1900s: Disorder, Inequality, and Social Change*. Durham, NC: Duke University Press, 55–56.
9. Salzman, James. 2017. *Drinking Water: A History*. New York: Overlook Duckworth, 59.
10. Taylor 2009, 190; Goldman 1997, 55–57.
11. Salzman 2017, 65–68; Blake, Nelson M. 1956. *Water for the Cities: A History of the Urban Water Supply Problem in the United States*. Syracuse, NY: Syracuse University Press, 115, 120, 138–142.
12. Melosi, Martin V. 2008. *The Sanitary City: Environmental Services in Urban America from Colonial Times to the Present*. Pittsburgh: University of Pittsburgh Press, 21.
13. Watershed Agricultural Council. n.d. "Croton and Catskill/Delaware Watersheds." Accessed September 3, 2021, www.nycwatershed.org.
14. Seneca Village. n.d. "The Croton Aqueduct." Accessed September 3, 2021, www.nyhistory.org.
15. Blake 1956, 164–166; Goldman 1997, 67–68.

16. Blake 1956, 135–142.
17. Blake 1956, 149–151; Goldman 1997, 69, 71; Taylor 2009, 261–262.
18. Goldman 1997, 163–164.
19. Finnegan, Michael C. 1997. "New York City's Watershed Agreement: A Lesson in Sharing Responsibility." *Pace Environmental Law Review* 14: 598–599.
20. Finnegan 1997, 606.
21. Finnegan 1997, 606n183. New Jersey v. New York, 283 U.S. 805 (1931). New Jersey v. New York, 347 U.S. 995 (1954). In 1961, the Delaware River Basin Commission was established pursuant to an interstate compact among Delaware, New Jersey, New York, Pennsylvania, and the federal government. Friedman, Gayle. 2017. "Delaware River Basin Commission." *The Encyclopedia of Greater Philadelphia*, https://philadelphiaencyclopedia.org.
22. Finnegan 1997, 609.
23. New York City Water and Sewer System. 2020. *New York City Water and Sewer System: Comprehensive Annual Financial Report for the Fiscal Year Ended June 30, 2020*, 6, www1.nyc.gov.
24. Finnegan 1997, 578.
25. Finnegan 1997, 609.
26. Finnegan 1997, 612.
27. Finnegan 1997, 613.
28. US Environmental Protection Agency (EPA). 2000. *The History of Drinking Water Treatment*, 2, https://archive.epa.gov. However, the passage of the act was not the first time that the federal government had taken an interest in water quality. In 1914, the federal Public Health Service established Drinking Water Standards for bacteria for interstate water carriers; state and local water systems were not required to comply with these standards, although many states adopted the standards, which were updated in 1925, 1946, and 1962. However, a 1969 survey found that only 60 percent of water-supply systems "met all the Public Health Service standards." This survey probably helped to lay the groundwork for the passage of the Safe Drinking Water Act. EPA 2000, 2–3.
29. Finnegan 1997, 617.
30. Nessen, Stephen. 2015. "Nearly 30 Years and $3.5 Billion Later, NYC Gets Its First Filtration Plant." *WNYC News*, June 17, www.wnyc.org; Finnegan 1997, 583.
31. Finnegan 1997, 580, 625; New York State Environmental Facilities Corporation. 1997. "New York City Watershed Memorandum of Agreement." January 21, https://dos.ny.gov.
32. Finnegan 1997, 626–629; Hu, Winnie. 2018. "A Billion Dollar Investment in New York's Water." *New York Times*, January 17, www.nytimes.com.
33. Finnegan 1997, 618.
34. Finnegan 1997, 623. *See also* Hu 2018.
35. *See* Church, Jennifer. 2009. "Avoiding Further Conflict: A Case Study of New York City Watershed Land Acquisition Program in Delaware County, NY." *Pace Environmental Law Review* 27: 393–410. New York City has also financed construction

jobs, loans for development, and reimbursement for property owners in watershed areas. *See* Hu 2018.

36. Finnegan 1997, 609; Office of the New York State Comptroller. 2017. *Federal and New York State Regulation of Drinking Water Contaminants*. www.osc.state.ny.us.
37. EPA. 2019. "City of New York to Comply with the Federal Safe Drinking Water Act and Prevent Contamination of the City's Drinking Water Supply." March 18, https://archive.epa.gov. The city had been resisting covering the Hillsview Reservoir for many years. Gelinas, Nicole. 2019. "Why Trump Should Call Off the EPA's Latest Assault on NYC." *New York Post*, March 25, www.nypost.com.
38. Tarr 1996, 132–133. The wealthy acquired indoor plumbing first. Taylor 2009, 200.
39. Cromwell, n.d.
40. Goldman 1997, 109.
41. Spiegel-Feld, Danielle, and Lauren Sherman. 2018. "Expanding Green Roofs in New York City: Towards a Location-Specific Tax Incentive." *NYU Environmental Law Journal* 26: 108. In other areas, the city has separate systems for domestic wastewater and stormwater. Spiegel-Feld and Sherman 2018, 111.
42. *See, e.g.*, Goldman 1997, 163; Miller 1994, 712.
43. Cromwell, n.d.; Andreen 2003, 226. The federal government provided general support in the form of loans and grants to states and local governments. Andreen 2003, 226.
44. Cromwell, n.d.
45. Miller 1994, 712–713. The federal government paid 75 percent of the $1.1 billion cost of building the plant. Miller 1994, 713.
46. Miller 1994, 712. New York City currently has fourteen sewage-treatment plants. Cromwell, n.d.
47. Miller 1994, 712.
48. Kensinger, Nathan. 2020. "NYC Has a Plan to Clean Its Sewage-Filled Waterways. Does It Go Far Enough?" *Curbed*, February 20, https://ny.curbed.com.
49. *See* New York v. New Jersey, 256 U.S. 296 (1921) (New York State tried to sue New Jersey for sewer discharges).
50. West Harlem Environmental Action v. New York City Department of Environmental Protection, Complaint, June 22, 1992, at para. 23.
51. Miller 1994, 710–711; Checker, Melissa. 2020. *The Sustainability Myth: Environmental Gentrification and the Politics of Justice*. New York: New York University Press, 59.
52. Miller 1994, 713.
53. Checker 2020, 70–72; Miller 1994, 720.
54. Miller 1994, 721.
55. West Harlem Environmental Action v. New York City Department of Environmental Protection, Complaint, June 22, 1992, at paras. 27–28.
56. Sze, Julie. 2007. *Noxious New York: The Racial Politics of Urban Health and Environmental Justice*. Cambridge, MA: MIT Press, 92–93.

57. Holloway, Caswell F., Carter H. Strickland, Michael B. Gerrard, and Daniel M. Firger. 2014. "Solving the CSO Conundrum: Green Infrastructure and the Unfulfilled Promise of Federal-Municipal Cooperation." *Harvard Environmental Law Review* 38: 356. *See also* Flowers, Catherine, and Mitchell Bernard. 2021. "When Environmental Racism Causes Hygienic Hell." *New York Times*, August 25, www.nytimes.com.
58. Holloway et al. 2014, 359–360.
59. Grumbles, Benjamin. 2007. "Using Green Infrastructure to Protect Water Quality in Stormwater, CSO, Nonpoint Source and Other Water Programs." Memorandum, EPA Asst. Admin. for Water, to EPA Reg'l Adminis., 1, www.epa.gov. *See also* Holloway et al. 2014, 360; NYC Department of Environmental Protection. n.d. "Green Infrastructure." City of New York. Accessed September 4, 2021, www1.nyc.gov.
60. Holloway et al. 2014, 361; Strickland, Carter H. 2012. "Case Study: The NYC Green Infrastructure Plan." C40 Cities, April 12, www.c40.org.
61. *See, e.g.*, Spiegel-Feld and Sherman 2018, 113–119; Holloway et al. 2014, 361.
62. Holloway et al. 2014, 365.
63. NYC Department of Environmental Protection. 2016. *Green Infrastructure Contingency Plan*, 1, www1.nyc.gov.
64. Valderrama, Alisa, John Lochner, and Marianna Koval. 2017. *Catalyzing Green Infrastructure on Private Property: Recommendations for a Green, Equitable and Sustainable New York City*, 8, www.nrdc.org ("More than 50 percent of the land area that the City has targeted for GI projects is in private hands.").
65. Crownhart 2021.
66. Valderrama et al. 2017, 6 (drawing parallels to the Watershed Protection Program).
67. Following the state's enactment of the New York City Municipal Water Finance Authority Act of 1984 (Public Authorities Law § 1045), the New York Water Board has authority for rate setting, and the New York City Municipal Water Finance Authority is responsible for financing capital improvements. Giuliani v. Hevesi, 681 N.E.2d 326, 328–329 (N.Y. 1997); Prometheus Realty v. New York City Water, 92 N.E.3d 778, 780, 782 (N.Y. 2017).
68. Prometheus Realty v. New York City Water, 92 N.E.3d 778, 782 (N.Y. 2017) (citing Public Authorities Law § 1045-g [4]). "The Water Board leases the infrastructure of the water supply and wastewater systems from the city," and the city has the option under the lease to require the board to pay the city rent. Prometheus Realty, 92 N.E.3d at 780. In recent years, the city has opted not to require the board to pay rent to the city. New York City Water and Sewer System 2020, 5; New York City Independent Budget Office. 2016. "Mayor Plans to Sink Rental Payment for Water System: But Only One- to Three-Family Homeowners Will See Savings This Year," *Focus On: The Executive Budget*, May, 1, https://ibo.nyc.ny.us.
69. NYC Department of Environmental Protection. n.d. "Billing FAQs." City of New York. Accessed September 4, 2021, www1.nyc.gov; NYC Department of

Environmental Protection. n.d. "Multi-Family Conservation Program." City of New York. Accessed September 4, 2021, www1.nyc.gov. The city requires that property owners of "all newly constructed or renovated properties" install water meters "at their own expense" and protect and replace those meters. For "most small properties," "DEP is responsible for" initially installing, replacing, and repairing water meters. NYC Department of Environmental Protection. n.d. "Water Meter FAQ." City of New York. Accessed September 5, 2021, www1.nyc.gov.

70. NYC Department of Environmental Protection. n.d. "Home Water Assistance Program." City of New York. Accessed September 5, 2021, www1.nyc.gov (program for low-income homeowners). NYC Department of Environmental Protection. n.d. "Multi-Family Water Assistance Program." City of New York. Accessed September 5, 2021, www1.nyc.gov (program for affordable multifamily housing projects).
71. Valderrama et al. 2017, 20. New York City Water Board. 2021. *Water and Wastewater Rate Schedule: Effective July 1, 2021*, 12, www1.nyc.gov.
72. Valderrama et al. 2017, 19. The city does charge certain parking lots a fee that might be considered a stormwater fee. Valderrama et al. 2017, 20 n.35.
73. New York City Water and Sewer System 2020, 5, 42.
74. New York City Water and Sewer System 2020, 6.
75. Miller 1994, 722; Riverkeeper. 2019. "How's the Water? New York City Water Qualify Factsheet." *Water Quality Reports*. Accessed September 5, 2021, www.riverkeeper.org. *See also* Ocampo, Josh. 2018. "The Hudson River Isn't as Gross as You Think: What a Huge Oyster Says about the Quality of NYC Water." *Mic*, August 14, www.mic.com.
76. NYC Department of City Planning. 2011. "Goal I: Expand Public Access." In *Vision 2020: New York City Comprehensive Waterfront Plan*, 22–33. Major new waterfront projects included the Hudson River Park, Brooklyn Bridge Park, Governors Island, Freshkills Park in Staten Island, Harlem River Park Greenway, and East River Esplanade South, among others. NYC Department of City Planning 2011, 24. For a critical analysis of the impact of the amenities such as parks on low-income communities and people of color, see Checker 2020, 49–83.
77. *See* Minelli, chapter 17 in this volume. *See also* Orton, Philip, Ning Lin, Vivien Gornitz, Brian Colle, James Booth, Kairui Feng, Maya Buchanan, Michael Oppenheimer, and Lesley Patrick. 2019. "New York City Panel on Climate Change 2019 Report, Chapter 4: Coastal Flooding." *Annals of the New York Academy of the Sciences* 1439 (March): 95–114.
78. *See* Center for an Urban Future. 2019. *Caution Ahead: Five Years Later—Assessing Progress and Challenges for New York City's Aging Infrastructure*, 4, 12–14, https://nycfuture.org. *See also* Office of the New York State Comptroller. 2017. *Drinking Water Systems in New York: The Challenges of Aging Infrastructure*, 6, 19, www.osc.state.ny.us.

PART III

Reducing Local Air Pollution

7

Clearing Delhi's Air

Hits and Misses in the Past Three Decades

TANUSHREE GANGULY

Delhi, India's capital city and home to over twenty million people, is ranked among the world's most polluted cities.[1] Throughout the past decade, the city has reported an annual-average concentration of particulate matter ($PM_{2.5}$) greater than one hundred micrograms per cubic meter ($\mu g/m^3$)—ten times the safe limit stipulated by the World Health Organization (WHO).[2] Following severe smog episodes in the winter of 2018 and 2019, the state government was compelled to declare a state of public health emergency in the city.

Delhi, officially named the National Capital Territory (NCT) of Delhi, is both a state and a Union Territory, which is a subdivision of the country that is governed by the central Union government of India. It differs from other Union Territories in the country, however, because it also has a state legislature that is independent of the central government. In other words, Delhi is a diarchy, where both the central and state governments share administrative responsibilities in the region. It is further divided into three municipalities, the New Delhi Municipal Council, and the Delhi Cantonment Area. The Union Ministry of Home Affairs (MHA) is responsible for managing law and order in the city, and the state government, headed by the chief minister and the Council of Ministers, is responsible for managing the state's finances and matters regarding health, education, transport infrastructure, and the environment. The municipalities are responsible for civic issues.

Delhi's unique system of governance is evident in the fact that it was the chief minister of Delhi, who is elected by the State Assembly and appointed by the president of India, who represented the city at the C40 World Mayors Summit held in Copenhagen in 2019. Delhi's chief minister, Arvind Kejriwal, signed the C40 Clean Air Cities Declaration in October 2019. By signing the declaration, he committed Delhi to, among other things, setting "ambitious pollution reduction targets within two years that meet or exceed national commitments, putting them on a path towards meeting World Health Organization guidelines," and "implementing substantive clean air policies by 2025 that address the unique causes of pollution in their cities."[3]

The chief minister has attributed Delhi's pollution woes to emissions from sources outside the city limits, misaligned government structures, and multiple governments' presence in the city. Other experts have raised similar critiques as well.[4] This chapter delves into these reasons and more, telling the story of Delhi's struggle with air pollution and the lessons learned from the clearing of skies during the pandemic lockdown.

What Is Polluting Delhi's Air?

Delhi's air quality problems are "seasonal and complex."[5] And while multiple efforts have been made to evaluate the contribution of emissions from different sources, there is significant variation in the results of these assessments.[6] The estimated contribution of $PM_{2.5}$ emissions from the transport sector, for example, ranges from 17.9 percent to 39.2 percent across different assessments. Similarly, the contribution from industry has been estimated to be as low as 2.5 percent and as high as 28 percent.[7] Despite these inconsistencies, it is clear that the major sources of emissions impacting Delhi's air quality include transport, industries, biomass burning (crop waste and solid fuel for cooking and heating), open waste burning, construction, and road dust. The erstwhile Badarpur thermal power plant, which is now completely shut down, was estimated to contribute 11 percent of Delhi's $PM_{2.5}$ concentration while still in operation.[8]

While emissions from seasonal or occasional sources like crop burning and bursting of firecrackers during Diwali result in elevated pollution levels during particular times of the year, emissions from vehicles, road resuspended dust, power plants, and brick kilns contribute to Delhi's particulate pollution year-round. For example, emissions from stubble burning in the neighboring states of Punjab, Haryana, and Uttar

Pradesh result in elevated $PM_{2.5}$ concentration in Delhi in the months of October and November.[9] The bursting of firecrackers during the Indian festival of Diwali, generally celebrated in October, also contributes to the substantial increase in particulate concentration around that time.[10] And all of Delhi is exposed to hazardous levels of $PM_{2.5}$ concentrations during the winter months.[11] By contrast, power plants and brick kilns contribute to Delhi's particulate pollution throughout the year.[12]

There is a lack of consensus regarding the relative contribution of sources outside of Delhi to the city's pollution problem. However, it is clear that the contribution is significant, with estimates ranging from 20 percent to 50 percent.[13] In general, the contribution of Delhi's own pollution to $PM_{2.5}$ concentrations has been estimated at less than 40 percent, though this amount varies across the city.[14] Notably, there is considerable spatial variation in $PM_{2.5}$ concentration in Delhi due to differences in the population size, the pattern of land use, the vehicular and traffic density, and the presence of small-scale industries.[15] Mapping of spatial variations is therefore critical for identifying vulnerable populations within the city.

Whatever the precise source of the pollution, an assessment of the annual $PM_{2.5}$ concentration trend for Delhi reveals that the monthly averages rarely touch 40 $\mu g/m^3$, the national standard for annual-average $PM_{2.5}$ concentration, though there are notably higher concentrations during certain months that significantly raise the city's annual average.

Who Manages Delhi's Air? Regulation versus Implementation

There are several governmental bodies at the state and central levels that are responsible for managing Delhi's air quality. At the highest level, the Union Ministry of Environment, Forest and Climate Change (MoEF&CC) is the nodal agency for planning, promoting, coordinating, and overseeing the implementation of India's environmental policies and programs across different states. The Central Pollution Control Board (CPCB), also a Union entity, provides technical services to the Union Ministry. Industry standards and motor-vehicle standards are set at the federal level by the MoEF&CC and Union Ministry of Road Transport and Highways (MoRTH), respectively.

At the state level, the State Pollution Control Boards (SPCBs) are responsible for improving environmental quality in India's states and advising the state governments on any matter concerning the prevention, control, and abatement of air pollution. Among other things, the SPCBs lay down industrial and automobile emission standards in consultation with the Central Board. The federal CPCB has also delegated authority to pollution control committees in Union Territories like Delhi. There, it is called the Delhi Pollution Control Committee (DPCC), and it implements all of the environmental pollution control laws for the government of Delhi. The functions of the DPCC as specified under the Air (Prevention and Control of Pollution) Act of 1981 include,

1. advising the Delhi Government on matters related to air pollution;
2. collecting and publishing data on the state of air quality in Delhi;
3. organizing awareness campaigns to disseminate information on air pollution-related issues;
4. setting standards for emissions from automobiles and industries; and
5. developing economically viable methods of air pollution control.[16]

In addition to the DPCC, the Department of Environment, Government of NCT of Delhi, is responsible for improving Delhi's air quality. This state-level entity shares responsibility with the DPCC for "protecting and improving" the environmental quality of the National Capital Region (NCR) of India.

It is worth noting at this point that air pollution mitigation warrants interventions across not only different levels of government but also different sectors, like transport, urban planning, road infrastructure and maintenance, waste management, and industrial pollution control. In Delhi, urban planning falls under the purview of the Delhi Development Authority, which is headed by the lieutenant governor of Delhi, a Union government appointee. Multiple agencies are also responsible for implementing transport policies in Delhi. The Delhi Transport Corporation is the main public-transport operator and operates buses in Delhi, but the Delhi-integrated Multimodal Transit System Ltd. also operates buses in Delhi. The Delhi Metro Rail Company is a central-state public-sector company and operates the Delhi metro. The municipalities are responsible for solid-waste management in Delhi. Maintenance of roads falls under the purview of both the municipal and state governments.

In other words, multiple agencies at multiple levels of government play a crucial role in ensuring clean air for Delhi. Coordinating this highly fragmented regime poses considerable challenges for the city.

Three Decades of Air Pollution Control in Delhi

The Delhi administration started taking steps to address air pollution in the 1980s, when the Supreme Court began to assert authority in environmental lawmaking in response to a public-interest litigation filed in the court in 1985.[17] In response, the court pushed the government to implement existing policies and formulate new policies to abate air pollution.

The central government introduced the Environmental Protection Act in 1986 and amended the Air Act of 1981 in 1987, empowering the central and state pollution control boards with authority to address the grave emergencies of air pollution.[18] The Motor Vehicles Act of 1988 and the Central Motor Vehicle Rules of 1989 added authority to set standards for vehicular emissions for manufacturers and users as well.[19] Notably, the Supreme Court has also played a sizeable role in catalyzing the government's efforts to rein in Delhi's air pollution.[20] This section discusses the decades of work on air pollution control and the efforts made at each level of government to address the issue.

Judicial Activism

Judicial action has guided environmental pollution control in Delhi over the past three decades. One landmark case that has shaped the discourse and action on pollution in the city is the "Delhi Vehicular Pollution Case," filed by the lawyer and activist M. C. Mehta in 1985.[21] Interestingly, this case is still in progress three decades later.

The Delhi Vehicular Pollution Case led to a complete transformation of public and private transport in Delhi. Supreme Court orders in the matter led to the phasing out of leaded gasoline, the introduction of premixed fuels for two-stroke-engine vehicles and removal of fifteen-year-old commercial vehicles in mid-1990, the conversion of the city's diesel bus fleet into compressed natural gas (CNG) by 2001, and the constitution of the Environmental Pollution (Prevention and Control) Authority (EPCA).[22] More recent orders in the matter have led to a ban on the import and use of pet coke and furnace oil in industries situated in Delhi, the adoption of a Graded Response Action Plan and Comprehensive Action Plan for pollution control in the city, the construction of peripheral expressways to divert non-Delhi-destined truck traffic, and the adoption of Delhi's parking rules. Supreme Court directives have also been instrumental in restricting the bursting of firecrackers during Diwali.

In parallel to the Supreme Court orders, the National Green Tribunal (NGT), which was established for the speedy disposal of cases related to environmental protection, has also helped steer air pollution mitigation measures in Delhi. In 2015, the NGT banned gasoline vehicles older than fifteen years and diesel vehicles older than ten years in the NCR. It also prohibited the parking of fifteen-year-old vehicles in any public area.[23] The state government was directed to implement the order.

Central Government Interventions

The government of India has also taken several concrete steps to address air pollution in Delhi, including by designating both short-term and long-term plans of action for responding to and mitigating the pollution and creating institutional bodies to guide actions on improving, among other things, Delhi's air quality.

GRADED RESPONSE ACTION PLAN

For immediate action, the MoEF&CC adopted the Graded Response Action Plan in 2017, which stratifies emergency measures according to air quality index (AQI) levels.[24] The measures are cumulative, which means that when there is a severe air quality concern, the measures taken will include all of those listed for the current AQI level, as well as all of the ones listed for lower levels. Response measures for emergency air quality conditions include banning trucks' entry into Delhi, pausing construction activities, introducing the odd-even scheme for private vehicles, and shutting down schools, the last of which is an optional measure. In November 2019, Delhi's air quality hit emergency levels, and the state government declared a public health emergency in response. The Delhi government decided to shut down schools after the declaration.

COMPREHENSIVE ACTION PLAN

The central government has also taken steps to create a long-term strategy to address air pollution. In 2018, the Union Ministry adopted the Comprehensive Action Plan, which identifies timelines and implementing agencies for executing long-term solutions—such as augmenting the city's air quality monitoring infrastructure or transitioning from polluting power plants to cleaner energy sources—for the prevention, control, and mitigation of air pollution in Delhi and the NCR.[25]

ENVIRONMENTAL POLLUTION CONTROL AUTHORITY

When the Environmental Pollution Control Authority was constituted in 1998, its sole objective was to assist the Supreme Court in protecting and improving the environment's quality. Its mandate was revised in 2015 to include powers to give directions on matters regarding the setting of environmental standards for emission limits and the closure, prohibition, or regulation of any industry, operation, or process.[26] The authority was instrumental in shutting down the highly polluting Badarpur thermal power plant, introducing environmental compensation charges for polluting trucks entering Delhi, and introducing clean fuels into Delhi.[27]

State Government Interventions

The state government in Delhi has also taken several actions to address local air pollution, from replacing fossil fuels with renewable energy to taking steps to transition to electric vehicles.

In 2016, the state adopted the Delhi Solar Policy to "reduce the state's expenditure of energy, strengthen its energy security, and reduce its reliance on unsustainable fossil fuels." Under the policy, the state promotes the "development of grid-connected solar plants on rooftops for meeting its own electricity needs and injecting surplus electricity into the distribution grid."[28] The state offers a limited-time generation-based incentive for existing and future net-metered connections in the residential segment only.

In the winter of 2019, the state government introduced its Winter Pollution Plan to combat the expected rise in air pollution due to stubble burning in the neighboring states of Punjab and Haryana. In addition to actions like planting trees, distributing masks, and increasing monitoring in pollution hot spots, the plan saw the resurgence of the odd-even scheme. The government first introduced the odd-even scheme, a car-rationing scheme that allows vehicles with number plates ending in even numbers to run on even dates and those with odd numbers to run on odd dates, in 2016. While the state government has experimented with odd-even schemes multiple times, there is still a lack of consensus on their impact on air pollution levels.[29]

The state also adopted the Guidelines for Scrapping of Motor Vehicles in 2018 to ensure that the process of scrapping of vehicles does not cause any harm to the environment and that hazardous components of scrapped vehicles are disposed of in a safe manner.[30] As mentioned earlier, use of diesel vehicles more than ten years old and gasoline vehicles more than fifteen years old is prohibited by the NGT in Delhi. The scrappage guidelines were thus introduced to facilitate the implementation of this decree.

The state government has taken steps to promote electric vehicles (EVs) as well. As a cornerstone of this policy, the state provides financial incentives in the form of purchase, scrapping incentives, and subvention of interest on loans, which will be provided to potential buyers of electric vehicles.[31] Road taxes and registration fees are also waived for EV owners, and a wide network of charging and battery-swapping stations will be established. For effective administration of the policy, a state EV Board has also been constituted, and a dedicated EV cell will be established.

Joint Interventions

Finally, some efforts to address air pollution span across both central and state governments. In anticipation of worsening air quality in the winter of 2020, for example, the Union government, the state government, and the Environmental Pollution Control Authority initiated discussions to ensure proper management of air pollution in the NCR region before the onset of winter.

Has Air Quality Improved?

In 2019, both the state and central governments made multiple claims that Delhi's air quality had significantly improved in the past three years. However, several independent researchers argue that the government has not properly substantiated these claims. In particular, these researchers argue that continuous, unbroken data from a consistent set of monitors is needed to draw year-to-year comparisons and that Delhi does not have sufficient data to satisfy this requirement.[32] Importantly, the number of continuous monitoring stations has increased significantly in recent years; compared to one continuous monitoring station in 2011, there are thirty-eight in Delhi today. Moreover, the data availability at every fifteen-minute interval has almost doubled in recent years.[33] Thus, in the near future, Delhi should be able to better assess changes in air quality.

The governments' own assessments, read in their entirety, also give mixed messages about the true state of Delhi's air quality. For example, a recent report released by the Central Pollution Control Board, based on data from manual monitoring stations in the city, revealed that the number of days with good air quality (when $PM_{2.5}$ is less than 60 μg/m³) had gone up from twenty-five in 2016 to fifty-nine in 2019, showing a positive trend upward. However, the report also estimated that Delhi's annual-average $PM_{2.5}$ concentration was 141 μg/m³, and further analysis of data from Delhi's manual monitoring stations reveals that the average $PM_{2.5}$ concentration in the city was 122 μg/m³ in 2018 and 120 μg/m³ in 2016, showing a negative upward trend.[34] While one metric suggests an improvement in air quality, the other points to a deterioration—hinting that while there may in fact be more days with good air quality, the bad days have been getting worse.

Given that high-frequency data for Delhi has only been available since 2018 and given the disparities in what the data show, depending on what metric is the focus of the assessment, it is difficult to ascertain whether Delhi's air quality has indeed improved since 2016. Yet answering this question will be critical, as Delhi continues to work to improve its air quality and makes decisions about its next steps.

What Is Ailing Air Quality Management in Delhi?

In addition to understanding the current status of air pollution in the city, it is also important to understand what factors may be hampering Delhi's efforts to improve its air quality and what challenges may need to be addressed. Despite a slew of mitigation measures and a large number of government entities working toward improving Delhi's environmental quality, it continues to grapple with air pollution. From manpower shortages to regime fragmentation, a range of issues plague air quality management in Delhi.

Lack of Coherence in Understanding Sources of Pollution in Delhi

It is important to understand the precise contribution that different sectors are making, because knowing what sectors are contributing the most will determine what actions Delhi will prioritize. Yet, as noted at the outset, while much is known about the sources that contribute to Delhi's pollution, there still is a lack of consensus on the relative contribution from each of the various polluting sources. For example, the relative contribution of regional pollution ranges from 20 to 50 percent across various estimates. While the lower estimates would suggest introducing extensive efforts to control local sources, the higher estimates would suggest a different, and more challenging, path forward—as Delhi does not have direct control over the main sources of pollution, it might instead need to focus on coaxing its neighboring states to reduce their own pollution levels.

Multiplicity of Political Parties

As explained earlier, Delhi is a diarchy, and different political parties control the central and the state governments of Delhi. This has fostered an unfortunate situation in which the different political levels of government often blame each other for pollution problems, thus complicating mitigation efforts further. For instance, the central and state governments have pointed fingers when addressing the issue of straw burning, which the state government often refers to as one of the primary causes of Delhi's poor air quality. In September 2020, right before the onset of the rice-harvesting season, the chief minister of Delhi wrote to the Union environment minister, highlighting the inadequacy of the central government's scheme for subsidizing straw-management machinery, which, according to the state government, has not successfully curbed the practice of stubble burning.[35] This is one of the many examples of the lack of coordination between the levels of government with regard to the formulation of mitigation strategies.

Multiplicity of Institutions

Delhi also faces challenges with coordinating among its various implementing and enforcement agencies at the same level of government, as well as across central and state governments, which hinders implementation efforts. Multiple agencies, for example, are responsible for implementing both the Graded Response Action Plan and the Comprehensive Action Plan. And while implementing agencies like the Department of Transport report to the state government, enforcement agencies like the traffic police report to the Union government. Moreover, the priorities of municipal agencies tend to differ from those of the pollution-regulating agencies, further complicating their coordination efforts. An interesting example of this is the recent pedestrianization of the Ajmal Khan Road, a busy market road in Delhi. The implementation of this project called for the participation of at least four different agencies. Due to the large number of stakeholders involved in the conceptualization and implementation of the project, it took nine years for the project to materialize.[36]

Manpower Crunch

In addition to the challenges of coordinating across multiple agencies, individual agencies also have internal challenges that further limit their ability to properly address Delhi's air pollution problem. Of particular import, while these institutions may hold authority to act on issues of air pollution, many of them suffer from a lack of manpower, severely impeding their ability to do so. The Delhi Pollution Control Committee, for example, which is the state's executive arm on matters related to pollution control, has been operating at three-fourths of its sanctioned strength. And the Environmental Pollution Control Authority reportedly has "very few active members and lacks dedicated central funding."[37] Thus, while it is empowered to conduct field inspections to oversee implementation of its directions on the ground, it has been unable to effectively exercise this power.

Lockdown, Blue Skies, and Lessons Learned

In the context of air pollution in India, Delhi is the most studied city and has been the focus of both judicial and policy action to reduce air pollution over the past three decades. And while the city has taken some significant steps to mitigate the problem, several notable challenges must still be overcome for the city to effectively contain the problem.

It is also important to note the impact that recent events have had on both Delhi's air quality and the recent actions of cities in addressing the problem. The COVID-19 pandemic in 2020 significantly impacted the city's air quality and its response to air pollution, as it is recognized that an increase in pollution levels could exacerbate respiratory ailments caused by viruses like SARS-CoV-2.[38] The first three months of the COVID-induced national lockdown saw a drastic improvement in Delhi's air quality, and AQI values went down by over 50 percent during the first phase of the lockdown alone. The drop was even higher in industrial areas within the city.[39] Transportation-related pollution decreased as well; Google mobility data shows that trips to retail and recreational areas, workplaces, and transit hubs in the city each dropped significantly, leading to a decrease in particulate concentration in areas of Delhi that have traditionally had high vehicular traffic.[40]

In addition to improving air quality in the city, the lockdown and COVID-related concerns have encouraged people to explore alternative commuting options, like biking. Bicycle manufacturers and retailers across the county have reported a surge in demand in the post-lockdown period, as commuters are wary of traveling in crowded public transport systems.[41] Interestingly, 25 percent of the commuters using buses, Intermediate Public Transport (IPT), and metro-rail have short commutes (less than five kilometers) that could easily transition into biking trips. In fact, one-third of everyday commutes in Delhi already happen on foot or by bicycle.

While Delhi has made multiple attempts in the past to develop bicycle tracks, no efforts are currently being made to cater to pedestrians or bikers and capitalize on the COVID-induced increase in bike ownership in the city. The state government is, however, taking steps to encourage bus ridership by ensuring that minimal contact between individuals occurs during such rides, and it began to launch contactless e-ticketing for Delhi buses in November 2020.[42]

These local efforts, such as promoting biking and extending the benefits of reduced vehicular trips beyond the lockdown, will undoubtedly have an impact on Delhi's air pollution, if they endure. However, while the state government needs to play a pivotal role in battling air pollution, without support from the central government and the neighboring states, Delhi may continue to grapple with poor air quality in years to come.

NOTES

1. World Health Organization. 2018. "Ambient Air Quality Database Application." https://whoairquality.shinyapps.io.
2. *Economic Times*. 2015. "$PM_{2.5}$ Level in Delhi 10 Times More than WHO Limits: Greenpeace." February 16, https://economictimes.indiatimes.com.
3. C40 Cities. 2019. "35 Cities Unite to Clean the Air Their Citizens Breathe, Protecting the Health of Millions." October 11, www.c40.org.
4. Chandra, Shekhar. 2018. "Why Delhi's Air Pollution Problem Never Gets Solved." *Mint*, June 20, www.livemint.com; Harish, Santosh, and Navroz K. Dubash. 2020. "Three Ways by Which Delhi's Kejriwal Govt Can Fulfil Its Promise to Curb Air Pollution." *The Print*, February 27, https://theprint.in.
5. UrbanEmissions.info. 2015. "Delhi's Air Quality—Is It Emissions or Meteorology?" December 14, https://urbanemissions.info.
6. These variations in estimated contributions can be attributed to several factors, including the "domain area of the study, number of sampling stations, time period of sampling, season of sampling, quality of surveys, emission factors, assumptions, and data on emission abatement technologies and efficiency of control." Jalan, Ishita, and Hem H. Dholakia. 2019. *What Is Polluting Delhi's Air? Understanding Uncertainties in Emissions Inventory*. New Delhi: Council on Energy, Environment and Water. www.ceew.in.
7. Jalan and Dholakia 2019, 10–12.
8. Kulkarni, Santosh, Sachin D. Ghude, Chinmay Jena, Rama K. Karumuri, Baerbel Sinha, V. Sinha, Rajesh Kumar, V. K. Soni, and Manoj Khare. 2020. "How Much

Does Large-Scale Crop Residue Burning Affect the Air Quality in Delhi?" *Environmental Science and Technology* 54(8) (March): 4790–4799. https://doi.org/10.1021.acs.est.0c00329.

9. Twenty percent of postmonsoon $PM_{2.5}$ concentration in Delhi has been attributed to nonlocal fire emissions from agricultural residue burning. Santosh et al. 2020.
10. UrbanEmissions.info. 2017. "Delhi's Air Quality—Understanding Emission Loads during Diwali 2017." Accessed March 15, 2021, https://urbanemissions.info.
11. Gorai, Amit K., Paul B. Tchounwou, Shanti S. Biswal, and Francis Tuluri. 2018. "Spatio-Temporal Variation of Particulate Matter ($PM_{2.5}$) Concentrations and Its Health Impacts in a Mega City, Delhi in India." *Environmental Health Insights* 12 (January): 1–9. https://doi.org/10.1177/1178630218792861.
12. UrbanEmissions.info. 2016. "What's Polluting Delhi's Air?" https://urbanemissions.info.
13. UrbanEmissions.info 2016; Purohit, Pallav, Markus Amann, Gregor Kiesewetter, Vaibhav Chaturvedi, Peter Rafaj, Hem H. Dholakia, Poonam Nagar Koti, Zbigniew Klimont, Jens Borken-Kleefeld, Adriana Gómez Sanabria, Wolfgang Schöpp, and Robert Sander. 2019. *Pathways to Achieve National Ambient Air Quality Standards (NAAQS) in India*. New Delhi: Council on Energy, Environment and Water. www.ceew.in.
14. ARAI and TERI. 2018. "Source Apportionment of $PM_{2.5}$ & PM_{10} of Delhi NCR for Identification of Major Sources." www.teriin.org.
15. Garg, Anchal, and N. C. Gupta. 2020. "The Great Smog Month and Spatial and Monthly Variation in Air Quality in Ambient Air in Delhi, India." *Journal of Health and Pollution* 10(27) (September): 1–14. https://doi.org/10.5696/2156-9614-10.27.200910.
16. Act 047 of 1987: Air (Prevention and Control of Pollution) Amendment Act, 1987.
17. Narain, Urvashi, and Ruth G. Bell. 2005. *Who Changed Delhi's Air? The Role of Courts and the Executive in Environmental Policymaking*. Resources From the Future, https://media.rff.org.
18. Act 047 of 1987: Air (Prevention and Control of Pollution) Amendment Act, 1987.
19. Act No. 59 of 1988: Motor Vehicles Act, 1988.
20. Urvashi and Bell 2005.
21. M. C. Mehta v. Union of India, 1991 SCR (1) 866.
22. Urvashi and Bell 2005.
23. UrbanEmissions.info. 2015. "Delhi's Air Quality—Benefits of Banning Older Vehicles." April 20, https://urbanemissions.info.
24. Central Pollution Control Board, Ministry of Environment, Forest and Climate Change (Government of India). 2016. "Graded Response Action Plan for Delhi & NCR." https://cpcb.nic.in.
25. DPCC. 2018. "Action Plan for Air Pollution and Its Control by Air Quality Monitoring Committee (AQMC)." https://cpcb.nic.in.
26. Ministry of Environment and Forests Order, S.O. 93 (E) of 1998, 16–22, https://parivesh.nic.in.

27. The authority has since been dissolved and has given way to the Commission on Air Quality Management in Delhi NCR and adjoining areas. Press Information Bureau, Government of India. 2020. "Commission for Air Quality Reviews Air Quality Scenario in National Capital Region and Adjoining Areas. Identifies Ten Immediate Measures." Press release. https://pib.gov.in.
28. GNCTD. 2016. "Delhi Solar Policy, 2016," 7, http://ipgcl-ppcl.gov.in.
29. Dantewadia, Pooja, and Vishu Padmanabhan. 2019. "Will Delhi's Odd-Even Policy Work?" *Mint*, November 4, www.livemint.com.
30. GNCTD. 2018. "Guidelines for Scrapping of Motor Vehicles in Delhi, 2018." August 24, https://transport.delhi.gov.in.
31. GNCTD. 2020. "Delhi Electric Vehicle Policy, 2020." August 7, https://transport .delhi.gov.in.
32. Harish, Santosh, and Kurinji Selvaraj. 2019. "AQI Tracking Needs Sound Analysis, Transparent Process." *Hindustan Times*, September 13, 2019, www.hindustantimes. com.
33. UrbanEmissions.info 2016.
34. Central Pollution Control Board (CPCB). 2020. "National Ambient Air Quality Status and Trends 2019."
35. *Indian Express*. 2020. "Need Time to Discuss Stubble Burning: Kejriwal Writes to Javadekar." September 27, https://indianexpress.com.
36. Patel, Shivran, and Surita Baruah. 2019. "What It Takes to Make a Road Pedestrian Friendly." *Indian Express*, May 20, https://indianexpress.com.
37. Nandi, Jayashree. 2019. "Lack of Funds, Staff Hamper SC-Backed Pollution Monitoring Agency's Work Powers." *Hindustan Times*, November 8, www.hindustan-times.com.
38. On October 5, 2020, the chief minister of Delhi launched a seven-point plan for controlling pollution in the city. Measures in the plan include dust management, targeted interventions in pollution hot spots, creation of a phone app to facilitate registration of pollution-related complaints, and creation of a control room for monitoring all air-pollution-related activities. *Hindustan Times*. 2020. "Delhi Launches Plan for War on Pollution." October 6, www.hindustantimes.com.
39. The first phase ran from March 25, 2020, to April 19, 2020. Mahato, Susanta, Swades Pal, and Krishna G. Ghosh. 2020. "Effect of Lockdown amid COVID-19 Pandemic on Air Quality of the Megacity Delhi, India." *Science of the Total Environment* 730 (August): 139086, 5. https://doi.org/10.1016/j.scitotenv.2020.139086.
40. Trips to retail and recreational areas in the city went down by more than 80 percent, trips to workplaces declined by 77 percent, and trips to transit hubs dropped by 78 percent. Susanta et al. 2020.
41. Sharma, Nisant. 2020. "India's Bicycle Craze Spikes, Then Hits a Bump during Pandemic." *Bloomberg Quint*, August 11, www.bloombergquint.com.
42. *Hindustan Times*. 2020. "Contactless E-ticketing on Delhi Buses from November: Gahlot." September 23, www.hindustantimes.com.

8

How to Fight Air Pollution

The London Way

FRANK J. KELLY

Air pollution has been a serious problem in London since the twelfth century, when bituminous coal was introduced into the city. This affordable and abundant form of energy was initially used for manufacturing before becoming a domestic fuel. By the late 1700s, coal was fueling the Industrial Revolution in addition to millions of domestic fires. As a consequence, London's appalling mixture of fog, smoke, and sulfur dioxide emissions became world famous. In 1905, during a public-health lecture, Harold Des Voeux, a London physician and prominent advocate for smoke abatement, termed the lethal mixture "smog."[1] Concern over the health effects of London's poor air quality also dates back over many centuries. In 1661, the diarist John Evelyn presented King Charles II with a treatise on the problem, suggesting that smoke pollution was shortening the lives of Londoners.[2]

Nearly two hundred years later, an article in *The Lancet* stated that "the air of this great city is, as all know too well, polluted by a variety of noxious gases and vapors diffused or held in solution."[3] The devastating health effects brought about by smog became all too evident in December 1952, when a toxic combination of London's cold, motionless, smoggy air brought about the worst pollution disaster in history. This great killer-smog episode claimed an estimated four to twelve thousand premature deaths and increased morbidity from cardiorespiratory causes.[4] It was also the impetus for the 1956 Clean Air Act, the major focus of which curtailed domestic coal burning in London and other major cities in the United Kingdom. The implementation of smokeless zones, controls imposed on industries, increased availability and use of natural gas, and changes in the industrial and economic structure of the UK led to a considerable reduction in concentrations of smoke and sulfur dioxide between the 1950s and the present day.

Air Pollution in London in More Recent Times

In December 1991, a severe wintertime air pollution episode occurred in London, characterized by unprecedented levels of benzene, carbon monoxide (CO), oxides of nitrogen (NO_x), and in particular, nitrogen dioxide (NO_2)—all components of motor-vehicle exhaust. In response to this, new air quality monitoring sites were established in and around London, and the equipment base of existing sites was extended beyond their then black smoke (BS) measurements. Continuous monitoring of particles with an aerodynamic diameter of 10 μm or smaller (PM_{10}) began to replace the BS measurements. During the 1990s and early 2000s, the airborne particulate and lead concentrations declined steadily following the phase-out of lead in gasoline, while CO, benzene, and 1,3-butadiene also fell dramatically, owing to the mandatory implementation of three-way catalysts and evaporative canisters in gas-engined vehicles. In turn, the reduction in volatile organic carbon (VOC) emissions produced a decline in the peak intensity of photochemical episodes. By contrast, the annual percentage reductions in NO_x levels (again achieved through the implementation of three-way catalysts on gas-engined motor vehicles) were substantially less than those achieved for CO and VOCs. This was due to the substantial and growing contributions to NO_x emissions from diesel-engined motor vehicles. The increasing use of diesel-engined vehicles also means that particulate matter (PM) is still of major concern, despite the reduction in black-smoke levels. As mentioned earlier, concentrations of PM_{10} declined during the 1990s, but these trends have slowed down; and during the 2000s, concentrations have remained constant. So, despite the air quality gains achieved in previous decades, like many other large cities around the world, London continues to experience high levels of air pollution, owing to a combination of mobile sources and the influence of regional background pollution.

Figure 8.1. London smog. (Photo by F. J. Kelly)

Air Quality Strategy

Following the reestablishment of the mayor of London office in 2000 and in view of widespread public concern about the health effects of air pollution, the new mayor, Ken Livingston, launched an air quality strategy in 2002, titled "Cleaning London's Air."[5] This strategy set out policies to move toward the point where pollution would no longer pose a significant risk to human health. The main aim of the strategy was the reduction of pollution from road traffic, which was the largest source of the pollutants of concern—$PM_{2.5}$ and NO_2—in the city. $PM_{2.5}$ refers to particles that pass through a size-selective inlet with a 50 percent efficiency cutoff at 2.5 μm aerodynamic diameter, which means that they are small enough to penetrate deep into the lungs. To this end, the goals were to reduce the number of vehicles on the road and to lower emissions from individual vehicles through modernization of the fleet vehicle stock.

In 2003, emissions from road transport contributed to approximately 40 percent of NO_x emissions and to 66 percent of PM_{10} emissions in the Inner London area. A reduction in London's road-traffic emissions was to be achieved in two ways: first, through a decrease in the number of vehicles on the road, and second, through reduced emissions from individual vehicles (i.e., modernization of the fleet vehicle stock). To help achieve the first aim, the mayor introduced a congestion charging scheme (CCS) in central London in February 2003.[6] One of the measures to tackle the second aim was a London-wide Low Emission Zone (LEZ), which was introduced in February 2008.

The Congestion Charging Scheme in London

The congestion charging scheme in London is a vehicle-charging scheme covering approximately twenty-two square kilometers, or 1.4 percent of the Greater London Area, and containing some of the most congested conditions in the capital. Vehicles crossing a cordon line to enter the congestion charging zone (CCZ) between the weekday hours of 7:00 a.m. and 6:00 p.m., termed the "congestion charging hours" (CCH), must pay a daily charge of what was originally £5 but was increased in July 2005 to £8 and then in July 2013 to £10. Assisted by revenue from the CCS, parallel improvements in the fleet of public transport vehicles and traffic management have been implemented to accommodate the shift in travel patterns following the introduction of the charging scheme, as well as the continued growth in demand. The principal traffic and transport objectives have been met, mirroring the effectiveness of similar schemes in Singapore, Stockholm, and Norway.[7] Similar road-pricing schemes have been considered for other UK cities, and it is likely that traffic zonal payment schemes will become more common elsewhere in the world. Indeed, Milan introduced such a scheme on a test basis at the beginning of 2007 to address the city's severe air pollution and traffic problems, and New York became the first US city to introduce a traffic-congestion charge in 2019. The CCS in London is therefore a likely forerunner in what may become a powerful and widely adopted approach to traffic demand management.

The CCS and Air Quality in London

There was considerable interest in determining whether the reduction in congestion and traffic achieved in London following the

Figure 8.2. Pollution monitoring on Marylebone Road, London. (Photo by F. J. Kelly)

implementation of the CCS has had a positive impact on the air quality. In principle, the CCS, in reducing congestion and the numbers of vehicles crossing the cordon, should reduce emissions and improve air quality both inside and outside the CCZ. However, one would not expect the CCS to elicit more than a small effect on air quality within the zone, considering that the charge has had a relatively moderate effect on traffic (approximately 20 percent reduction in vehicles entering the CCZ) in a small (1.4 percent of Greater London) area of London. This is particularly the case when one considers the host of other contributors to emissions both within London and from a regional background. For example, changes associated with the CCS in traffic flows and vehicle speeds had the potential to produce both increases and decreases in PM and NO_x emissions. In addition, improvements in public transport vehicles (and with this, an increase in the number and flows of diesel-powered buses and taxis entering the CCZ), the introduction of traffic-management measures, and the magnitude and location of road works were also likely to have air quality impacts. In fact, following appropriate analysis, the CCS was found to result in no or only a very small improvement in air quality.[8]

The London-Wide Low Emission Zone (LEZ)

The key driver for the introduction of the LEZ was the need to improve the health and quality of life of people who live and work in London by improving air quality.[9] In addition, the scheme was designed to move London closer to achieving national and EU air quality objectives for 2010. The LEZ was designed to tackle these objectives by restricting the entry of the oldest and most polluting vehicles across Greater London, an area of 2,644 square kilometers.[10] The scheme operates 24 hours a day, 365 days a year, using cameras to identify the registration numbers of vehicles and the Driver and Vehicle Licensing Agency (DVLA) database to identify a vehicle's emissions. The LEZ applies to diesel-engine heavy-goods vehicles (HGVs), buses and coaches, larger vans, and minibuses. The LEZ initially targeted vehicles with disproportionately high emissions (lorries over twelve tons), obliging them to meet specific modern Euro emission standards, which set limit values for exhaust emissions for new vehicles. This first phase of the LEZ was followed by the inclusion of lighter HGVs, buses, and coaches in July 2008, while large vans and minibuses were to be included from October 2010. However, the mayor at the time, Boris Johnson, decided to delay this third phase by eighteen months due to poor economic conditions. Specifically, the LEZ required heavy-duty vehicles to meet the Euro III emission standard for PM_{10} and then the Euro IV standard in 2012. A standard for NO_x was not included as part of the scheme, as there were too many unresolved issues about certification and testing of NO_x abatement equipment.

In order to deter the use of high-polluting vehicles and provide an incentive for operators to upgrade their vehicles, operators of vehicles that did not meet the LEZ standards were required to pay a charge of £200 (lorries, coaches, and buses) or £100 (vans and minibuses) for each day they are driven into the zone. Furthermore, should an operator of a non-compliant vehicle not pay the daily charge, then following the service of a penalty charge notice, penalty charges of £1,000 (lorries and buses) or £500 (vans and minibuses) apply; these penalties are reduced by 50 percent if paid within fourteen days or increased by the same increment if not received within twenty-eight days.

Potential Health and Environmental Impacts of the LEZ

It is well established that poor air quality has marked effects on certain, sensitive individuals. Because the LEZ was implemented across the whole of Greater London, an area in which more than eight million people live and work, the scheme had the potential to bring about a range of health benefits for London residents. When introduced, the LEZ was predicted to reduce the emission of PM_{10} in London by sixty-four tons (or 2.6 percent) in 2008 and, in turn, to decrease the area of Greater London that exceeds the EU air quality limit values for PM_{10} of 40 $\mu g/m^3$.[11] In practice, however, the benefits were smaller than expected and difficult to identify, as Euro emissions databases were subsequently found to be inaccurate in reflecting real-world driving conditions.

Between 2008 and 2014, evidence was found for only small decreases in NO_x and NO_2 concentrations at roadside and background locations within London's LEZ, and there was no evidence of improvements in PM_{10}.[12] These modest changes were not associated with improvements in children's respiratory health.[13] Notably, at the rate of improvement observed for NO_2 following implementation of the LEZ, London would not have reached full compliance with its legal requirement for decades. With poor air quality attaining increasing public and professional concern, London's current mayor, Sadiq Khan, made improving air quality a cornerstone of his policy agenda, leading to the early introduction of the Ultra Low Emission Zone (ULEZ), an exemplar Clean Air Zone (CAZ), in April 2019. The ULEZ prescribes the permissible levels of vehicle emission with the aim of lowering exposure to NO_2 (as a proxy for diesel tailpipe emissions) and PM.

The ULEZ was implemented by Transport for London (TfL) with the specific goal of improving air quality and, as a consequence, population health. It was configured to deliver major impacts on London's air quality rapidly: modeling by the Environmental Research Group (then at King's College London) for TfL predicted that the ULEZ would reduce total NO_2, PM_{10}, and $PM_{2.5}$ in central London by 47, 16, and 11 percent, respectively. The ULEZ covers the existing CCZ and operates 24 hours a day, 7 days a week, 365 days a year. It requires vehicles entering the zone to either meet emission requirements or pay a daily charge; the requirements are based on Euro standards that limit the release of NO_x and PM from engines. Specifically, the ULEZ initially stipulated Euro 3 standard for motorcycles and mopeds, Euro 4 (NO_x) for petrol cars and vans, Euro 6 (NO_x and PM) for diesel cars and vans, and Euro VI (NO_x and PM) for heavy vehicles. The ULEZ is being implemented in three phases: (1) in April 2019, restrictions were placed in the CCZ area; (2) beginning in April 2021, restrictions were tightened for the existing London-wide LEZ to include charges for heavy-duty vehicles that exceed NO_x or NO_2 emissions (in addition to PM); and (3) beginning in October 2021, the ULEZ area boundary was extended to the North and South Circular Roads in Greater London. Initial results from this effort have demonstrated a marked improvement in London's air quality.[14]

How the Public Transport Network Is Also Changing

London has the largest bus fleet in Europe, with over nine thousand buses, of which around 70 percent are double-deck buses. Since 2018, the current mayor has spent more than £300 million to transform London's bus fleet by retrofitting thousands of buses and phasing out pure diesel double-deck buses. All buses in central London now meet or exceed the cleanest Euro 6 standards. In October 2020, TfL achieved its goal of all buses London-wide meeting or exceeding the cleanest standard, reducing emissions from the bus fleet by 80 percent compared to 2016 levels. As a consequence, the whole of London is now a Low Emission Bus Zone.

Since January 2018, all newly registered taxis in London have been required to be zero-emission capable (ZEC). Taxi drivers were supported in the move to cleaner vehicles by TfL's Taxi Delicensing Scheme, which launched in 2017. The scheme provided payments of up to £5,000 to retire the oldest taxis from London licensing. In 2019, this scheme was restructured to provide top-level payments of £10,000. There are now over thirty-five hundred ZEC taxis and over seventy dedicated taxi rapid-charge points. Improvements to the taxi fleet are expected to reduce NO_x emissions from taxis by around 65 percent in 2025, compared to 2013. To address congestion related to the increase in the number of private-hire vehicles, since April 2019, they are no longer exempt from the congestion charge.

Looking to the Future

The experiences in London clearly show that more fundamental changes in urban transportation systems are required for air quality improvements to be achieved and for city streets to become more pleasant environments. Although London has introduced a ULEZ, banning all but the cleanest vehicles and incentivizing the use of zero-emission cars and taxis in a small area of central London will be insufficient to achieve compliance with NO_2 standards until at least 2025. However, it seems that attitudes toward car ownership are changing. Many urban youths of today, it seems, would rather use taxis and car-share than own their own car. Polls in London associated with the mayoral elections in 2016 put air quality issues high on the political agenda and indicated that there was popular support for a further tightening of vehicle use in the city. Cars promote a sedentary lifestyle, and a shift to cycling and walking has health benefits for urban dwellers.[15]

The car industry has also recognized the need for change, and companies such as BMW and Ford have launched car-share schemes. Car-sharing has obvious benefits to cities. Zipcar estimates that every shared vehicle replaces up to twenty private cars, thus reducing total vehicle miles and land devoted to parking. The car industry is also pushing forward the development of city-friendly plug-in electric vehicles. Worries over battery life and the lack of coordinated charging networks have, until recently, held back electric vehicle sales, even with generous government subsidies. In the UK, however, the electric vehicle fleet has grown rapidly in recent years—to more than 400,000 plug-in and 165,000 pure electric vehicles—suggesting that the tide is turning in favor of zero-emission vehicles. In contrast, goals for (hydrogen) fuel-cell-powered cars have not been achieved because of limited fueling networks and high initial vehicle costs. However, as cities like London have seen recent growth in both size and population, even zero-emission vehicles are not the solution. Moving increasing numbers of people efficiently around a city can only be achieved by expanded mass transit systems. The increased use of bus networks and traditional subways is providing city dwellers with more flexible, cheaper, and less polluting options. Rapid transit systems offer only a partial solution to city congestion and poor air quality, however, because many users may not live within easy walking distance of the transit points. To be effective, rapid transit systems must be linked with other transport options at the start and end of the journey. The creation of Autonomous Mobility-on-Demand networks may solve this problem.[16] Modeled after bicycle-share programs, users would have access to a network of lightweight electronic vehicles (LEVs) distributed at charging stations throughout the city. Customers would be able to summon a vehicle through a smartphone app and abandon it upon reaching their destination, from where the vehicle would find its way back to a charging station or new user. Trials of this technology are ongoing in Milton Keynes, just north of London.[17] In the city of the future, mass public transport systems may lead to many more car-free roads, transforming the landscape and enhancing the urban experience. Moving toward such an environment depends, however, on continued and exacting scientific research, translating such research into realistic and effective policies, and successfully encouraging members of the public to use public transport, value exercise, and not drive for short journeys.[18] The reward will be improved health and quality of life for the growing urban populations around the world.

NOTES

1. Wise, William. 1968. *Killer Smog: The World's Worst Air Pollution Disaster*. iUniverse, 42–48.
2. Evelyn, John. (1661) 2013. "Fumifugium; or, The Inconvenience of the Aer and Smoake of London Dissipated." In *Historical Perspectives on Preventive Conservation*, 262. Brighton: National Society for Clean Air.
3. *The Lancet*. 1856. 68(1718): 139–140.
4. Bell, Michelle L., and Devra Lee Davis. 2001. "Reassessment of the Lethal London Fog of 1952: Novel Indicators of Acute and Chronic Consequences of Acute Exposure to Air Pollution." *Environmental Health Perspectives* 109: 389–394. https://doi.org/10.1289/ehp.01109s3389.
5. The Greater London Council, the elected government for Greater London, was abolished in 1986 by the Local Government Act 1985. Strategic functions were split up across central government departments. In 1998, Londoners voted in a referendum to create a new governance structure for Greater London, leading to the appointment of a directly elected mayor, the first in the UK. Greater London Authority (GLA). 2002. "The Mayor's Air Quality Strategy, Cleaning London's Air."

6. GLA 2002.
7. Chin, Anthony T. H. 1996. "Containing Air Pollution and Traffic Congestion: Transport Policy and the Environment in Singapore." *Atmospheric Environment* 30(5): 787–801; Seik, Foo Tuan. 2000. "An Advanced Demand Management Instrument in Urban Transport: Electronic Road Pricing in Singapore." *Cities* 17(1): 33–45; Victoria Transport Policy Institute. 2007. *Online Transport Demand Management Encyclopaedia*. www.vtpi.org.
8. Kelly, Frank, Ross Anderson, Ben Armstrong, Richard Atkinson, Ben Barratt, Sean Beevers, Dick Derwent, David Green, Ian Mudway, and Paul Wilkinson. 2011. "The Impact of the Congestion Charging Scheme on Air Quality in London. Part 1. Emissions Modeling and Analysis of Air Pollution Measurements." Research Report 155, Health Effects Institute, Boston, MA, 5–71.
9. GLA 2002.
10. London Mayor. 2007. "Statement by the Mayor on the London Low Emission Zone." www.tfl.gov.uk.
11. Transport for London. 2006. "Proposed London Low Emission Zone: Scheme Description and Supplementary Information." Mayor of London, Greater London Authority, London.
12. Mudway, Ian S., Isobel Dundas, Helen E. Wood, Nadine Marlin, Jeenath B. Jamaludin, Stephen A. Bremner, Louise Cross, Andrew Grieve, Alex Nanzer, Ben M. Barratt, Sean Beevers, David Dajnak, Gary W. Fuller, Anna Font, Grainne Colligan, Aziz Sheikh, Robert Walton, Jonathan Grigg, Frank J. Keyy, Tak H. Lee, and Chris J. Griffiths. 2019. "Impact of London's Low Emission Zone on Air Quality and Children's Respiratory Health: A Sequential Annual Cross-Sectional Study." *The Lancet Public Health* 4(1): e28–e40. https://doi.org/10.1016/S2468-2667(18)30202-0.
13. Mudway et al. 2019; Wood, Helen E., Nadine Marlin, Ian S. Mudway, Stephen A. Bremner, Louise Cross, Isobel Dundas, Andrew Grieve, Jonathan Grigg, Jeenath B. Jamaludin, Frank J. Kelly, Tak Lee, Aziz Sheikh, Robert Walton, and Christopher J. Griffiths. 2015. "Effects of Air Pollution and the Introduction of the London Low Emission Zone on the Prevalence of Respiratory and Allergic Symptoms in Schoolchildren in East London: A Sequential Cross-Sectional Study." *PLoS One* 10(8): 1–12. https://doi.org/10.1371/journal.pone.0109121.
14. For example, in 2016, monitoring sites in London recorded over four thousand hours above the short-term legal limit for NO_2. In 2019, this reduced to just over one hundred hours, a reduction of 97 percent. Greater London Authority. 2020. "Air Quality in London 2016–2020: London Environment Strategy: Air Quality Impact Evaluation." www.london.gov.uk.
15. Woodcock, James, Phil Edwards, Cathryn Tonne, Ben G. Armstrong, Olu Ashiru, David Banister, Sean Beevers, Zaid Chalabi, Zohir Chowdhury, Aaron Cohen, Oscar H. Franco, Andy Haines, Robin Hickman, Graeme Lindsay, Ishaan Mittal, Dinesh Mohan, Geetam Tiwari, Alistair Woodward, and Ian Roberts. 2009. "Public Health Benefits of Strategies to Reduce Greenhouse-Gas Emissions:

Urban Land Transport." *The Lancet* 374(9705): 1930–1943. https://doi.org/10.1016/S0140-6736(09)61714-1.

16. Chin, Ryan. 2013. "Solving Transport Headaches in the Cities of 2050." *BBC Future*, June 17, www.bbc.com.
17. *BBC News*. 2013. "Driverless Cars to Be Introduced in Milton Keynes." November 7, www.bbc.com.
18. Kelly, Frank J., and Julia C. Fussell. 2015. "Air Pollution and Public Health: Emerging Hazards and Improved Understanding of Risk." *Environmental Geochemistry and Health* 37(4): 631–649. https://doi.org/10.1007/s10653-015-9720-1.

9

The Authority and Experience of the City of Beijing with Regulating Air Pollution

ALVIN LIN

As the capital of China and the center of national political authority, the city of Beijing holds a special place in China's political system and its development of environmental policy. A mega-city of 21.54 million people, Beijing's experience addressing its air pollution challenges in the four decades since the country's opening up in 1978, particularly during its rapid growth period between 2000 and 2020, can provide important insights for other global mega-cities that are facing similar challenges due to rapid development and increasing emissions from industrialization, transportation, and other sources.

Beijing's experience with development and air pollution parallels the experiences of other mega-cities in the United States and Europe, such as London, New York, and Los Angeles, as well as in Asia, such as Tokyo, during their rapid-development phases. However, given the size and scale of China's development, the dominance of coal in its energy and industry structure, and the magnitude and rapidity of the response measures needed to address its severe air pollution challenges, Beijing's experience has been notable.

Background of Beijing's Legal Status and Authority to Adopt Environmental Laws

China has a top-down governance structure, which means that the central government in Beijing provides the overall legal and policy direction for provinces and cities, including by setting five-year plans, developing national laws and policies, and establishing overall guidance and minimum standards. Provinces and cities will then implement these laws and policies by adopting their own lower-level plans, targets, laws, and policies that are adapted to their local circumstances but that also support and mirror the central, top-level precedents.[1] In essence, the central government generally leads, while the provincial and city governments follow; in cases where provincial or city governments do take the lead in policy making, such as with Shenzhen's piloting of a market economy, this is done with the explicit support of the central government.[2]

At the national level, China enacted a framework Environmental Protection Law on a trial basis in 1979, which was subsequently revised in 2015. Looking specifically at national air pollution controls, the national Air Pollution Prevention and Control Law was first enacted in 1987 and revised most recently in 2015.[3] Provinces and cities have developed their own local air pollution laws based on this national law. Beijing, for example, enacted local measures to implement the national air pollution law in 1988, which paralleled the national law and addressed the key sources of air pollution emissions, including coal burning, vehicles, industrial emissions, and construction, and introduced fines for violations.[4]

Beijing's Air Pollution Challenge and Pressures to Improve Air Quality

Beginning with China's ascension to the World Trade Organization in 2001, which initiated a wave of foreign direct investment in the country and an acceleration of the country's economic development, and continuing with the preparation for hosting the Summer Olympics in Beijing in 2008, which triggered a massive construction boom, the city of Beijing has experienced a period of supercharged growth. During this time, the city saw its traffic transformed from predominantly bicycles to cars; the construction of roads, skyscrapers, shopping malls, and multistory apartment buildings replace the traditional one-story "hutong" residences; and the rapid expansion of iron and steel, cement, and coal power plants to supply the building materials and energy for this growth and construction.

Beijing's environmental challenges also skyrocketed during this time. Between 1998 and 2017, China's coal consumption increased by 2.7 times, from 1.35 billion tons of coal in 1998 to 3.86 billion tons by 2017, and the number of vehicles on Beijing's roads grew by 335 percent, from less than

2 million to nearly 6 million.[5] All the while, Beijing's GDP grew by 1,078 percent, from less than 250 billion RMB to more than 2.5 trillion RMB. Its population grew by 74 percent as well, from about 12.5 million people to 22 million people.

This growth in economic activity and the resulting increase in emissions, combined with a growing public and governmental awareness about the health and environmental impacts of air pollution, have led to the passage of increasingly stronger laws and policies to reduce air pollution emissions. Particularly in the years after the 2008 Olympics, which included severe winter smog episodes in the winters of 2011–2012 and 2012–2013, policy makers have paid much greater attention to the need to strengthen air pollution measures.

Beijing's status as the national capital has contributed to its role as a national leader in setting the most aggressive air pollution policies and standards. A number of events of national importance, such as the 2008 Summer Olympics, the 2014 Asia-Pacific Economic Cooperation (APEC) Leaders' Meeting, and the 2022 Winter Olympics, have increased pressure on Beijing to tighten air pollution controls. But even independent of these events, as a city surrounded by many of the most industrial and highest coal-consuming provinces in the country, Beijing has had strong incentives to develop effective control policies. And, as described in the following sections, a number of these policies were pioneering in the Chinese context, including higher emission control standards for industry and vehicles, the conversion of power and heat from coal to natural gas, and the regional coordination of air pollution efforts. These policies have become models for China's national approach to air pollution.

The Progressive Strengthening of Beijing's Air Pollution Policies

As a UN Environment and Beijing Municipal Ecology and Environment Bureau review of Beijing's air pollution policy and progress notes, Beijing enacted a series of air pollution policies and actions between 1998 and 2017. The policy evolution can be divided into three stages, with the regulations becoming progressively more stringent in each phase. Critically, as has already been alluded to, a central theme that runs across these periods of local action is that Beijing's air pollution policies have been closely coordinated with the central government, which provides fiscal support to the city. As others have observed, due to Beijing's unique political importance, "as early as 1998, with direct instructions from the central leadership, air pollution control in Beijing became the top priority in environment management in China."[6]

Phase 1: 1998–2008

Beijing launched the first local-government air pollution program in China in 1998 in response to worsening air pollution. These policies were significantly strengthened as part of the preparation for the 2008 Beijing Summer Olympics.[7] Beijing adopted a number of important air pollution measures during these early years, including replacing coal with clean fuels; applying new techniques in power plants regarding desulfurization, dust collection, and NO_x combustion; increasing green land covering in the city; managing dust at construction sites; and establishing pollution control measures for vehicles, including setting new vehicle emission standards, improving the quality of the fuel supply, and retiring heavy-duty and old vehicles.[8] However, despite these efforts, the policies were ultimately unable to significantly improve the city's air quality, due to the increase in emissions from industrializing neighbor provinces and an increase in the number of vehicles on the road. Nevertheless, the introduction of these air pollution measures proved that vehicle-exhaust controls "contribute[d] the most to concentration abatement potential of NO_x and $PM_{2.5}$ in Beijing."[9]

Phase 2: 2009–2012

During the 2008 Olympics, Beijing piloted a number of short-term administrative measures to rapidly reduce air pollution. Some of the most prominent measures included shutting down factories in neighboring provinces, limiting vehicle travel (with odd- and even-day restrictions based on license-plate numbers), and strengthening end-of-pipe controls on coal power plants. Many of these measures subsequently became recurring short-term tools used to reduce air pollutant emissions during other politically significant events in Beijing and other cities—such as the World Exposition in Shanghai and the Asian

Games in Guangzhou in 2010, the APEC Conference in Beijing in 2014, and the Military Parade in Beijing in 2015—and some became a permanent part of the city's air pollution control efforts. Efforts to jointly coordinate pollution control efforts across provinces were the predecessor for the current regional air pollution management mechanism in the Beijing-Tianjin-Hebei region, as well as other key air pollution regions such as the Yangtze River Delta.[10]

Phase 3: 2013–2017

The third stage of Beijing's air pollution policy developed in response to the severe air pollution episodes of autumn and winter 2012–2013, when coal-based winter heating and atmospheric inversion conditions combined to cause severe air pollution episodes in much of the country. These episodes hit particularly hard in the north, including in Beijing, and sharpened public attention on air pollution and its public health impacts. In response, the State Council, China's cabinet, issued the nation's first "Air Pollution Action Plan" in September 2013. This plan, which covered the five-year period between 2013 and 2017, contained the strongest pollution controls yet. The Air Pollution Action Plan was a signal of the seriousness of the government's intent to address China's air pollution challenges. For the first time, the plan set quantitative targets for annual average $PM_{2.5}$ concentration reductions in key air pollution regions—a 25 percent reduction by 2017 compared to 2012 for the Beijing-Tianjin-Hebei area, a 20 percent reduction for the Yangtze River Delta area, and a 15 percent reduction for the Pearl River Delta area—and an even stricter target for Beijing as the nation's capital: to reduce the average annual $PM_{2.5}$ concentration to 60 micrograms per cubic meter ($\mu g/m^3$) or less by 2017, from 89.5 $\mu g/m^3$ in 2012.[11] By 2017, strong city and regional efforts enabled Beijing to reduce its average annual $PM_{2.5}$ concentration to 58 $\mu g/m^3$ and enabled the Beijing-Tianjin-Hebei area to reduce its annual average $PM_{2.5}$ concentration by 25 percent.[12]

Just as with the national air pollution law, provinces and cities took their cue from the first national air pollution action law to develop and establish their own local air pollution action plans. Beijing took the lead once again, releasing its own Clean Air Action Plan at the same time as the national Air Pollution Action Plan in August 2013. The plan was accompanied by a "key responsibility allocation document," which set forth eighty-three measures to comprehensively reduce air pollution emissions from key sectors, including coal combustion in power and residential heating, vehicle emissions, and industrial and construction activities, as well as supporting policies including financial measures, information transparency, monitoring, emergency response systems, and so on to support these efforts. Notably, the document lists the specific agencies and officials responsible for implementing each measure as a way to ensure accountability.[13]

Beijing's first local law on air pollution, Ordinances of Beijing Municipality for Prevention and Control of Air Pollution, passed in 2014 as well and was the first local regulation in the nation that targeted $PM_{2.5}$ pollution and aimed to comprehensively treat and control a wide range of $PM_{2.5}$ sources. The passing of this law marked a shift in air pollution control, as the United Nations has described, "from end-of-pipe treatment to process-wide control; from setting concentration limits to controlling both the concentration and total emission amount; and from government control to social governance, with equal emphasis put on corporate governance, sectoral governance, and regional coordination."[14]

A Sectoral Review of Beijing's Actions to Reduce Key Air Pollutant Emissions

Given Beijing's national significance and abundant financial and technical resources, the city has been able to lead the country in reducing air pollutant emissions in key sectors, including from coal combustion in the power sector, industry, residential heating, and vehicle emissions. This section briefly reviews key policies that Beijing has adopted in each of these sectors.

In the power sector, Beijing tightened the SO_2 and NO_x reduction standards of its coal power plants, part of a nationwide trend toward stricter SO_2, NO_x, and $PM_{2.5}$ emissions standards that led to the retrofitting of most of the country's coal power plants to meet "ultra-low emissions standards" comparable to natural gas plants. Beijing eventually phased out its four remaining coal power plants by 2017, replacing them with natural gas power plants and electricity imported from surrounding provinces—which, while it has continued to be generated mainly from coal power, is also being generated from increasing amounts of nonfossil sources such as wind and solar.[15] Beijing has phased out coal power within its jurisdiction well ahead of the national trend, in which coal power continues to be a major, albeit decreasing, share of power generation (a 68 percent share in 2020).

Beijing has also changed its economic and industrial structure, relocating heavy industry outside the city; for example, eight million tons of steelmaking capacity for the steelmaker Shougang Group, the city's largest polluter, were relocated to the coastal city of Caofeidian in the lead-up to the 2008 Beijing Olympics.[16] For the remaining coal boilers in the city, necessary for winter district heating, Beijing also increased the size and emission standards of coal boilers before eventually replacing them with more expensive but cleaner natural gas boilers. Since 2017, the city has also focused on replacing dispersed household coal burning with no emissions controls in the rural suburbs and villages surrounding the city with gas or electric heating, providing huge subsidies for heating equipment and to make up for more expensive operating costs.

In order to address air pollution from rapidly growing vehicle ownership (6.2 million vehicles, responsible for some 45 percent of $PM_{2.5}$ air pollution in the city), Beijing has led the country—along with other major cities, such as Shanghai, Shenzhen, and Guangzhou—to implement the highest fuel quality and vehicle emission control standards in China and is currently implementing the China 6 vehicle air pollutant emission standards, which are among the toughest in the world.[17] In addition, the city has built an extensive public transportation system, including one of the largest subway systems in the world, and placed limits on vehicle ownership, through a lottery system with an annual cap on new vehicles, and on vehicle use, with limits on driving based on license plates, as mentioned previously, such as restricting 20 percent of vehicle license plates on any one day.

Air Pollution Monitoring and Enforcement

In conjunction with the aforementioned regulatory tools, another key tool of Beijing's air pollution policy, and of environmental governance in China generally, has been the implementation of an extensive ambient air quality monitoring network in the city and across the country. The data gathered by this monitoring network is reported to the public and is used to measure progress—or the lack thereof—in tackling air pollution. Such air quality reporting has kept air pollution progress and policy high on the public's and the local government's agenda and has led to continual efforts to improve on progress.

Reasons for Beijing's Success and Lessons for Other Cities

While Beijing continues to face air pollution episodes and will need further efforts to reach China's air quality standard of 35 $\mu g/m^3$ $PM_{2.5}$ and the even lower WHO standard of 10 $\mu g/m^3$, it has come a long way since the Olympics and the 2012–2013 smog episodes. Beijing's progress may be attributable in large part to the following factors:

- As a capital city with key political importance, Beijing has had financial, political, and technical resources far beyond most cities in China—and around the world—to devote to fighting air pollution. Key political events such as the 2008 and 2022 Olympics, political meetings, and so on have kept progress on air quality high on the agenda.
- Along with political will, Beijing has also continuously developed stronger air pollution laws and plans for air pollution regulation—in line with strengthening national air pollution laws and plans—and continuously raised emission standards, promoted industrial structure shift, and strengthened complementary policies such as air pollution monitoring and emergency response measures.
- The plans have set specific air quality improvement targets based on science-based planning and continual improvements by sector. They have also set clear responsibilities for which agencies are accountable for reaching each target and policy.
- Beijing has benefited greatly from international cooperation on air pollution policy, including through exchanges with United Nations Development Programme, United Nations Energy Programme, and US Environmental Protection Agency experts to share experiences and best practices on addressing various sources of air pollution.
- Beijing has also focused on implementation and enforcement of standards, including through environmental inspections and campaigns to

tackle big and small polluters (such as industry facilities and cars).

- Given that Beijing's air quality is very much affected by neighboring provinces, it has also led efforts on joint prevention and control of air pollution, working and coordinating with the wider region of provinces in Beijing's airshed.
- In line with the preceding points, Beijing and China have used air pollution information transparency, both of ambient air quality through a nationwide air quality monitoring network and continuous emissions monitoring systems from specific industrial sources, such as power plants and factories, to raise public awareness of air pollution and to hold polluters accountable for violating emissions standards.

These factors have enabled Beijing to meet its goals to continuously improve air quality. In the next decade, it will have the chance to deepen these policies and make more progress in neighboring provinces, in order to address the severe air quality challenges in northern China. Given China's goal of achieving carbon neutrality by 2060 and the need for some cities and provinces to be forerunners in achieving carbon peaking and then neutrality, it is also quite likely that Beijing will turn its future policy focus toward reaching net-zero emissions well before 2060, just as it led the nation in setting stronger air pollution control targets and policies. The Beijing Winter Olympics in February 2022, which aimed to be net zero, are one showcase for this effort. With Beijing having accumulated substantial expertise and experience in addressing air pollution over the past two decades, the next decade and more will continue to demonstrate how the city can lead in reducing both air pollution and greenhouse gas emissions.

NOTES

1. Other scholars prefer to use the term "quasi-federal" in describing China's system of governance and the separation of powers between the central and provincial levels, noting, for example, that "in terms of institutional functions, China's central agencies have a great deal of power over economic planning, tax policy, some pricing, and standard-setting. Meanwhile, local governments control much permitting, other aspects of pricing, portions of production, and land policy." Compared to "more top-heavy forms of unitary government rule," China's governance structure is characterized by *tiao-kuai* relations, with "functional, vertical relationships (*tiao*)" dominating in areas of governance with more centralized

control and "territorial, horizontal (*kuai*) relationships" dominating in situations with greater local government responsibility. Davidson, Michael. 2019. *Creating Subnational Climate Institutions in China*. Discussion Paper, Harvard Project on Climate Agreements, www.belfercenter.org.

2. *Reuters*. 2020. "China Gives Shenzhen More Autonomy for Market Reform, Integration." October 18, www.reuters.com.
3. Library of Congress. 2020. "Regulation of Air Pollution: China." December 30, www.loc.gov.
4. *Measures of Beijing Municipality for Implementing the Law of the People's Republic of China on Prevention and Control of Air Pollution*. 2001. www.asianlii.org.
5. National Bureau of Statistics of China. 2020. "China Statistical Yearbook: 2020." China Statistics Press, December 1, www.stats.gov.cn; UN Environment. 2019. "A Review of 20 Years' Air Pollution Control in Beijing." www.unep.org.
6. Jin, Yana, Henrik Andersson, and Shiqiu Zhang. 2016. "Air Pollution Control Policies in China: A Retrospective and Prospects." *International Journal of Environmental Res Public Health* 13(12): 1219. https://doi.org/10.3390/ijerph13121219.
7. UN Environment 2019, 19.
8. UN Environment 2019, 19.
9. Jin et al. 2016.
10. Jin et al. 2016.
11. Finamore, Barbara. 2013. "China Pledges to Tackle Air Pollution with New Plan." NRDC, September 13, www.nrdc.org.
12. UN Environment 2019, 4.
13. *Baidu Encyclopedia*. n.d. "Beijing 2013–17 Clean Air Action Plan." Accessed May 12, 2021, https://baike.baidu.com (in Chinese); Beijing Government. 2013. "Full Text of 'Beijing 2013–17 Clean Air Action Plan Key Responsibility Allocation.'" D1EV.com, September 3, www.d1ev.com (in Chinese).
14. UN Environment 2019, 27.
15. *See* Xinhua. 2018. "Beijing's Last Large Coal-Fired Power Plant Suspends Operations." *XinhuaNet*, March 18, www.xinhuanet.com.
16. Xinhua. 2005. "Steel Giant Shougang to Move Out of Beijing." *China Daily*, February 8, www.chinadaily.com.cn; Mysteel. 2019. "Relocation—Boon or Bane for China Steelworks?" April 29, www.mysteel.net.
17. Yan. 2019. "China Focus: China Starts Implementing Tougher Vehicle Emission Standards." *XinhuaNet*, July 2, www.xinhuanet.com; "China: Light-Duty Emissions." *TransportPolicy.net*. Accessed October 12, 2020, www.transportpolicy.net.

PART IV

Reducing Greenhouse Gas Emissions

10

Delhi's Journey to Reduce Greenhouse Gases

Initiatives and Learnings from the Electricity Sector

NEERAJ KULDEEP AND TIRTHA BISWAS

Global cities are major contributors to greenhouse gas (GHG) emissions. Studies indicate that cities consume over 78 percent of global total final energy consumption and are responsible for 60 percent of global GHG emissions.[1] For a developing country like India, an expanding economy leads to rapid urbanization, which in turn increases energy consumption—and its associated GHG emissions. This is especially true for the national capital of Delhi, which is growing at a faster rate than the rest of the country.[2]

Delhi is a city-state, which means the border of the city also represents the border of the state. Hence, it is the only state in India that has 97.5 percent of its population living in an urban area. Delhi is also a highly populous city. According to the latest census, in 2011, Delhi had a population of 16.7 million, making it the second most populous metropolitan area in the world.

The majority of Delhi's economic activity comes from the services sector, which contributed more than 80 percent of the gross state domestic product in both 2018 and 2019; Delhi's relatively small industrial sector contributed only around 14 percent.[3] As a result, the state primarily relies on interstate trade to cater to its population's increasing material demands.

Between 2011 and 2019, Delhi saw considerable growth in its per capita income, commensurate with the national average. However, when compared in absolute terms, Delhi's per capita income is nearly three times the average national per capita income. Along with a higher per capita income, Delhi citizens also have a higher consumption footprint; per capita electricity consumption in Delhi is nearly 40 percent higher than the national average.[4] Delhi also has the highest vehicle ownership of all major cities in India, at almost ten times the national average.[5] As a result, it has a greater GHG footprint than many of its neighbors.

Delhi's high economic growth rate, however, is threatened by the adverse consequences of climate change. In response, the state government has begun to implement some promising policies and programs to reduce GHG emissions, which may provide examples for other states to follow. The decision to shut down the thermal power plants within the state's boundary has been a key contributor to the localized reduction of emissions, alongside other policies promoting the use of decentralized renewables in the residential sector and compressed natural gas usage for transportation. These policies and the questions about their effectiveness are explored further in this chapter.

Delhi's Power Sector

Power demand in Delhi is increasing.[6] Every summer, news articles are published regarding the ever-growing power demand. Between 2009 and 2019, the peak power demand grew steadily, with a 5 percent compound annual growth rate. And between 2010 and 2019, electricity consumption increased from twenty billion units to thirty billion units. Notably, 2020 was an exception, due to COVID-19 and its impact on economic activities.[7]

As an urban center, domestic consumers constitute the largest share of the total electricity demand in Delhi, at 53 percent between 2018 and 2019. Commercial and industrial consumers are responsible for another 23 percent and 14 percent of the city's total electricity demand, respectively.[8] High domestic load results in a very peculiar demand curve in Delhi, with peak demand in the morning and evening hours. Air-conditioning load, in particular, is the key contributor to Delhi's peak demand, which presents challenges to efforts to decarbonize the grid; the domestic air-conditioning demand tends to peak during morning and late-afternoon hours, while solar energy production peaks in the middle of the day. This mismatch could impede the share of distributed solar due to higher integration costs.

In December 2019, the Central Electricity Authority, as part of its Nineteenth Electric Power Survey, released projections on electricity demand and peak power requirement for Delhi. Electricity demand is likely to rise from thirty billion units in 2019 to forty-seven billion units by 2030. Similarly, Delhi's peak power requirement is expected to be about twelve thousand megawatts (MW) by 2030.[9] The growing electricity demand primarily occurs in the domestic and commercial consumer categories, which can be attributed to population growth, rising income and appliance ownership, affordable electricity prices, and rising ambient temperatures.

Delhi's Energy Mix

As we will see, the power sector is a major contributor to Delhi's GHG emissions. Delhi relies heavily on coal for its electricity requirement—a significant source of emissions, with a 74 percent share in the final electricity generation. Given the substantial gap between peak and base load, gas is another preferred source of electricity to meet the peak demand, contributing an additional 14 percent. In contrast, renewable energy use is low. Hydro and nuclear-based generation make up only 9 percent and 2 percent, respectively, of the city's final electricity generation. And the share of other renewables collectively, including solar, wind, and biomass, makes up only 1 percent of the overall electricity demand in Delhi.[10]

Due to air pollution concerns and limited gas availability, Delhi has been increasingly relying on electricity from power plants located far away from the city. And because Delhi does not count emissions from these out-of-state plants in its GHG inventory, the shift toward out-of-state generation has deflated the state's reported emissions.

In fact, all of the coal-based power-generating stations that are supplying power to Delhi are now located outside the city boundaries. Two coal-based power plants, the Rajghat power station and the Badarpur thermal power station, both commissioned in 1989 and 1970 and located within the city boundaries, have since been shut down due to a combination of economic and air quality concerns. The motivations for the plants' closure are discussed in more detail later in the chapter.

While there are no longer coal power plants in Delhi, two gas-fired power plants remain within the city. Due to the scarcity of domestic gas, these two plants are predominantly used to meet peak demand. These plants are also part of Delhi's islanding scheme to first ensure or restore power to Delhi in the events of grid failure or unavailability of power from outside the city. However, the availability of gas has declined over the years, and the share of gas-based generation in Delhi's overall electricity mix has declined as well. There is a plan by Pragati Power Corporation Ltd. to set up an additional 750 MW gas power plant in Delhi, but the plan is currently on hold due to the unavailability of gas.[11]

GHG Emissions and Source Apportionment

Energy use across various end-use sectors remains the single largest contributor of GHG emissions in Delhi, and in 2015, the emissions from the energy sector represented approximately 89 percent of the city's total emissions (figure 10.1). The waste sector, which includes municipal solid waste and domestic wastewater, is the second largest contributor, at 10 percent. Both the agricultural and industrial sectors in Delhi have a negligible contribution to the city's total emissions.

Moreover, electricity generation within the city boundary is one of the key sectors contributing to Delhi's overall GHG emissions. Back in

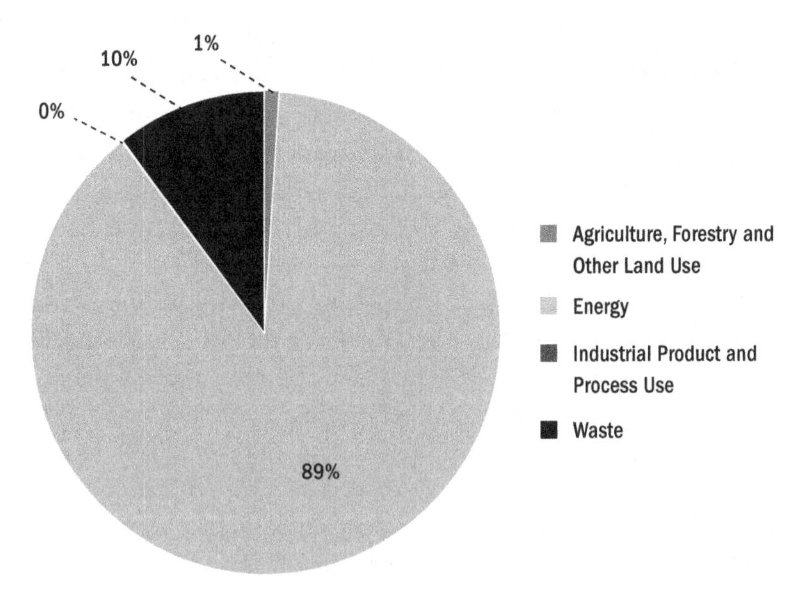

Figure 10.1. Sector-wise disaggregation of GHG emissions in 2014. (Sharma, Mukesh, and Onkar Dikshit. 2016. "Comprehensive Study on Greenhouse Gases (GHGs) in Delhi." Indian Institute of Technology Kanpur, October.)

TABLE 10.1. Share of Delhi's GHG Emissions by Sector

Economic activity	GHG emissions (CO_2eq million tons), 2014	Share in total emissions
Electricity consumption	16.38	43%
Transport	12.27	32%
MSW burning	3.12	8%
Residential sector	2.45	6%
Refrigerant use	1.25	3%
Livestock	0.57	2%
Industry	0.56	1%
Cropland	0.39	1%
Incinerator	0.27	1%
Green cover	0.18	0%
Restaurant	0.17	0%
Landfill	0.09	0%
Agriculture	0.08	0%
Diesel generator	0.05	0%
Cremation	0.04	0%
Aircraft	0.01	0%
Drain	0.01	0%
Wetland	0.01	0%
Total	37.9	

Source: GHG Platform. n.d. "GHG Platform India." Accessed October 15, 2020, www.ghgplatform-india.org.

2014, when Delhi was still heavily reliant on local thermal power plants, electricity generation and consumption made up about 43 percent of the total GHG emission load.[12] The contributions came primarily from the two coal-based power plants, referred to in the previous section, both of which had reached their useful operational life and have since been shut down. Since then, Delhi has increased imports of electricity from other states to cater to its soaring electricity demand. Between 2015 and 2016, Delhi imported approximately 80 percent of its electricity, resulting in a carbon emissions footprint of 24 million tons of CO_2—roughly five times the emissions from the sector in that year.[13]

Table 10.1 disaggregates emissions across key end-use activities. Electricity generation and use and transport activities contributed to the majority of Delhi's overall GHG emissions. As the table demonstrates, the second largest source of emissions is the transport sector. Between 2005 and 2015, emissions from the transport sector increased from 8 to 13 million tons of CO_2. Road transport remained the major source for transport emissions, representing a share of 61 percent of the total sectoral emissions. Aviation fuel consumption, used for domestic aviation, contributes the remaining 39 percent.

Despite the rise in transport-related emissions, the recent GHG inventory for Delhi indicates that between 2005 and 2015, overall GHG emissions from Delhi declined—dropping from 29.3 to 23.3 million tons of CO_2eq during a time when the economic growth of the city remained robust.

The decrease in emissions is partially due to the city's low-carbon initiatives, but it is also largely the result of the city's accounting methodology, which does not include emissions embedded in products, including electricity, that were produced outside the city. Because most of Delhi's gross state domestic product comes from the services sector, the production-based accounting method may significantly underestimate the GHGs attributable to Delhi's residents.[14] As a case in point, and as stated previously, coal generation was shifted out of the city, and Delhi relies heavily on imported electricity from other states to cater to its vast population base—yet the emissions expended in the production of such electricity are not attributed to Delhi in its GHG inventory.

Delhi's Journey to Reduce GHG Emissions

Delhi has taken actions that have resulted in a lower GHG emissions inventory for the city. However, many of these actions have been primarily motivated by concerns about air pollution and affordability rather than concerns about climate mitigation. As per the World Health Organization (WHO)'s survey of world cities, Delhi has been consistently ranked among the worst polluted cities.[15] The annual average $PM_{2.5}$ concentration in Delhi has been more than 100 $\mu g/m^3$, which is almost ten times higher than the safe limits stipulated by the WHO.[16] Average $PM_{2.5}$ concentrations during winter months are even higher, at more than 200 $\mu g/m^3$ on most days, which makes the air in Delhi more toxic than any other city.[17]

In November 2017, a thick layer of smog covered the entire city, and $PM_{2.5}$ and PM_{10} levels went above the AQI censors' measurement limit of 1,000 $\mu g/m^3$.[18] This event, referred as "The Great Smog of Delhi," alarmed all members of society, including policy makers and enforcing agencies. A democratic demand for clean air, alongside an increased awareness of air pollution and its problems, led to a widespread push for the government to take concrete action to curb air pollution in the city. The state government's proactiveness in shutting down coal plants and supporting solar deployment and electric vehicles through incentive schemes and energy-efficiency measures was essential in fulfilling the public demand for clean air.

The current ruling party in Delhi government, the Aam Aadmi Party (AAP), came into existence out of an India-wide anticorruption movement in 2011. The political party was officially formed in 2012 and won the Delhi Assembly election in December 2013. The core principles of the AAP, which include a commitment to pro-people governance and a rejection of VIP culture, resonated well with the masses in Delhi and across India. AAP was focused on improving the basic necessities—electricity, water, education, health, and so on—for the people of Delhi and won a clear mandate in the Delhi Assembly, with sixty-seven out of seventy seats. Since then, the AAP government has worked with agencies to bring reforms across numerous sectors, including the power sector, and has implemented a variety of measures such as a generation-based incentive scheme and virtual and group net-metering policy, which are described further in the following section, and the closure of Delhi's two remaining coal-fired generation plants.

The closure of Delhi's coal-fired plants was motivated by a mix of environmental and economic concerns. On the environmental front, a 2015 report by a team of officials from the Central Pollution Control Board, the Delhi Pollution Control Committee, and the Ministry of Environment, Forest and Climate Change concluded that both plants were also defying the emissions norms, with PM_{10} levels observed above 150 $\mu g/m^3$.[19] The Badarpur plant, in fact, was ranked as the poorest performing plant on environmental emissions control in multiple independent studies, contributing on its own between 80 and 90 percent of particulate matters, NO_x, and SO_x emissions from the energy sector.[20]

There were economic motivations for closing these plants as well. During the previous tenure of the current Delhi government, in 2013, it promised to provide electricity to domestic consumers at affordable prices. The promise was fulfilled with a 100 percent subsidy for the first two hundred units and a 50 percent subsidy for the next two hundred units.[21] However, providing cheaper electricity required a cheaper source of power. The age-old thermal power plants were operating with reduced efficiency, and the cost of generation was higher than other generating sources. Shutting down these plants saved about twenty paise per unit of electricity being generated, leading all government bodies and distribution companies (DISCOMs) to support permanent shutdowns.[22]

With the path to closure cleared, the Rajghat power station was permanently shut down in 2015. The Badarpur plant initially faced a temporary shutdown in 2017 to alleviate acute air pollution in Delhi after a thick layer of smog covered the entire city, but eventually, with a recommendation from the Environment Pollution Prevention and Control Authority, the plant was permanently shut down in October 2018.[23]

Initiatives Reducing GHG Emissions in Delhi

While the government has historically had varying motivations for acting, Delhi has pursued several actions in the power and electricity sectors that are specifically targeted toward reducing GHG emissions. However, the impact of these policies on Delhi's GHG emissions inventory is as of yet unclear. This section explores some of the actions that Delhi has taken, including actions to increase energy efficiency, promote renewable energy, transition to cleaner transport energy sources, and reduce electricity use in the buildings sector.

State Action Plan on Climate Change

In 2008, the government of India, with recommendations from the Prime Minister's Council on Climate Change, launched the National Action Plan for Climate Change (NAPCC). The program aims to identify measures on climate change mitigation and adaptation, which aid in advancing India's climate and development-related objectives.[24] States were asked to prepare their individual—but more robust—State Action Plans on Climate Change (SAPCC), in line with the strategy outlines in the NAPCC. Delhi was among the twenty-two states and Union Territories in India to put such a plan in place.

The first SAPCC for Delhi was prepared in 2009.[25] The plan had sixty-five agenda points for four years, beginning in 2009, and thirty-six priority actions, including both national and subnational goals, with points on providing clean air and water, promoting a noise-free environment, managing municipal waste, increasing green cover, and mitigating climate change. The State Action Plan became the foundational piece for more focused strategies, including ones to reduce reliance on coal, to promote solar energy in the state, to promote energy efficiency, to restrict diesel generators, to promote compressed natural gas (CNG) fuel, to promote electric vehicles, and so on.

Delhi's commitment to climate change has remained in the years since, with several key policies and programs enforced post-2012. In 2019, Delhi submitted its revised State Action Plan on climate change, with revised targets on mitigation and adaptation for the following ten years. The revised plan echoes the key principles of the previous SAPCC and focuses on six vulnerable areas, including energy, transport, and urban development.[26]

Energy Efficiency

Separately, Delhi has also worked to improve energy efficiency in buildings. In 2006, the government of Delhi established, within the state Department of Power, the Energy Efficiency and Renewable Energy Management Centre (EEREM) pursuant to the powers granted under clause (d) of Section 15 of the Energy Conservation Act of 2001.[27] EEREM is the state-designated agency for the implementation of state and central schemes on energy efficiency and renewable energy. Since 2006, when the focus was on energy efficiency, the EEREM has been instrumental to the success of various energy-related programs, including the Unnat Jyoti by Affordable LEDs for All (UJALA) scheme and the Street Light National Programme (SLNP), which targets to replace conventional streetlights with energy-efficient LED lights.

Under the UJALA scheme, each household is provided with four LED lamps at the cost of INR 70 (approximately US$1), payable in monthly installments of INR 10 along with the household's electricity bills. As of October 2020, about 13.2 million LED bulbs and tube lights have already been distributed in Delhi, which is expected to save about 345 MW of power demand over time, as well as 1.7 billion kilowatt-hours (kWh) of electricity and 1.4 million tons of CO_2 every year.[28] The EEREM is also implementing similar schemes with DISCOMs to distribute efficient air conditioners, which are the key contributors in Delhi's growing electricity demand.

Delhi has also been promoting energy efficiency in new buildings through the adoption of the Green Building Code.[29] The target is to achieve 250 green buildings by 2020 and 500 by 2030. Furthermore, the government has implemented the Building Energy Efficiency Programme for retrofitting existing buildings with energy-efficiency solutions; however, the government has not yet undertaken significant activities to implement these goals.[30] Some of the 2020 targets include retrofitting 50 percent of the existing commercial and industrial buildings and achieving a demonstration of the "Net Zero Energy Building" concept.

Finally, as part of Delhi's climate change agenda, the government of Delhi has also promoted several demand-side measures to reduce electricity consumption through mandatory audits of large public and commercial institutions and deployment of energy-efficient street lamps, as described earlier.[31] Additionally, the government has supported the state DISCOMs in implementing several DSM schemes; notable ones include the Renewable Energy Assisted Pumps, Bachat Lamp Yojana, Appliance Replacement Program, and Automatic Demand Response program with Smart Meter.

Renewable Energy

Reducing the GHG emissions associated with electricity consumption has been one of Delhi's key focus areas, and apart from improving energy efficiency, it has primarily done so by promoting a shift to solar energy. Solar installation in Delhi has expanded in recent years, and cumulative solar installations in Delhi have increased at an exponential rate over the past five years.[32] Much of this action occurs at the state level, though DISCOMs have also taken on a key role, particularly in overcoming barriers for consumers.

STATE SOLAR POWER POLICY

Historically, the state has taken action to promote solar energy. In 2001, the government of Delhi mandated use of solar water heaters for all industries, hotels, education institutions, and other commercial and residential buildings with an area of more than five hundred square meters, including both existing and new buildings. More recently, the state's solar power policy has set a pathway for expanding solar installations in Delhi.

India's solar mission aims to achieve 100 gigawatts (GW) of solar generation capacity nationwide by 2022, with a 40 GW target for rooftop solar installations.[33] In furtherance of this goal, states were asked by the Ministry of New and Renewable Energy to formulate policies to boost solar deployment in their respective states. Delhi announced its rooftop solar policy in 2016, with the aim to reduce Delhi's reliance on conventional energy and to develop solar energy to provide affordable, reliable, continuous power to the citizens of Delhi and lower energy prices in the long run.

The policy, established initially with a five-year operative period, aims to achieve 1 GW of solar capacity by 2020 and 2 GW by 2025, within city boundaries.[34] This would translate into 4.2 percent and 6.6 percent of total energy consumed in 2020 and 2025, respectively. The policy also provides an annual deployment trajectory for solar installations and emphasizes the creation of market-based mechanisms, with supportive policies, regulatory regimes, and incentive structures to drive the demand and adoption of solar with minimal use of state government subsidies. The regulatory and incentive schemes include the following:

- *Group and virtual net-metering.* As Delhi is an urban settlement with multistory buildings, for the first time in India, the policy proposed implementing innovative group and virtual net-metering to cater to the specific needs of the potential solar owner.[35] Group net-metering allows a consumer to install a larger solar system and use the excess solar electricity export to offset electricity consumption for multiple buildings across Delhi (within the same distribution licensee's area of supply), whereas virtual net-metering allows consumers without roof ownership to access solar electricity from existing solar plants. Both metering mechanisms were officially notified by the Delhi Electricity Regulatory Commission in 2019.[36]

- *Generation-based incentives.* A generation-based incentive was proposed to further improve the economic viability of solar for consumers. The scheme provides an additional INR 2 per unit of gross solar energy generated. The incentive is only applicable for the first three years, during the operative period of the scheme.[37]
- *Renewable purchase obligation (RPO).* Under the RPO obligations, the policy recommends that DISCOMs prioritize at least 75 percent of their RPO targets through sourcing renewable energy power from within the state, a higher target than is currently set by the state's electricity regulator.[38] This would support rooftop solar deployment, since the generation from net-metered rooftop solar systems is accounted toward RPO fulfillment. The policy proposes various other incentives in the form of exemptions, including electricity tax and fees, open-access charges, wheeling and banking charges, and transmission charges. The policy also amends the building bylaws for Delhi, allowing solar systems to be installed over and above the maximum permissible building height without any special permissions from local authorities.

The incentives provided as part of the Delhi solar policy are in addition to the capital subsidy provided under the Central Finance Assistance scheme from the Ministry of New and Renewable Energy. Under this scheme, residential consumers are given 40 percent of the system cost for 1 to 3 kW of solar capacity and 20 percent for 3 to 10 kW of solar capacity.[39] Thus, residents will be able to take advantage of both state and national subsidies for solar power.

Separately, another special scheme, the Agriculture-cum-Solar Farm Scheme of National Capital Territory (NCT) of Delhi, was notified for the agricultural consumers with the objective of providing additional income to farmers with solar power plants. As part of the scheme, a farmer can install a minimum of one megawatt of solar capacity on a Rural Electric Supply Cooperative model.[40]

OVERCOMING BARRIERS TO SOLAR DEPLOYMENT

The early success of solar deployment in Delhi could be attributed to EEREM, which led the efforts on demonstration pilots and mobilized various departments to install solar on government buildings. Gradually, as the economic viability improved for solar and more consumers started to adopt solar, the role of DISCOMs became central. The Ministry of New and Renewable Energy's Rooftop Solar Phase 2 program, for example, recognizes DISCOMs as the lead implanting agency for rooftop solar, and DISCOMs in Delhi have been implementing innovative approaches to further accelerate rooftop solar deployment.[41]

Yet despite the efforts by the state to promote solar energy, there have been several barriers that have hampered its efforts. A large share of electricity connections in Delhi are in the domestic category, and domestic consumers face significant barriers in adopting rooftop solar in particular due to high up-front cost, less consumer awareness, lack of access to exclusive roof ownership, lack of access to affordable finance, and a low quality of vendors, among others.

To overcome these barriers, BSES Rajdhani, one of the DISCOMs in Delhi, launched a solar city initiative, Solarise Dwarka, primarily to create demand aggregation through consumer awareness and provide the Rural Electric Supply Cooperative model (no up-front investment by the consumer). The program became successful in a short span, creating about 2.6 MW of rooftop solar capacity by high-rise residential societies within six months. Similarly, BSES Yamuna, another DISCOM, is in the process of testing business models, including community solar for shared roof owners, on-bill finance for those who see up-front cost as a barrier, and solar partners for those who live in rented properties without roof ownership.[42]

A lack of consumer awareness has been identified as another significant impediment to solar adoption, as most consumers are still unaware of the true benefits of a rooftop solar system. Those who are aware of the benefits and are interested in rooftop solar do not necessarily have information on the quality of systems and vendors, DISCOM processes, financing options, government schemes, and so on. However, utilities are taking measures to remove informational barriers to the installation of solar panels.

For example, Delhi DISCOMs are actively engaging with consumers and Resident Welfare Associations to fill the information gap. Even during the COVID-19 pandemic, DISCOMs have been regularly conducting workshops and interactions to increase consumer awareness. DISCOMs have also created a centralized portal with information on solar and with net-metering applications, providing a transparent process for consumers and reducing the lead time on processing applications.

Transportation

In the transport sector, Delhi has taken steps to transition toward cleaner sources of energy, and the city is one of the few early movers to CNG. While the primary objective was to address the rising air pollution levels, a transition to CNG also significantly reduced the carbon emissions footprint of the sector. The state government used a series of regulatory measures to increase the penetration of CNG within the transport energy mix, and in 2001, the Supreme Court enforced a switch-over to CNG for all public transport vehicles.[43] This was also supported by a fiscal measure that exempted CNG from the value-added tax.[44]

Recently, the state government also began to incentivize vehicle electrification and set a goal for 25 percent of new vehicle registrations by 2024 to come from battery electric vehicle sales.[45] Some other supporting incentives include providing interest subvention on loans, waiver of road tax and fees, and infrastructure support through the establishment of charging stations and battery-swapping facilities.

Conclusion

While Delhi has adopted a wide variety of measures to reduce its GHG emissions, there are reasons to wonder whether its commitment will continue. Peak power demand in Delhi is steadily increasing as the population and economy continue to grow. And, as noted previously, Delhi's population has increased threefold in the past thirty years, and the monthly per capita expenditure in the city is one of the highest in

India. These factors are contributing to a higher adoption of electric appliances and a higher electricity demand.

In addition, as discussed, Delhi faces particular challenges in addressing its increasing power demand. As an urban center, air conditioners are the key contributors to the overall power demand in Delhi, which results in two peaks; other states, in comparison, usually face only a single peak in a day. Additionally, the large difference between base load and peak load imposes significant stress on the distribution and transmission infrastructure.[46] Though growing peak demand leads to more revenue and a more efficient utilization of distribution assets for DISCOMs, it also requires a higher investment in upgrading the infrastructure.

In order to overcome challenges associated with growing peak demand and lower base load, DISCOMs in Delhi are promoting rooftop solar adoption and supporting various energy-efficiency measures, and the state has introduced its own initiatives to promote solar and electric vehicle adoption. However, it is unclear to what extent these policies have contributed to meaningful reductions in GHGs and how much more progress can be achieved with these measures alone.

NOTES

1. United Nations. 2021. "Cities and Local Action: Cities and Pollution." United Nations Climate Action, January 18, www.un.org.
2. As the gross state domestic product of Delhi grew at a compound annual growth rate of 7.42 percent during the past five years, the national GDP increased with a compound annual growth rate of only 5 percent during the same period. RBI. 2020. "Handbook of Statistics on Indian States 2019–2." October 13, https://m.rbi.org.in.
3. Planning Department, Government of National Capital Territory of Delhi. 2019. "Economic Survey of Delhi, 2018–19," 3, http://delhiplanning.nic.in.
4. Data cited are based on authors' own assessment.
5. Energy and Resources Institute. 2014. "Proliferation of Cars in Indian Cities: Let Us Not Ape the West," 2, www.teriin.org.
6. Electricity distribution in Delhi is divided into five regions, each served by a separate entity. A large part of Delhi is served by three distribution companies (DISCOMs): BSES Yamuna, BSES Rajdhani, and Tata Power Delhi Distribution Limited. Apart from these three DISCOMs, the New Delhi Municipal Council supplies power to ensure uninterrupted power to the administrative district area, and the Military Engineering Services is responsible for electricity supply to the cantonment area. All three DISCOMs are based on a public-private partnership model, in which the Delhi government holds a minority stake. Private DISCOMs

came into existence during a major power-reform phase in India between 1997 and 2002, which also shaped Delhi's power sector as we know it today. In 1997, the Delhi Vidyut Board replaced the Delhi Electric Supply Undertaking, which was responsible for electricity generation, distribution, and transmission in Delhi. The Delhi Electricity Regulatory Commission was formed in 1999 as an independent entity to regulate electricity tariffs and to reform the Delhi power sector. However, the Delhi Vidyut Board still failed to provide a reliable power supply, and due to both a rampant increase in electricity theft and frequent power outages, the entity struggled to remain financially viable. Eventually, the unbundling of Delhi Vidyut Board, which was merely a change in the legal status of the Delhi Electric Supply Undertaking, took place in July 2002. Six companies, including three private DISCOMs, one transmission company, one power holding company, and one generation company, were formed to bring efficiency and improve the overall health of the power sector. Nosal, Leah. n.d. "LAD Case Study: New Delhi Electricity." Stanford University Center on Democracy, Development and the Rule of Law.

7. In 2020, the peak demand stood at 6,314 MW, compared to 7,409 MW in 2019. Data cited are based on authors' own assessment.
8. Department of Environment, Government of Delhi. 2017. "Delhi State Action Plan on Climate Change." October 12, http://moef.gov.in.
9. Central Electricity Authority, Government of India. 2019. "Report on Nineteenth Electric Power Survey of India (Volume II)—(National Capital Region)." December.
10. Data cited are based on authors' own assessment.
11. Government of Delhi. 2019. "Economic Survey of Delhi—2018–19." January 17, http://delhiplanning.nic.in.
12. Sharma, Mukesh, and Onkar Dikshit. 2016. "Comprehensive Study on Green House Gases (GHGs) in Delhi." Indian Institute of Technology Kanpur, October, 25, www.indiaenvironmentportal.org.in.
13. Delhi Transco Limited. 2017. "Electricity Statistics 2015–16 and 2016–17." October 12, http://dtl.gov.in.
14. Between 2019 and 2020, Delhi generated a gross state domestic product of US$117 billion. However, the economic output from the industry sector only remained at a meager US$15 billion, representing only 13 percent of the gross state domestic product. The Intergovernmental Panel on Climate Change 2006 guidelines on National GHG Inventories is the widely accepted methodology and is used in developing national GHG emissions inventory submissions to the United Nations Framework Convention on Climate Change. However, as mentioned, a drawback of such a methodology is that it does not consider embedded emissions. This drawback seems particularly noteworthy when assessing the GHG emissions of economic areas where the service sector dominates.
15. Rice, Doyle. 2018. "90% of People Breathe Polluted Air: New Delhi Is World's Most Polluted Big City." *USA Today*, May 2, www.usatoday.com.

16. PTI. 2015. "$PM_{2.5}$ Level in Delhi 10 Times More than WHO Limits." *Economic Times*, February 16. https://economictimes.indiatimes.com.
17. Dodge, Eric, and Kevin Rowe. 2015. "Why Delhi's Air Worsens during Winters." *Pollution Watch* (blog), *Scroll.in*, October 20, https://scroll.in.
18. Garg, Anchal, and N. C. Gupta. 2020. "The Great Smog Month and Spatial and Monthly Variation in Air Quality in Ambient Air in Delhi, India." *Journal of Health & Pollution* 10(27) (September): 2. https://doi.org/10.5696/2156-9614-10.27.200910; Central Pollution Control Board. 2009. "National Ambient Air Quality Standards, Central Pollution Control Board Notification." Report No. B-29016/20/90/PCI-L. *Gazette of India*, pt. 3, sec. 4, November 18, 4, http://cpcb.nic.in.
19. *Indian Express*. 2015. "Badarpur, Rajghat Thermal Power Stations to be Shut Down." December 5, https://indianexpress.com.
20. *Economic Times*. 2016. "The Badarpur Plant's Effect on Air Pollution and Why It Needs to Be Shut Down." August 12, https://economictimes.indiatimes.com.
21. Bhaskar, Uptal. 2020. "Delhi Shows Way for Other States as Electricity Subsidies Light Up Electoral Fortunes." *Mint*, February 12, www.livemint.com; PTI. 2019. "Delhi Govt's Free Electricity Scheme an Example of Smart Governance—Kejriwal." *Energyworld.com*, *Economic Times*, October 10, https://energy.economictimes.indiatimes.com.
22. Goswami, Sweta. 2017. "Delhi Pays Huge Price for Power from Pollution-Heavy Badarpur Plant." *Hindustan Times*, March 16, www.hindustantimes.com.
23. TNN. 2018. "Badarpur Thermal Plant Shut for Good." *Times of India*, October 16, https://timesofindia.indiatimes.com.
24. Prime Minister's Council on Climate Change, Government of India. 2018. "National Action Plan on Climate Change (NAPCC)." October 15, http://moef.gov.in.
25. ICLEI. 2014. "Delhi State Action Plan on Climate Change—How It All Began." November 17, http://southasia.iclei.org.
26. Department of Environment, Government of National Capital Territory (NCT) of Delhi. 2019. "Delhi State Action Plan on Climate Change." http://moef.gov.in.
27. Energy Conservation Act. 2001. http://legislative.gov.in.
28. Ministry of Power, Government of India. 2020. "National Ujala Dashboard." October 28, www.ujala.gov.in.
29. Energy Efficiency and Renewable Energy Management Centre. 2018. "Delhi Energy Conservation Building Code 2018." http://web.delhi.gov.in.
30. Ministry of Power, Government of India. 2020. "Indo-Swiss Building Energy Efficiency Project (BEEP)." www.beepindia.org.
31. *See generally* Mehta, Rakesh. 2009. "Climate Change Agenda for Delhi 2009–2012." http://web.delhi.gov.in.
32. The total installed capacity stood at 122 MW in July 2020, compared to 12 MW in December 2015. PTI. 2020. "Delhi Power Discoms See Surge in Residential Rooftop Solar Power Connections." *Energyworld.com*, *Economic Times*, August 2, http://energy.economictimes.indiatimes.com.

33. *InsightsIAS*. 2020. "RSTV: The Big Picture—India's Solar Energy Push." July 27, www.insightsonindia.com.
34. Department of Power, Government of NCT of Delhi. 2016. "Notification: Delhi Solar Policy 2016." http://ipgcl-ppcl.gov.in.
35. Department of Power 2016, 7, 8.
36. Delhi Electricity Regulatory Commission. 2019. "Delhi Electricity Regulatory Commission (Group Net Metering and Virtual Net Metering for Renewable Energy) Guidelines, 2019." www.derc.gov.in.
37. Department of Power 2016, 10.
38. Department of Power 2016, 5.
39. *Economic Times*. 2019. "How Much Subsidy Can Individual Residential Households Avail for Installing Rooftop Solar Systems under the Phase-II of Grid-Connected Rooftop Solar Programme." October 23, https://economictimes .indiatimes.com.
40. Government of Delhi. 2018. "Agriculture-cum-Solar Farm Scheme in NCT of Delhi." http://web.delhi.gov.in.
41. Ministry of New and Renewable Energy, Government of India. 2019. "Guidelines on Implementation of Phase-II of Grid Connected Rooftop Solar Programme for Achieving 40 GW Capacity from Rooftop Solar by the Year 2022." August 20, https://solarrooftop.gov.in.
42. Kuldeep, Neeraj, Selna Saji, and Kanika Chawla. 2018. *Scaling Rooftop Solar: Powering India's Renewable Energy Transition with Households and DISCOMs*. New Delhi: Council on Energy, Environment and Water, June, xiii, www.ceew.in.
43. Aijaz, Rumi. 2018. "The Herculean Task of Improving Air Quality: The Case of Delhi and NCR." ORF Issue Brief 267, Observer Research Foundation, New Delhi, November, 12, www.orfonline.org.
44. Dutta, Sanjay. 2020. "Draft Policy to Push CNG, PNG in Cities." *Times of India*, January 23, https://timesofindia.indiatimes.com.
45. Transport Department, Government of NCT of Delhi. 2020. "Delhi Electric Vehicles Policy, 2020." August 7, https://transport.delhi.gov.in.
46. State Load Dispatch Centre Delhi. 2019. "Annual Report 2018–19." www.delhisldc .org.

11

Energy Cities

The Case of London

YAEL R. LIFSHITZ

This chapter unpacks the complementarities between two key trends: the rising importance of cities as subnational actors and the rise of distributed generation. The central premise here is that there are potential synergies between the two.

Whereas in the past our electricity was produced almost entirely by a few larger facilities and then transmitted to our homes, these days, with a solar panel on our roof, we can all be part of the energy production cycle. This is known as "distributed generation." Solar panels are probably the most intuitive example, but distributed generation includes a much wider range of technologies, from diesel-powered generators to smart meters and even tiny wind turbines that can plug into your phone. Importantly, distributed generation is inherently about producing electricity in close proximity to where it is consumed.

Cities, in turn, are particularly well suited to develop policies that can enable the flourishing of distributed generation, given the jurisdictional powers they typically hold over things like buildings, transport, streets, and more. Cities' potential in this regard is even greater, given that they house about half of the world's population, which means many more energy users lie within the city realms.[1] The chapter thus examines whether, and to what extent, cities and local communities advance the transition toward carbon-neutral energy by adopting or enabling distributed generation. It does so by drawing an example from one of the most vibrant global cities: London.

The chapter makes two main scholarly contributions. First, it offers a rich description of London's distributed generation efforts. Second, more broadly, it analyzes the role that cities can play in the transition toward distributed generation, which is particularly significant in an era when more than half of the world lives in cities. By calling attention to the intersection between urban hubs and distributed production of energy, the chapter also contributes to the conceptual discussion regarding the nature of "distributed" governance.

The discussion proceeds as follows. The first section offers a conceptual analysis of the nexus between cities and distributed generation and its potential. The second section illustrates by drawing on the case of London. The third section offers concluding remarks.

Synergies of Proximity

Cities are increasingly important to our global economy and to our cultural and social fabric.[2] Yet 2020 has brought with it a particular challenge to cities. The same traits that made cities so attractive to so many people, such as high connectivity and a high concentration of people and business in a relatively small area, also made them more susceptible to the effects of a pandemic.[3] For example, between March and May 2020, London and a few other major cities in the UK recorded a higher pandemic-related death rate per capita as compared to rural areas.[4] At the point of writing this, it is probably too soon to tell how exactly the global pandemic will impact cities. But even if the global rise in urbanism slows down, our twenty-first-century cities are not likely to dissipate so quickly.[5] The allure of an educated workforce and the innovative industries will probably persist.[6] Cities, which already house over half of the world's population, will probably remain a force to reckon with.

The benefits of cities arise out of proximity. Economically, lots of skilled workers are clustered together, which facilitates the transfer of knowledge and spurs creative industries. Socially and culturally, the proximity allows for social interactions and makes restaurants, pubs, and theatres viable.

Cities are also large consumers of electricity. According to one estimate, cities account for over 75 percent of global energy consumption and greenhouse gas emissions.[7] Energy in cities can mean many things, from transportation to buildings, industry, and more.[8] This chapter focuses on *distributed* generation. The terms "distributed generation" (DG) or "distributed energy resources" (DERs) generally refer to energy that is produced in proximity to where it is consumed.[9] This can be understood in contrast to more centralized models, under which electricity is produced in large facilities (think of a huge wind farm or a gas-powered power plant). The paradigmatic example of distributed generation is often a rooftop solar panel. But distributed generation is not limited to rooftop solar. It encompasses a wide variety of technologies and resources, including diesel generators, reuse of runoff heat, and more.[10] The benefits of distributed generation include improving resiliency, potentially reducing greenhouse gases, and potentially improving access to energy and providing financial benefits to participants.[11]

The global COVID pandemic has also impacted energy use. More people working from home also means a rise in domestic energy consumption (and, accordingly, a decline in office energy consumption). If you are now working from home more often, it makes more sense to invest in your at-home infrastructure. The pandemic might also have increased the need for a sense of independence with regard to, inter alia, energy. This too could feed into the growth of decentralized energy. It also simultaneously increases the importance of cities in this regard: with more than half of the population in urban areas, more residential energy use may also mean more energy use within the city realms.

The argument here is that with at least half of the world's population in cities, and given cities' jurisdictional capacities over key aspects such as housing, transport, and streets, cities are in a position to promote distributed energy.

On functional grounds, in a sense, this is another illustration of the "scale principle." Dean Lueck has shown that the scale of the resource can or should match the scale of the jurisdiction.[12] The same could be true with regard to the scale of the most effective solution or governance mechanism. Katrina Wyman and Danielle Spiegel-Feld apply this principle specifically to cities. Wyman and Spiegel-Feld make a case, on functional grounds, for including cities as actors on the international stage. Using the example of marine plastic waste, they show that, even if the problem is ultimately global, sometimes subnational actors are actually best placed to address the problem.[13]

I suggest that the scale principle applies in the context of distributed generation as well. The production of distributed generation and its consumption are often within the city's realm. The example of a solar panel is helpful again: let us assume that you want to install a solar panel on your roof. For the sake of this discussion, let us assume that you not only own your roof but also have secured all the necessary finances for the panel.[14] In order to install a solar panel on your roof, you still need some kind of planning approval or permit. That, typically, is controlled by the city or local authority. The same would be true for on-site batteries, on-street electric charging, reuse of runoff heat, and more. When both the generation and the use of the energy are local, the local authority is almost inevitably involved. As such, the city is well positioned to either facilitate or inhibit the growth of distributed generation.

To be sure, advancing distributed generation at the city level does not imply that national policy has no role to play. In fact, given the way the electric grid operates and the way infrastructure is set up, certain aspects of our energy operations will inevitably go beyond the scale of a specific city. Consider the electric grid: in the UK, for example, the grid is split, roughly, into four parts, all of which exceed the scope of one particular city. Likewise, roads and pipelines are set up in a way that encompasses areas larger than any particular city. Indeed, infrastructure like pipelines and grids tend to be regulated at a national or at least regional scale, not a local scale.[15] There are also useful economies of scale in utility-scale (nondistributed) generation. Moreover, national policies like carbon pricing or feed-in tariffs could greatly impact the adoption of distributed generation.[16]

The complementarities between urban hubs and distributed generation could also be justified on nonfunctional grounds. As Wyman and Spiegel-Feld point out, cities could also have a stronger claim to "democratic legitimacy" as compared to other nonstate actors.[17] The same could be true for distributed generation and the participatory aspect that goes along with it.[18]

To illustrate the synergies and the challenges of urban energy, the following section presents a case study, which is focused on distributed generation in London.

The Case of London

The UK has set an ambitious, legally binding target of net-zero emissions by 2050.[19] To date, the UK has already cut roughly 40 percent of its emissions, relative to a 1990 baseline.[20] According to one analysis, "almost all" of the recent emissions reductions came from the electricity sector.[21] The UK transitioned, within a decade, from a largely coal-based energy sector to one that is increasingly based on renewables. In 2010, fossil-fuel-based electricity accounted for roughly three-quarters of the UK consumption. In 2019, for the first time, renewables overtook fossil-fuel-based production.[22] On a city level, London has committed to net-zero carbon emissions by 2030. Importantly for our purposes, London aims to achieve 15 percent of its energy from local sources by 2030.[23]

A brief primer on the structure of local governance in London is in order. Local governance in London is divided into two "layers." The city-wide layer, known as the Greater London Authority (GLA), is made up of the Greater London Assembly and the mayor of London. Much of GLA's work is concerned with its "special responsibility for police, fire, strategic planning and transport."[24] The city is also divided into thirty-three districts (thirty-three borough councils and the City of London Corporation). This layer of governance is responsible for most of the day-to-day services of its residents, for example, "education, housing, social services, environmental services, local planning and many arts and leisure services."[25] This division of governance becomes significant in the context of distributed energy, as some of the activities that relate to distributed activities are in the realm of the boroughs, while others are more directly governed by the GLA.

This case study offers a snapshot of local-government activities with regard to distributed generation. Rather than reviewing the activities on the basis of the technology they support or the scale of the projects, the discussion offers a taxonomy based on the type of local-government action or activity. Doing so sets the stage for further discussion, both for the city itself moving forward and, importantly, for drawing broader insights regarding local-government activities and distributed governance.

Mandates

Introducing mandates is one way that local governments can promote the installation of distributed technologies.[26] These can often be thought of as demand-side measures, meaning that they aim to increase the "pull" (typically from the consumer side) for distributed generation (while other measures, such as funding, discussed later, aim to boost supply). London has not directly mandated the implementation of distributed technologies, yet there is a range of policies that indirectly promote or encourage the proliferation of certain technologies. These include requirements for new buildings, transportation, and informational mandates.

Consider the mandates for new buildings. The London Plan lays out several energy policies that must be satisfied for all new major developments as part of the planning and approval process.[27] First, it requires that all new major developments be zero-carbon. This target has applied to *residential* development since 2016 and, under the 2021 London Plan, will also cover nonresidential development. As part of this target, first, planning proposals are expected to meet an on-site 35 percent reduction in CO_2 emissions, going beyond what is required by national standards.[28] Second, where on-site reductions have been maximized, either any shortfall must be made up through carbon-offset payments to the local council's carbon-offset fund or all major development proposals will need to have an energy strategy that demonstrates how this zero-carbon target will be achieved.[29] As part of this energy strategy, developers need to: show plans for energy efficiency and a preference for local energy resources; say that they are seeking to "maximize opportunities for renewable energy by producing, storing and using renewable energy on-site"; and make suggestions on how to utilize demand-side response.[30] Finally, the London Plan requires certain "strategic" developments to undertake "whole life-cycle carbon emission assessments."[31]

More specifically, major developments that are located in Heat Network Priority Areas (located by the London Heat Map) are required to incorporate systems that rely on "existing or planned heat networks" where possible.[32] This provision is meant to encourage developers to use to runoff heat as a source of available, and local, energy.

London also operates transportation-related mandates. The Ultra Low Emission Zone, established in April 2019, places a tax on vehicles that enter central London, ranging from £12.50 to £100.[33] While this zone is not directly intended to reduce greenhouse gas emissions or to change the energy mix, it has nonetheless been influential on those accounts. Mandating vehicles to reduce emissions or pay reduced the proportion of "noncompliant" vehicles by 30 percent between March and

July 2019.[34] The Ultra Low Emissions Zone (ULEZ) will be expanded to the north and south circular in October 2021, which the GLA expects, in turn, will encourage commuters to use existing and newly expanded bicycle lanes.[35]

In addition to the citywide scheme, the mayor has also supported smaller-scale borough-led initiatives that make certain streets accessible only to ultra-low-emissions vehicles (ULEVs).[36] Two notable examples are as follows. First, as part of Camden Council's goal to "reduce pollution in the vicinity of 23 schools in Frognal/Fitzjohns," it decided to implement a Healthy School Street Zone in the area from September 2020.[37] For this to be successful, the council will close off certain streets and restrict entry to electric vehicles (EVs).[38]

Finally, local governments may also introduce softer, information-providing mandates. These are designed to overcome an otherwise market failure for parties to monitor, collect, and report useful data. For example, the newest version of the London Plan introduced a five-year reporting requirement on the "actual operational energy performance of major developments" within the city.[39] The emphasis on actual performance, rather than estimated or projected performance, is meant to encourage the identification and reporting of any differences.

The mayor also monitors and reports on the collective progress made by the implementation of the energy policies laid out in the 2016 version of the London Plan through the publication of annual Energy Monitoring Reports. Under this information-collecting scheme, in 2019, the GLA requested that all thirty-five of London's Local Planning Authorities (LPAs) provide information on the development and usage of their carbon-offset funds (funds that are then used for furthering carbon reductions, as discussed shortly).[40] The collection of this information culminated in a report published by the GLA, which contained recommendations to LPAs for the "next phase of the zero carbon and offset policy."[41] Under the new London Plan, this process will be repeated annually.[42]

Direct Funding

One of the key ways in which London supports distributed generation is through financing. In London, funding for distributed energy projects takes various forms, ranging from small "seed" grants awarded to local community energy schemes to large-scale renewable-energy and district heating projects that require more sizeable injections of capital.

Early-stage or "seed"-type funding is especially crucial for local community projects, in which a lack of early-stage funding is often a key challenge.[43] Although many of the community-scale projects will not rely on one source of funding (but rather combine private and public funding from various sources), city and local-authority funding does seem to play an important role in community projects in London.[44]

The London Community Energy Fund (LCEF) is a notable example.[45] The fund provides either small grants or capital funding for local community energy projects across the city.[46] This type of support is necessary, for example, for organizations like the Schools Energy Co-Op, which targets schools for the installation of solar paneling.[47] The LCEF mostly supports projects that generate clean energy, although it also supports a few projects that reduce demand for energy. The majority of projects involve solar photovoltaic (PV) panels (though some other notable examples include LED installations or EV charging points). As of 2021, the LCEF has gone through four funding cycles, supporting a total of eighty-six local energy projects in London. The typical recipients are nonprofits, co-operatives, and charities that act as the intermediary between the community and the various other public and private bodies involved (such as local authorities or investors). The projects generally look to install panels on community buildings or schools, where they can provide a community-facing educative role.

With regard to larger-scale energy projects, the London Energy Efficiency Fund (LEEF) operated between 2011 and 2015 specifically to support small to medium enterprises (SMEs) and the public sector.[48] Although this fund is no longer active, its significance is twofold: First, it is notably "the first dedicated energy efficiency fund in the UK."[49] Second, it also acted as the blueprint for subsequent funds that followed, and as such, it may be illustrative of the kinds of projects that may come under the current (newer) funds.[50]

The (much larger) successor to LEEF, launched in 2018, is the Mayor of London's Energy Efficiency Fund (MEEF), established through the European Regional Development Fund. It totals £500 million to be invested in low-carbon projects by May 2023, provided mostly in the form of loans. It is notable, in particular, for involving private-sector investors.[51] The MEEF considers investments in a range of projects, from decentralized energy and energy storage to EV charging and more.[52] Importantly, it requires that at least 70 percent of investments are made to the public sector, with the remainder of investments available for the private sector. In addition to this, "for every £7000 invested, one tonne of carbon needs to be saved, and energy efficiency projects need to achieve a 20% energy saving."[53] An example of one such recipient is the London borough of Enfield, for the construction of a large-scale heat network of insulated pipes, which will utilize and distribute low-carbon waste heat from a local incineration plant to heat the buildings and homes in the new development.[54] City Hall also offers grant funding for the "technical, commercial, legal and strategic work" involved with the implementation of large-scale decentralized energy projects.[55]

The first of these initiatives, the Decentralised Energy Project Delivery Unit (DEPDU), was a £2.8 million scheme (90 percent funded by the European Commission) that operated between 2011 and 2015.[56] Its successor was the £3.5 million Decentralised Energy Enabling Project (DEEP), launched in 2017, which was half financed by the European Regional Development Fund.[57] It allocated all of its funding by the end of 2020.[58] The latest initiative is the £6 million Local Energy Accelerator (LEA), a follow-on and expansion to DEEP, which is also half funded by the European Regional Development Fund, in partnership with the GLA. Agreed in November 2020, the scheme will run until July 2023 and aims to be more ambitious than its predecessors with regard to contributing toward London's zero-carbon target. Whereas both DEPDU and DEEP focused primarily on district heating that utilized gas-fired combined heat and power, the LEA stipulates that supported networks should utilize "low carbon heat sources."[59] Furthermore, its ambit will be broader, so as to encompass projects that touch on, for example, demand-side response and vehicle-to-grid technology.[60]

Project funding is also available at the council level. The Camden Climate Fund, set up by Camden Council, is an illustrative example. The fund is able to distribute grants to householders, businesses, and community groups that are looking to install renewable energy or heating systems on their properties. The funding is based on money from the carbon-offset payments that are collected when new developments fail to comply with the London-wide carbon-reduction targets, mentioned earlier.[61] The sums available tend to be smaller than citywide funding schemes, but the eligibility criteria and the purposes to which funding can be applied are more varied. For example, in Camden Council, a household fund provides funding for the installation of solar PV or solar thermal. Homeowners, private tenants, and private landlords are eligible to apply, and applications are assessed based on carbon-reduction potential, cost, and feasibility of delivery.[62] Similar funding is available under the Community Energy Fund and the Business Fund, with variations on the size of the funding pool and the eligibility criteria.[63]

Another example of council-level funding comes from the borough of Islington, with the Islington Community Energy Fund (ICEF).[64] Like the grants available in Camden, the Islington fund makes use of carbon-offset funds collected by the council. It targets "innovative energy projects," with a specific focus on tackling fuel poverty in the borough. However, unlike the Camden Climate Fund, the scope of the Islington fund is narrower, focusing on *community* projects. Thus far, Islington Council has gone through four funding rounds, totaling £360,000.[65] In previous rounds, some notable examples included a local housing cooperative, which aims to be "energy self-sufficient" and uses the funds to install a solar PV array; a church that was awarded a grant to install battery storage, so that it could use all of the energy generated by a solar PV array; and "radiant heat panels," designed to provide heating to small communal areas.[66]

Notably, the funding, both citywide and at the borough level, is typically tied to a particular project or technology, rather than being available to hire staff.[67] This seems to be one of the challenges for community-scale projects, which, in addition to project-specific finance, also need to sustain human resources and incur ongoing maintenance costs.

Self-Consumption

The city is also a large consumer and thus can have a significant impact by adopting DG within its own buildings and facilities. City Hall already operates with 100 percent "green" energy, which includes renewables.[68] The London Fire Brigade has installed solar capacity in over half of its buildings and is pursuing further installations. The Metropolitan Police

Service currently has twenty-two solar PV systems, and two further installations are in progress.[69]

A key area in which the city's own consumption can impact its energy strategy is transportation. In London, the transportation and infrastructure sectors currently consume most of the city's energy.[70] The subway, London's vast subterranean train network, derives its energy almost entirely from the (central) national grid.[71] Transport for London (TfL), the agency responsible for running London's transportation systems, suggests that there has been an overall reduction in their CO_2 emissions, although this is attributed to reductions in the UK grid more generally.[72]

More progress has been made aboveground. The double-decker bus, one of London's most famous hallmarks, is going electric. London currently has over 485 electric buses, making it the largest electric fleet in Europe.[73] A few bus routes operate exclusively on electric buses or within "Low Emission Bus Zones," and the city has contracted for dozens of more double-deckers, with additional financing on the way.[74] The aim, ultimately, is to electrify London's entire bus fleet.[75]

Aggregation

City governments are well placed to coordinate and consolidate the demand of their inhabitants. The idea is that by aggregating demand for DG technologies, cities can facilitate their implementation by enabling individuals to access economies of scale that would not be available to purchasers acting alone. Other aggregative measures may include meter aggregation, which allows for a single DG system to offset several meters, or group billing, which allows electricity customers to consolidate multiple accounts on one bill.

In London, the mayor's "Solar Together" initiative leverages group buying power to drive down the unit price of solar paneling, which lowers the financial barrier to entry into the renewable-energy market for consumers.[76] Reducing the costs of installation can open up the market to those who might otherwise be turned off by the up-front costs and long-term nature of the investment. This has typically been a problem for London, with its transient population.[77]

Essentially, the way the Solar Together program works is that homeowners, renters (with permission), SMEs, or Commonhold Associations register their interest in the scheme. City Hall then hosts a reverse auction with approved providers so as to attain the lowest price for the bulk package of solar panels. A personalized offer is sent through to those who registered, which customers may choose to accept or not. Importantly, though, the buyers must pay the fully quoted price for the panels, unless other avenues for funding are available. For example, at the borough level, the Camden Climate Fund may be able to provide up to 50 percent of the quoted costs of an accepted offer (up to £1,500).[78] The average savings on the cost of installing solar panels was 35 percent.[79]

The mayor of London has also set up London Power, a green energy company that provides Londoners with electricity and gas. London Power does not produce or deliver electricity directly. Instead, it partnered with a private energy company, Octopus Energy, which is the energy company behind the London Power operations. London Power essentially sells to its own customers the services that are provided by Octopus Energy. The point is that, by taking advantage of City Hall's bargaining power, which is achieved by aggregating the demand of individual city dwellers for affordable energy, the mayor has been able to offer tariffs that are both unique to London Power and cheaper than those available to regular customers of Octopus Energy. As of March 2021, 5,521 homes were powered by London Power.[80]

In January of 2018, the GLA launched a twelve-month pilot (using nPower as its third-party licensed supplier), which allowed the city to use its procurement power to purchase "excess electricity generated by London boroughs, public bodies and others and sell it to consumers," which, in this first instance, was Transport for London (TfL).[81] If successful, the plan was to roll out the scheme across other public sector bodies, such as the National Health Service, and even to private consumers. However, the GLA has confirmed that there are currently no plans to extend the pilot.[82]

Lastly, in addition to *providing* funding (as discussed earlier), local authorities (whether citywide or council or otherwise) can use their power to help *aggregate* funds from individuals, which can be put toward local energy. For example, a local authority in West Berkshire (west of London) has issued a "Green Bond" (the first UK local-government authority to do so), with the aim of raising funds for solar panels on buildings owned by the local authority.[83] The scheme, developed by the ethical investment platform Abundance Investment, will allow both residents and nonresidents to crowdfund local renewable-energy projects through a Community Municipal Investment. In this way, the scheme provides cash-strapped local authorities with a mechanism by which to fund local green projects.[84] A few other local councils intend to follow this model.[85]

Providing Supportive Infrastructure

The city can also provide supportive physical, legal, and legislative infrastructure and guidance in order to create an environment that is more conducive to the promotion of DG technologies (or at least one that does not inhibit their success). An intuitive example of *physical* infrastructure could be the provision of EV chargers. The borough of Camden, for example, facilitates chargers on public streets.[86]

Consider, next, city permitting. Permitting has the power to shape the local landscape in very dramatic ways, and it has been described by some scholars as one of the most influential government programs.[87] In the UK, as a matter of *national* policy, solar panels are a form of permitted development.[88] In other words, the default is that their installation does not require planning permission (subject to certain technical and practical requirements). Lifting this legal requirement and, therefore, making the installation of DG technology less of an onerous process may widen overall participation in the DG space. However, councils can limit this freedom to install solar panels.[89] Camden Council (mentioned earlier), for example, placed certain limitations on the installation of solar panels in conservation areas. Although this may seem at first glance like an insignificant limitation, it actually impacts a large area within the council.[90]

Other regulatory hurdles, however, tend to be outside the city's purview, given the way that electric grids are regulated. Ofgem is the UK regulator for both gas and electricity markets. Ofgem supports energy innovators through its Innovation Link Programme, which notably (in addition to general feedback) provides a regulatory sandbox.[91] The sandbox service allows for a temporary suspension of specific regulations that inhibit the trials of various products and services and provides real-world evidence and support for the viability of innovative energy products and services that are currently ruled out by regulation, thus informing future potential regulatory change. Ofgem ran two application windows for the original sandbox in 2017, which resulted in seven approved sandbox trials. What is particularly notable for our purposes is that almost every trial sought "to maximise the benefits of locally produced (and sometimes stored) electricity for local consumers" and some sought "the use of platforms to facilitate peer-to-peer energy trading."[92] Thus, while this is not an example of a local-led initiative, it does serve to show that the regulatory innovation on this front is being pursued with regard to the local-level activities.[93]

Another function that local authorities can serve is capacity building. One example of this type of activity in London relates to heat load and utilization.[94] The GLA has published a number of studies and reports to provide assistance to boroughs, developers, and planners. For example, the GLA oversees the publication and periodic update of the London Heat Network Manual and the London Heat Map.[95] These are designed to provide the comprehensive practical guidance needed for the development of networks and to help borough councils to identify potential heat network opportunities, as well as to locate opportunities for expansion in existing networks. The GLA has also researched the overall decentralized energy capacity in London, as well as the potential use of secondary/waste heat in district heating systems.[96]

Facilitating Private-Public Partnerships

Lastly, the city and its boroughs interact and collaborate with different community or business organizations in various ways.[97] For just a few examples, consider the Community Energy London, which partners with the GLA and specific boroughs such as Camden and Islington to support local projects such as installing solar panels on state schools (publicly funded schools).[98] Community Energy London provides support in the form of helping to secure finances, capacity building, and assistance in permitting. A similar example is the Camden Climate Change Alliance, which works closely with Camden Council.[99] A recent project that grew out of this collaboration is a solar project on a high school in Hampstead (on the crowd-sourcing model).[100] The Green Finance Institute, for example, aims specifically to develop the market for financing energy efficiency in buildings. It is supported, inter alia, by the City of London Corporation.[101]

Concluding Remarks and Thoughts on Distributed Governance

The central premise of this chapter is that cities are creatures of proximity. They thrive precisely because people, ideas, and businesses are clustered together. There is also a *conceptual* proximity to users, which stems from the city being closer to them in the "sovereignty food chain." Things like the planning of the homes we live in, our daily commute, and the heat that keeps us warm are often in the realm of cities. Proximity is also a defining feature of distributed generation. These proximities can be complementary.

Cities can harness the synergies of proximity with regard to distributed generation. The paradigmatic example is that of a solar panel on a residential roof. Since the city or local authority typically controls the permitting of residential homes, it also has the power to make it very easy for rooftop solar—or, alternatively, very difficult. Cities can also support (or hinder) activities such as charging stations for EVs, dedicated lanes for bicycles or EVs, use of runoff urban heat, and more. All of this takes place within the city realms.

This chapter illustrates the existing (and potential) synergies by offering a snapshot of London's activities with regard to energy, with a specific focus on distributed generation. It does so by organizing the discussion into categories based on the type of activity or the type of involvement on the part of the local authority (rather than the technology or its scale). This taxonomy also aims to set up the discussion regarding potential activities moving forward and to facilitate a comparative discussion, as this book sets out to do.[102]

The first category pertains to *mandates*. As an example, London has set requirements on new buildings to meet particular standards with regard to energy use and carbon emissions. These requirements, although primarily focused on new buildings, could serve to boost demand for distributed generation, in particular, solar PV and reuse of runoff heat. The second category is *direct funding*. An example in this category is the London Community Energy Fund, which provides funding to local community energy projects across the city. Funding is also available at the borough level; for example, the Camden Climate Fund offers funding for both individuals and small business. The third category focuses on *self-consumption*. Examples in this category include the use of distributed generation by the city itself, like solar panels on the Fire Brigade buildings. Another important example is that of transportation, which the city controls. London currently has a few dozen electric buses. It plans, ultimately, to electrify its entire fleet.

The fourth category stems from the city as a natural *aggregator*. London, for example, uses its bargaining power to obtain competitive rates for potential solar buyers through its Solar Together program. The fifth category focuses on providing *supportive infrastructure*. This could relate to physical infrastructure, such as chargers for EVs, placed on city streets, or it could be legal and administrative support, for instance, clearing regulatory hurdles to permitting. The sixth and final category is *private-public partnerships*, which underscores the significance of community-based and business organizations and their key role in facilitating distributed generation in cities. An example is a fruitful cooperation of Community Energy London with the GLA and specific boroughs such as Camden and Islington to support local projects, such as installing solar panels on state schools.

Of course, the claim is not that policies aim to promote distributed generation, or more broadly the energy transition, need to be limited to cities. First, distributed generation could also flourish in rural areas. Second, given the way energy grids are set up and the location of many large-scale renewable resources, other (nonlocal) levels of government almost necessarily need to be involved. For example, one of the hurdles to setting up micro-grids is the current regulation by Ofgem (which regulates electricity across the UK). This regulatory hurdle was explicitly mentioned by industry members who were interviewed as part of this case study. While the city can assist micro-grids in many ways, it cannot overturn Ofgem regulations.

Cities could also benefit from policies on the national level. The UK has adopted legislation that aims to decarbonize the energy sector.[103] This in turn helps the city achieve its own net-zero target. For example, the subway system has decreased its carbon footprint over the past decades; yet the reductions are attributed not to particular actions on the part of the city but to an overall reduction in the UK grid. Recently, as part of the COVID recovery package, the UK government included a program that aims specifically at homes and local buildings: a £1 billion program to make public buildings greener, including schools and hospitals, "to fund both energy efficiency and low carbon heat upgrades."[104] Although these programs are not intended to target cities specifically, given that over 80 percent UK residents live in urban areas, they are likely to help cities in furthering their carbon-reduction targets and to advance distributed generation.[105] The claim, then, is not that cities are the sole actors but rather that a fruitful cooperation between multiple levels of government is helpful and that cities have a large role to play in this space.

Finally, the rise of distributed generation in urban locations presents an opportunity to begin unpacking distributed governance. The question in this context is, more broadly: What happens when decisions are dispersed, and how do we govern such interactions? This form of governance, which is growing in our times, presents new challenges. In some sense, we have had these distributed mechanisms in private law for centuries. But we are seeing more of these types of mechanism due to various technological advancements and social changes.[106] The rise of distributed generation is one example. Others include the impact of Facebook likes, Amazon reviews, and individual transportation choices. Maybe, in fact, this represents more about a push and pull between central and decentral. And maybe it is redefining what "central" and "decentral" mean.

In any case, the rise of urban distributed generation presents an interesting example worth studying further. Cities are likely to continue to be at the forefront of these transitions, and the nexus between cities and distributed governance will continue to impact our global cities and our everyday lives.

NOTES

For helpful comments and conversations at various stages of this project, I am grateful to Danielle Spiegel-Feld and Katrina Wyman and participants of the Global Sustainable Cities Workshop, the King's College Research Seminar, Society for Legal Scholars annual conference, Sustainability Conference of American Legal Educators, ASU, and KPLUS Alliance, Nexus Governance in the Context of Climate Change Workshop. Chloe Gershon, who was instrumental to the research process, and Nafisa Eshan provided excellent research assistance.

1. To be sure, the claim is not that rural areas cannot promote distributed generation. There are certainly many economic and technical advantages to distributed generation in rural areas. The claim here is that only cities have the capacity and the potential to promote distributed generation, regardless of developments in rural areas.
2. *But cf.* Frey, William H. 2020. "Even before Coronavirus, Census Shows U.S. Cities Growth Was Stagnating." Brookings, April 6, www.brookings.edu.
3. Urban Living. 2020. "Great Cities after the Pandemic." *The Economist*, June 11, www.economist.com ("The virus has attacked the core of what makes these cities vibrant and successful" because cities "cram together talented people who are fizzing with ideas"); London Intelligence. 2020. "How Coronavirus is Impacting London." *King's College London*, May 8, www.kcl.ac.uk ("Aspects of London, such as its high global connectivity and dense concentration of jobs and economic activity in the central area, should now perhaps be seen as vulnerabilities as well as assets.").
4. Caul, Sarah. 2020. "Deaths Involving COVID-19 by Local Area and Socioeconomic Deprivation: Deaths Occurring between 1 March and 31 May 2020." Office for National Statistics, June 12, www.ons.gov.uk ("Between March and May 2020, London had the highest age-standardised mortality rate, with 137.6 deaths per 100,000 persons involving COVID-19"; "The highest age-standardised mortality rate involving the coronavirus (COVID-19) was in urban major conurbations, with 123.5 deaths per 100,000 population.").
5. Urban Living. 2020. "Great Cities after the Pandemic." *Economist*, June 11, www.economist.com.
6. London Intelligence. 2020. "How Coronavirus Is Impacting London." King's College London, May 8, www.kcl.ac.uk; Centre for London. 2020. *The London Intelligence: Your Quarterly Review of the State of the City*. http://kcl.ac.uk ("London's strengths remain—its diverse economic base, highly educated workforce, institutional depth and local, national and international roles.").
7. Evolving Cities. n.d. "About ICEC." Accessed October 15, 2020, https://evolvingcities.org.
8. *See, e.g.*, LSE Cities. 2014. *Cities and Energy: Urban Morphology and Heat Energy Demand*. www.lse.ac.uk (offering a comparative study of energy efficiencies created by the spatial configuration of cities in each of the four largest European cities: London, Paris, Berlin, and Istanbul); IRENA. 2016. "Renewable Energies in Cities." www.irena.org.
9. *See* Lifshitz, Yael. 2019. "Private Energy." *Stanford Environmental Law Journal* 38: 131.
10. *See, e.g.*, Nolden, Colin. 2019. "The Governance of Sustainable City Business Models." *ECEEE Summer Study Proceedings*, 2019, 987–996.
11. Community solar many be more resilient to economic turmoil. Foehringer Merchant, Emma. 2020. "Residential Solar Is Hurting under Coronavirus: Community Solar May Be More Resilient." *GTM*, April 15, www.greentechmedia.com. To

the extent that DG replaces high-emitting sources, it can also potentially reduce greenhouse gas reliance. *See* Shrader, Jeffrey, Burçin Ünel, and Avi Zevin. 2018. *Valuing Pollution Reductions: How to Monetize Greenhouse Gas and Local Air Pollutant Reductions from Distributed Energy Resources*. Institute for Policy Integrity. https://policyintegrity.org. Lifshitz 2019, 133–134. DeWeerdt, Sarah. 2020. "There's a Good Chance Your Home Could Produce All the Electricity You Need by 2050." *Daily Science*, March 10, https://anthropocenemagazine.org.

12. Studying various wildlife regimes, Lueck shows how the range of the species' habitats correlates with the range of the jurisdiction charged with managing the wildlife. Lueck, Dean. 1989. "The Economic Nature of Wildlife Law." *Journal of Legal Studies* 18(2): 291–324.
13. Wyman, Katrina, and Danielle Spiegel-Feld. 2020. "Cities as Environmental Actors: The Case of Marine Plastics." *Arizona Law Review* 62(2): 487–506.
14. Whether you can install that solar panel could depend, as I have argued elsewhere, on whether it is *your* roof, that is, whether you have the necessary property rights. Lifshitz 2019, 138–142.
15. In the UK, Ofgem regulates the grid. In the United States, FERC regulates pipelines; regional operators manage the grid.
16. *See* Ofgem. n.d. "About the FIT Scheme." Accessed October 15, 2020, www.ofgem.gov.uk. The feed-in tariff program operated in the UK between April 2010 and April 2019.
17. Wyman and Spiegel-Feld 2020, 597 (citing Porras). Frug, Gerald E. 1982. "The City as a Legal Concept." *Harvard Law Review* 93: 1069, 1072, 1096, 1106. Frug describes states as "intermediate" levels of government between local and national governments in the U.S. Frug 1982, 1105n188.
18. This also ties into energy democracy.
19. The Climate Change Act 2008 (2050 Target Amendment) Order 2019, art. 2(2).
20. Evans, Simon. 2020. "Analysis: UK Low-Carbon Electricity Generation Stalls in 2019." *Carbon Brief*, January 7, www.carbonbrief.org.
21. For a review of UK energy policy, *see* data on sources in the UK energy sector: *Carbon Brief*. n.d. "Energy Policy Archive." Accessed October 15, 2020, www.carbonbrief.org.
22. Evans, Simon. 2019. "Analysis: UK Renewables Generate More Electricity than Fossil Fuels for First Time." *Carbon Brief*, October 14, www.carbonbrief.org.
23. Mayor of London and London Assembly. 2018. *London Environment Strategy*, 263, www.london.gov.uk. This goal is scaled back from a previous goal of 25 percent. According to Daniel Barrett, lead of smart energy and innovation at Greater London Authority, the revision of this target is a result of "the pace at which markets are maturing, national policy and regulatory environments, capacity and capability across the sector." Barrett, Daniel. 2020. Email to the author, July 16.
24. Local Government Association. 2010. *Local Government Structure Overview*, 4, www.local.gov.uk.

25. London Councils. n.d. “The Essential Guide to London Local Government.” Accessed July 14, 2020, www.londoncouncils.gov.uk.
26. For example, California became the first US state to mandate the installation of solar panels in new residential dwellings. Lifshitz 2019, 148n123 and accompanying text.
27. Policy SI 2 of the London Plan 2021. Mayor of London. 2021. *The London Plan: The Spatial Development Strategy for Greater London*. Greater London Authority, 342, http://london.gov.uk. A “major development” is defined, generally, based on the size (for residential development, when “the site area is 0.5 hectares or more”; for other development, “where the floor space is 1,000 square metres or more, or the site area is 1 hectare or more”) or the number of dwellings (more than ten). *See* London Plan 2021, 512 (referring to part 1 2(1) of the Town and Country Planning (Development Management Procedure) (England) Order 2015).
28. The national building requirements are set in Part L of the 2013 Building Regulations.
29. London Plan 2021, 346–347.
30. London Plan 2021, 342, 346–347.
31. London Plan 2021, 347. The same requirements are not mandated for minor developments. That said, the mayor does encourage boroughs both to require energy strategies for development proposals where it may be considered appropriate and to make sure that “opportunities for on-site electricity and heat production from solar technologies (photovoltaic and thermal)” are taken advantage of. London Plan 2021, 343.
32. London Heat Map. 2020. “Heat Map.” Accessed July 1, 2020, http://maps.london.gov; London Plan 2021, 349.
33. Transport for London. 2020. “Ultra Low Emission Zone.” https://tfl.gov.uk.
34. Noncompliant vehicles are defined based on whether they meet the Ultra Low Emissions Zone (ULEZ) standard, which includes NO_x and PM for diesel cars, lorries, buses, coaches, and heavy vehicles; NO_x for gasoline cars; CO, THC, and NO_x for motorcycles, mopeds, etc.
35. Mayor of London and London Assembly. 2019a. *Central London Ultra Low Emission Zone: Four Month Report*, 6, 3, www.london.gov.uk.
36. Mayor of London and London Assembly. n.d.-a. “The Mayor’s Ultra Low Emission Zone for London.” Accessed July 14, 2021, www.london.gov.uk.
37. Camden Council. 2019. *Appendix A: Healthy Streets, Healthy Travel, Healthy Lives: Camden Transport Strategy 2019–2041*, 15, www.camden.gov.uk.
38. A similar and possibly even more ambitious program is the City Fringe Low Emission Neighbourhood in the Councils of Tower Hamlets, Hackney, and Islington. One aspect of the scheme involves restricting the access of two groups of streets in Hackney during peak hours (between 7 and 10 a.m. and 4 and 7 p.m.) to ULEVs, cyclists, and pedestrians. Mayor of London and London Assembly. n.d.-b. “City Fringe Low Emission Neighbourhood.” Accessed July 14, 2021, www.london.gov.uk.

39. Transport for London. 2019. *Health, Safety and Environment Annual Report: 2018/19*, 29–30, http://content.tfl.gov.uk. London secured £50 million ($74.4 million) over eight years from Barclays Bank to sponsor the city's bicycle-sharing program and £37 million ($55.1 million) over ten years from Emirates Airlines to sponsor a cable car across the river Thames. McKinsey & Company. 2013. *How to Make a City Great*, 19, www.mckinsey.com.
40. The 2011 London Plan introduced a scheme to allow for carbon offsetting where development cannot meet the carbon dioxide reduction target (which is set out in policy 5.2 of the plan). This is to be paid to the local borough and "ring-fenced" to avoid leakage and secure delivery of carbon dioxide savings elsewhere.
41. Mayor of London and London Assembly. n.d.-c. "Energy Monitoring Reports." Accessed July 14, 2021, www.london.gov.uk; Mayor of London and London Assembly. 2019b. *Carbon Offset Funds Survey Results 2019: Greater London Authority Report on the Findings of the 2019 Carbon Offset Funds Survey*. www.london.gov.uk.
42. Policy SI 2(D), London Plan 2021, 342.
43. Edgar, James, Joe Ahern, and Mark Williams. 2020. *The Future of Community Energy: A WPI Economics Report for SP Energy Networks*. WPI Economics, 44, http://wpieconomics.com.
44. Smith, Nadia (SELCE director and project manager). 2020. Interview with the author. Community Energy England suggests that this is due to the fact that organizations are increasingly able to source funding "outside community energy-specific government support mechanisms, as well as the sector's increasing participation in larger-scale partnership-funded projects." Robinson, Sandy, and Dominic Stephen. 2020. *Community Energy: State of the Sector 2020*, 23, https://communityenergyengland.org.
45. The LCEF was set up to fill the gap when the Urban Community Energy Fund (UCEF), run by the UK government, came to a close in 2016. After significant lobbying, the mayor of London decided to introduce a London equivalent as part of his Energy for Londoners program. Mayor of London and London Assembly. n.d.-d. "London Community Energy Fund." Accessed June 22, 2020, www.london.gov.uk.
46. The provision of capital funding is new to the 2021–2022 cycle. Whereas in previous years, the fund only provided development grants of up to £15,000, it introduced, for the 2021–2022 cycle, a second stream (Stream B), which provides capital funding. The change came in response to a shift in national policy, the removal of the feed-in tariff, which had previously been in place and had provided solar producers a return on their investment. Absent the feed-in tariff, the challenge to community projects is not only in the setup stages but also in making the project financially viable. London Assembly. 2020. "London Community Energy Fund Phase 4 (2020/21) FAQ." www.london.gov.uk.
47. Rolfe, Andy. 2020. Interview with the author.
48. The London Green Fund (LGF) was established by the mayor of London, in conjunction with the European Commissioner for Regional Policy. The LGF initially

totaled £120 million and provided a source of investment for projects that would reduce the city's carbon emissions. From 2009 to 2015, the funds were disbursed into three separate urban-development funds (UDFs), the most relevant for our purposes being the London Energy Efficiency UDF (LEEF). Lawrence, Karen. 2011. "LEEF Funding for Public Buildings in London Launched." Local Energy, September 13, www.localenergy.org.uk.

49. Amber Infrastructure Group. n.d. "London Energy Efficiency Fund." Accessed July 20, 2020, www.amberinfrastructure.com.
50. The LEEF targeted various technologies, including small-scale renewables, as well as district heat networks. Lawrence 2011. City-level funding is also sometimes available for specific purposes—for example, funding for heat. Mayor of London and London Assembly. n.d.-e. "Energy Leap Project Pilots." Accessed October 15, 2020, www.london.gov.uk.
51. European Investment Bank. 2018. *Multi-Region Assistance Project—Revolving Investment for Cities in Europe (MRA-RICE): Case Study—London*, 5,www.fi-compass.eu.
52. Mayor of London and London Assembly. 2018. "Mayor's £500m Energy Fund to Help Cut Carbon Emissions." July 9, www.london.gov.uk.
53. Amber Infrastructure Group. n.d. "MEEF Is a £500m Low Carbon Fund for Londoners." Accessed July 2, 2020, www.amberinfrastructure.com.
54. Amber Infrastructure Group. 2020. "MEEF Funds Low-Carbon Heat Network to Supply Better Value Energy." June 22, www.amberinfrastructure.com.
55. Barrett 2020.
56. As a result, eighteen decentralized energy projects, ranging from the development of energy master plans to the construction of energy centers and heat networks, received active support from the scheme. Altogether, the schemes were estimated to have had a "carbon dioxide equivalent (CO_2e) reduction savings of 43,904 tonnes, electrical capacity of 47.5 MW and thermal capacity of 105.3 MW." Barrett 2020.
57. Mayor of London and London Assembly. 2019c. *London Environment Strategy: One Year on Report*, 5, www.london.gov.uk.
58. With the support of recent investment from the UK national government through the Heat Network Investment Project fund, it aims to achieve "CO_2e reduction savings [of] 17,400 tonnes and 3MW by 2023." Barrett 2020.
59. On Energy Islington. 2020. *Islington Community Energy Fund Prospectus*. www.islington.gov.uk.
60. Mayor of London and London Assembly. n.d.-f. *Local Energy Accelerator: Scaling Up Clean and Flexible Energy Systems*. Accessed October 15, 2020, www.london.gov.uk.
61. Collected through the planning system when new developments in Camden fall short of the mayor of London's carbon targets under Policy SI2 of the London Plan. Camden Council. n.d. "Camden Climate Fund." Accessed July 1, 2020, www.camden.gov.uk.

62. Funding of 50 percent of costs up to funding limit of £1,500. Camden Climate Fund. 2019a. *Camden Climate Fund: Domestic Terms & Conditions*. www.camden.gov.uk.

63. The Community Energy Fund (2019–2021) covers 50 percent of costs up to a funding limit of £25,000 to local community groups. The energy or heat that is produced must be for the benefit of the community. Camden Climate Fund. 2019b. *Camden Climate Fund: Community Energy and Heating Terms & Conditions Phase 2 (2019–2021)*. www.camden.gov.uk. The Business Fund (2019–2021) covers 50 percent of costs up to funding limit of £10,000 to SMEs that satisfy certain conditions, including that they are a member of the Camden Climate Change Alliance and employ less than 250 full-time-equivalent employees. Camden Climate Fund. 2020. *Camden Climate Fund: Business Terms & Conditions Phase 2 (2019–2021)*. www.camden.gov.uk. The duty for borough councils to ring-fence payments made by developers existed under Policy 5.2(E) of the London Plan. Mayor of London and London Assembly. 2016. *The London Plan: The Spatial Development Strategy for London Consolidated with Alterations since 2011*, 181, www.london.gov.uk. Policy SI2(D) of the London Plan 2021 both expands on and makes this duty clearer: "Boroughs must establish and administer a carbon offset fund. Offset fund payments must be ringfenced to implement projects that deliver carbon reductions. The operation of offset funds should be monitored and reported on annually." London Plan 2021, 342.

64. On Energy Islington 2020.

65. Community Energy England. 2020. *Setting Up a Local Authority Community Energy Fund*, 2, https://communityenergyengland.org.

66. On Energy Islington 2020; Islington Council. n.d. "Community Energy Fund Round Three." Accessed July 9, 2020, www.islington.gov.uk.

67. Smith 2020. Though on occasion there might be some limited scope for funding core costs (staff and office). *See, e.g.*, On Energy Islington 2020.

68. Mayor of London and London Assembly 2019c.

69. "TfL is currently progressing 1.1 MW of solar installations through the RE:FIT programme. The London Fire Brigade (LFB), with solar already installed in over 50 per cent of their buildings and a capacity of 0.82 MW, is currently reviewing the options for further installations. And the Metropolitan Police Service (MPS) has 22 Solar PV systems, having installed new systems at Kilburn and Hendon since 2016, with two further systems in progress to increase their total capacity to 0.68 MW." Mayor of London and London Assembly 2019c.

70. These sectors consumed 57 million metric tons and 22.7 million metric tons, respectively, in 2018. Department for Business, Energy and Industrial Strategy. 2018. *UK Energy in Brief 2019*, 9, https://assets.publishing.service.gov.uk. The "London Underground is responsible for most consumption" of "total grid electricity usage." Accordingly, recent electricity and gas consumption stands at 1.291 GWh and 79,062,005 kWh, respectively. Transport for London. 2019. *Health, Safety and Environment Annual Report: 2018/19*, 60, http://content.tfl.gov.uk.

71. Although it does have "an emergency back-up supply" at Greenwich. Mayor of London and London Assembly. 2015. "Mayor & TfL Launch Low Carbon Future for Greenwich Power Station." January 8, www.london.gov.uk.
72. Transport for London 2019, 59. There could be technical challenges to relying on DG in the context of the subway system, especially if the DG systems were renewable based (e.g., solar), given their intermittency. Further advancements in battery technologies might be able, in the future, to mitigate some of these technical difficulties.
73. Transport for London. 2021. *Bus Fleet Audit*. https://content.tfl.gov.uk; Mayor of London and London Assembly. 2019d. "London's Electric Bus Fleet Becomes the Largest in Europe." September 5, www.london.gov.uk.
74. Ambrose, Jillian. 2020. "UK Electric Buses Boosted by Innovative £20m Battery Deal." *The Guardian*, June 23, www.theguardian.com ("finance enough batteries to power about 100 electric buses").
75. Mayor of London and London Assembly 2019d. Double-deckers have another benefit: they carry more passengers per road space and per unit of emission. Department for Transport. 2018. *Transport Energy Model Report: Moving Britain Ahead*, 34, 2.87, https://assets.publishing.service.gov.uk.
76. This initiative falls under Action 3 of the Mayor's Solar Action Plan, which focuses on helping "Londoners to retrofit solar energy technology on their homes and workplaces through Mayoral programmes and funding." Mayor of London. 2018. *Solar Action Plan for London*. www.london.gov.uk.
77. London is the "lowest regional performer in terms of solar, with only 2.25 percent of the UK's overall solar output coming from the capital." Power Technology. 2019–2018. "Solar Together London: Will This New Strategy Light Up the Capital?" July 25, www.power-technology.com.
78. Camden Climate Fund 2019a.
79. Mayor of London and London Assembly 2019c, 7.
80. Mayor of London and London Assembly. n.d.-g. "London Power Quarterly Report: January–March 2021." Accessed October 15, 2020, www.london.gov.uk.
81. The pilot was launched as part of the License Lite program, operated by Ofgem. Mayor of London and London Assembly. n.d.-h. "Energy Supply." Accessed June 22, 2020, www.london.gov.uk.
82. Barrett 2020.
83. Holder, Michael. 2020. "West Berkshire Council Launches UK's First Local Government Green Bond." *Business Green*, July 16, www.businessgreen.com.
84. The Green Bond for West Berkshire reached its £1 million target in October 2020, hitting 640 investors. Business Green. 2020. "UK's First Local Government Green Bond Raises £1m for West Berkshire District Council." October 14, www.businessgreen.com.
85. Warrington Council in Cheshire announced that it will do the same. The Warrington scheme hopes to fund a solar farm and nearby battery-storage facility. Leeds Council (a separate UK city) is in the pipeline. Hill, Toby. 2020. "Council-

Issued Green Bond Enables Residents to Invest in Solar Farm." *Business Green*, August 25, www.businessgreen.com.

86. Some people argue against public support for EVs, however. *See* Klass, Alexandra B. 2020. "Regulating the Energy 'Free Riders.'" *Boston University Law Review* 100(2): 581–649 (reviewing the arguments made in this context).
87. Ellickson, Robert C. 2019. "Zoning and the Cost of Housing: Evidence from Silicon Valley, Greater New Haven, and Greater Austin." SSRN, January 13, https://doi.org/10.2139/ssrn.3472145.
88. The Town and Country Planning (General Permitted Development) (England) Orders of 2008 and 2012 are national statutory instruments that remove considerable legal barriers to the proliferation of DG technology by confirming that solar panels are a form of permitted development on domestic and nondomestic premises, respectively. The Town and Country Planning (General Permitted Development) (England) Order 2015 increases the allowable capacity on noncommercial solar installations that counts as permitted development.
89. For properties located in conservation areas, for example, solar paneling cannot be installed on a wall that fronts a highway but is otherwise permitted unless the borough council has removed permitted development rights for roof alterations by way of an Article 4 direction. The Town and Country Planning (General Permitted Development) (England) Order 2015, Schedule 2, Part 14, A.1(c) (2015).
90. Applies to Belsize Conservation Area, Hampstead Conservation Area, Swiss Cottage Conservation Area, and specific properties in Frognal Way. Camden. 2019. *Solar Together—Camden Council Planning Guidance—Non-Domestic*, 2, www.camden.gov.uk.
91. Ofgem. 2018a. *What Is a Regulatory Sandbox?* www.ofgem.gov.uk. One example of a project supported by Ofgem is Ripple Energy. The project is an example of community-based crowdfunding mode, which offers individuals—including city dwellers—an opportunity to financially invest in renewable energy, no matter where they live and whether they rent or not. Graig Fatha Wind Farm. 2020. *Co-Pilot Wind Project Ltd. Co-operative Society*. https://static.rippleenergy.com. In its advisory capacity, Ofgem supported Ripple Energy, by providing "bespoke" information, guidance, and feedback on Ripple Energy's intended approach. Ofgem. n.d. "Innovation Link Case Studies." Accessed October 15, 2020, www.ofgem.gov.uk.
92. Ofgem. 2018b. *Insights from Running the Regulatory Sandbox*, 1. www.ofgem.gov.uk. For example, Verv Energy, alongside British Gas, is trialing "a new arrangement that maximizes benefits from local generation and tests peer-to-peer electricity trading across a distributed ledger platform." The site for the trial is Banister House, Hackney, where Repowering London and Hackney Council supported the creation of Banister House Solar and the installation of solar panels to provide electricity for communal areas. PowerVault is supplying the batteries for storage in the trial. Since starting, the trial allowed for the UK's first energy trade on blockchain. Verv. 2018. "We've Just Executed the UK's First Energy Trade on

the Blockchain as We Look to Power a London Social Housing Community with Sunshine." *Medium*, April 12, https://medium.com.

93. Several regulatory challenges still remain, however. First, the regulatory sandbox is not a mechanism by which developers can expect a permanent regulatory change; the appropriate policy channels are available for this. Ofgem 2018a. Furthermore, the existing regulatory landscape, with its costs and complexities, favors incumbent energy suppliers. This has inhibited the growth of community-level renewable-energy projects and microgrid technology. Smith 2020. The Local Electricity Bill, introduced in Parliament earlier this year, brings attention to this issue and currently has the support of 201 members of Parliament. The main aim is to make it possible for electricity generators "to become local suppliers" and for this to be financially feasible. Power for People. n.d. "The Community Energy Revolution: Campaign for the Local Electricity Bill." Accessed October 15, 2020, https://powerforpeople.org.uk. Its application, however, remains to be seen.
94. Another example of capacity building could be the support for and provision of energy studies. *See, e.g.*, Mayor of London and London Assembly, n.d.-h.
95. GLA. 2014. *London Heat Network Manual*. www.london.gov.uk; GLA. 2011. *Decentralised Energy Capacity Study: Phase 3: Roadmap to Deployment*, ix. www.london.gov.uk.
96. GLA 2011.
97. On the potential of public-private partnerships in climate mitigation, *see generally* Grannis, Jessica. 2020. "Community-Driven Climate Solutions: How Public-Private Partnerships with Land Trusts Can Advance Climate Action." *William and Mary Environmental Law and Policy Review* 44(3): 701–744.
98. Community Energy London. 2019. *Annual Report 2019*. www.communityenergy.london.
99. Camden Climate Change Alliance. 2020. "Partnerships Delivering Community Energy in Camden." June 24, www.camdencca.org.
100. Camden Climate Change Alliance 2020. (discussing the details of the Hampstead High School solar project).
101. Green Finance Institute. 2019. "Green Finance Institute Establishes Coalition for the Energy Efficiency of Buildings." December 11, www.greenfinanceinstitute.co.uk.
102. For other categorizations and for various policy recommendations, see Rosenberg, Jonathan. 2019. *Remarkable Cities and Their Fight against Climate Change*. Environmental Law Institute; Mike Gerrard's Pathways project, available at LPDD. n.d. "Pathway 5. Electricity Decarbonization." Accessed October 15, 2020, https://lpdd.org; Vivian, Sabrina, Kanchan Swaroop, Matt Haugen, Samantha VanDyke, and Sydney Troost. 2020. *Investigating City Commitments to 100% Renewable Energy: Local Transitions and Energy Democracy*. https://ilsr.org.
103. Energy Act 2013.
104. UK Government, Her Majesty's Treasury. 2020. *Policy Paper: A Plan for Jobs 2020*. www.gov.uk.

105. UK Department of Environment Food and Rural Affairs. 2020. "Official Statistics: Rural Population 2014/15." August 27, www.gov.uk.
106. A technological shift, of course, does not necessarily result in a more decentralized model. To begin unpacking what distributed governance means, a more nuanced understanding is required. Is Uber really less centralized than the taxi industry, or has it just shifted power from one group of stakeholders to another?

12

Greenhouse Gas Emission Reduction in Beijing

Goals, Actions, and Recommendations

MAO XIANQIANG, HU TAO, HE FENG, XING YOUKAI, AND GAO YUBING

In June 2015, China submitted its "Enhanced Actions on Climate Change: China's Intended Nationally Determined Contributions" to the United Nations Framework Convention on Climate Change (UNFCCC) Secretariat and committed "to achieve the peaking of CO_2 emissions around 2030 and make best efforts to peak early." Furthering these efforts, President Xi Jinping announced at the seventy-fifth UN General Assembly in September 2020 that China will increase its nationally determined contributions, adopt more effective policies and measures, strive to reach the peak of CO_2 emissions by 2030, and achieve carbon neutrality by 2060.[1]

As the capital of China, Beijing is both a world-famous ancient capital and a modern international city. It stands as the cultural center, international communication center, and scientific and technological innovation center of the country. Beijing has also experienced substantial economic growth as of late: in 2018, Beijing achieved a regional GDP of 3,032 billion yuan, an increase of 6.6 percent over the previous year.

In recent years, the Beijing Municipal Government has attached great importance to addressing climate change and has regarded the co-control of carbon emissions and air pollutants as an effective means for promoting the construction of China's ecological civilization, improving air quality, and achieving high-quality development.[2] Following the global trend, Beijing has actively explored its options in pursuing low-carbon development and was selected as one of the national low-carbon pilot cities in 2012 by the central government of China.

This chapter explores the state of greenhouse gas (GHG) emissions in Beijing, as well as the goals that the city has set and the efforts taken thus far to achieve these goals. It also reviews the preliminary results of these efforts. The chapter concludes with a consideration of the existing challenges in furthering these efforts and describes the options available for Beijing to overcome the challenges and continue making progress.

GHG Emissions in Beijing

Beijing has not released any official statistical data on its GHGs emissions to date. However, the Chinese Academy of Environmental Planning—an agency affiliated with the national Ministry of Ecology and Environment—estimated that, in 2015, total GHG emissions in Beijing were 162.25 $MtCO_2$-eq, with a per capita emissions rate of 7.48 tons of CO_2-eq.

As tables 12.1 and 12.2 demonstrate, the total CO_2 emissions were around 159 Mt, including 106 Mt of direct emission and 53 Mt of indirect emission. Industrial energy consumption and transportation were the main sources of direct emissions, accounting for 54 percent and 27 percent of direct emissions, respectively. The indirect CO_2 emissions come from electricity imported from other provinces.

TABLE 12.1. Beijing's GHGs Emission in 2015 ($MtCO_2$-eq)

CO_2	CH_4	N_2O	HFCs	Forestry carbon sink	Total GHGs
159.04	3.64	0.50	1.80	−2.73	162.25

Source: China City Greenhouse Gas Working Group. 2015. *China City Greenhouse Gases Emissions Dataset.*

TABLE 12.2. Beijing's CO_2 Emissions in 2015

Emission sources		CO_2 emissions (Mt)
Direct emissions	Agricultural	0.67
	Industrial energy consumption	57.08
	Services	9.75
	industrial processes	2.21
	Household	8.37
	Transportation	28.24
	Subtotal	106.31
Indirect emissions		52.73
Total CO_2 emission		159.04

Source: China City Greenhouse Gas Working Group. 2015. *China City Greenhouse Gases Emissions Dataset.*

Beijing's GHG Emission Reduction Targets and Policy Instruments

The Beijing Municipal Government has been determined to play a leading and exemplary role in GHG emissions reduction and low-carbon development and has accordingly set a GHG emissions reduction target. In August 2016, the Beijing Municipal Government issued "Beijing's 13th Five-Year Plan of Energy Conservation and Tackling Climate Change," covering the period between 2016 and 2020, and committed the city to peaking its CO_2 emissions as early as possible—aiming for during the current fourteenth five-year plan period, which runs from 2021 to 2025.[3]

The five-year plan outlined several specific energy consumption and GHG emission reduction targets and goals, including,

1. *Continuing to lead the country in energy efficiency improvements.* By the end of 2020, the city aimed to limit the total energy consumption to below 76.51 Mtce and to drop the energy consumption per unit of regional GDP output (i.e., carbon intensity) by 17 percent, as compared to 2015.
2. *Achieving the intensity peaking of CO^2 emissions per yuan of economic output by 2020.* This means that the CO_2 emissions per 10,000 yuan of GDP will fall by 20.5 percent, as compared to 2015 levels.
3. *Continuing to increase the proportions of clean and low-carbon energy in total energy use.* In 2020, the total coal consumption was to be limited to below nine million tons, the proportion of noncoal energy in total energy use was to be over 90 percent, and the proportion of new energy and renewable energy was to increase to over 8 percent.

To achieve these carbon reduction targets, Beijing has used three major policy instruments: regulations, market-based instruments, and education and public-awareness campaigns.

Regulations

The Beijing Municipal People's Congress has passed a couple of local regulations on carbon reduction, including regulations addressing air pollution, promoting forestry carbon sink work, and establishing a pilot emissions trading program. In addition, each of the city's five-year plans, including its thirteenth five-year plan—which are drafted by the local government, separate from and in furtherance of China's own national five-year plan—must be reviewed and approved by the Beijing Municipal People's Congress, according to China's constitution.

The Beijing Municipal Administration has issued several regulations as well to regulate actions taken by people and firms. For example, one such regulation prohibits public buildings, including shopping malls, restaurants, libraries, auditoriums, and school classrooms, from having an indoor temperature lower than 26 degrees Celsius (78.8 degrees Fahrenheit) during the summer season, in order to avoid consuming too much electricity by overcooling. Supervisors hired by the Municipal Government often check temperatures in public buildings and report violation results to the government, which decides whether to give penalties to the building operators.

Market-Based Instruments

Beijing also uses market-based instruments to provide economic incentives or disincentives for people and firms to adjust their actions toward low-carbon options. For example, Beijing established a carbon emission trading program in December 2013, restricting the emissions of certain "key carbon emitters" but allowing all carbon emitters in the city, both industrial and nonindustrial, to participate in an exchange market and buy or sell their carbon allowances.[4]

Beijing has also applied a block pricing system to set up escalation tariffs for electricity, water, and other utilities, in order to ensure that households are able to meet their basic needs while also discouraging luxury use. The city applies a feed-in-tariff system for developing its solar and wind power as well, implementing its own initiative in furtherance of China's national policy.

Education and Public Awareness

Education is another tool that Beijing has used to achieve its carbon reduction targets, in order to raise public awareness of climate change.

Government-owned TV stations often have various programs on climate change, and one of the key components of the mayor's annual report to the Municipal People's Congress, which is normally broadcast to the public, is carbon reduction work.

Environmental education is being institutionalized by the government as well. All of the science textbooks used in the K–12 education system, for example, now include climate change content. And the Beijing Environmental Education Center, established decades ago under the Beijing Municipal Ecology and Environment Bureau, is one of the major organizations that advocates for climate change efforts.

GHG Emission Reduction Actions in Key Sectors

Using these three tools, Beijing has taken steps in various sectors to reduce the city's GHG emissions and to promote a low-carbon society, as part of its low-carbon development plan under its current fourteenth five-year plan and previous five-year plans. This section explores some of the specific actions taken in each sector.

Adjusting Beijing's Industrial Structure toward a Low-Carbon Economy

At the highest level, Beijing has been promoting a transition toward a low-carbon economy by encouraging its economic structure to move away from heavy industry. As has been demonstrated, carbon emissions are mostly a product of primary industries, such as agriculture and mining, and secondary industries, such as manufacturing.[5] The tertiary industry, which includes financial and information services, produces significantly fewer carbon emissions. For decades, Beijing has been adjusting its industrial structure and shifting toward a low-carbon economy, reducing emissions from its industries. The proportion of the primary and secondary industries in the economy continues to decline.[6] Meanwhile, the proportions of the tertiary and service industries continue to rise, and the service industry now accounts for 81 percent of Beijing's regional GDP.[7] This transition toward a service-oriented economy makes it easier to decouple Beijing's economic growth from GHG emissions.

Adjusting Beijing's Energy Structure toward Low-Carbon Energy Consumption

Beijing has also made substantial progress in reducing the carbon intensity of the city's energy mix. In the recent years, coal has dropped from nearly 14 percent of the total energy consumed in 2015 to nearly 2 percent in 2019, and the city aims to eventually eliminate coal entirely.[8] Meanwhile, the proportion of natural gas in the energy mix continues to rise, from roughly 29 percent in 2015 to 34 percent in 2018. Recently, Beijing has also made strides in increasing its production and consumption of new and renewable energy. In 2018, the total utilization of new energy and renewable energy amounted to about 7.9 percent of the total energy consumption in Beijing. This includes solar energy, geothermal and heat-pump systems, biomass, and wind energy.

Looking first at heat pumps, Beijing has extensively promoted the integrated development of heat pumps and urban heating networks. The Municipal Government has allocated quite a significant budget toward conducting this plan, and the utilization of geothermal and heat-pump systems has gradually developed from single small projects to regional complex projects. By the end of 2018, Beijing had completed more than thirteen hundred geothermal and heat-pump projects, with a total application area of 55.54 million square meters, which could replace about 710 Mtce per year of traditional energy and reduce CO_2 emissions by 1.9 Mt per year.

In addition, Beijing has used a national feed-in tariff to incentivize the development and construction of decentralized wind and solar power projects in enterprise facility sites, parks, and rural areas. By the end of 2018, Beijing's wind power projects had a cumulative installed capacity of 190 MW, with an annual power-generation capacity of 348 MWh, equivalent to the total electricity consumption of 125,000 households a year. Solar installations have been rising rapidly as well; in 2018, the installed capacity of photovoltaic power-generation projects in Beijing was 448 MW, which represents a 170 percent increase since 2015.[9]

The main biomass energy resources are household waste and biogas resources in Beijing. By the end of 2018, in order to meet national environmental requirements, Beijing had built nine waste-incineration power plants with a total installed capacity of 286 MW and an annual power generation of approximately 1,713 MWh and two biogas power plants with a total installed capacity of 15.8 MW and an annual power generation of 70 MWh. Although these plants were built to comply with national regulations, they served to help lower the carbon intensity of electricity used in Beijing.

Making Energy-Efficiency Improvements

In the period of the thirteenth five-year plan, between 2016 and 2020, Beijing's economy continued to grow. The total energy consumption increased from 68.53 Mtce in 2015 to 73.15 Mtce in 2018, and the growth rate has increased year by year. However, during this period, the energy consumption per unit of regional GDP simultaneously decreased year by year, from 0.298 tce per 10,000 yuan in 2015 to 0.254 tce per 10,000 yuan in 2018, which is less than half of the average level of the country. To achieve the energy-efficiency targets, the Beijing Municipal Government, together with the national government, has both heavily subsidized citizens to use energy-efficient electronic and electric appliances and raised electricity prices for industrial users, especially manufacturers.

Promoting Low-Carbon Transportation

In order to encourage the development of low-carbon transportation, Beijing has promoted electric vehicles, implemented a fleet sale restriction policy, sped up the construction of the metro system, and promoted green public transportation and bicycle-sharing options, making it more convenient for the public to make low-carbon travel decisions.

During the period covered by the thirteenth five-year plan, Beijing continued its policy of capping the total amount of internal combustion vehicles, permitting a very slow growth rate. The total number of vehicles in Beijing increased by around 8 percent since 2015, with an overall stable growth rate. By the end of 2018, there were 6.084 million vehicles in Beijing, an increase of only 175,000 vehicles over the previous year. However, despite the small increase in vehicles, there has been a substantial increase in new-energy vehicles on the road. While in 2015, there were only 28,000 electric passenger vehicles in Beijing, the number rose to 225,000 in 2018. In addition, the number of electric trucks in Beijing significantly increased as well: there were 17,753 electric trucks in Beijing in 2018, an increase of 49 percent over the previous year.

Developing and enhancing the public transportation system is always the priority of Beijing Municipal Government, in order to meet the growing demands of a rapidly increasing passenger population. By the end of 2018, there were 888 bus operating lines and 22 metro lines, with a length of 636 kilometers. The total number of passengers' ridership was 3.85 billion in 2018, with an average of more than 10 million rides per day.[10]

Beijing has also promoted alternative forms of transportation, such as biking. In 2018, Beijing started the reconstruction of special bicycle roads and completed an expansion of 928 kilometers of biking and hiking systems.[11] In 2019, a 6.5-kilometer long, 6-meter-wide, 20-kilmoter-per-hour-speed bike highway was completed for bikers who commute from the Xierqi IT industrial park to the Huilongguan residential area. According to the Municipal Governmental Plan, the bike highway is to be extended farther to connect with the broader biking system.[12] By the end of 2018, Beijing had established 3,575 bicycle-rental service points, an increase of 9 percent over the previous year. The number of public bicycles has increased as well, reaching 104,000—an increase of 2 percent over the previous year. Shared bicycles have grown rapidly as well since their introduction in Beijing in September 2016. By the end of 2018, there were nine shared bicycle operating companies, with a total of 9.91 million bicycles, used in total around 610 million times.[13]

Promoting Low-Carbon Construction

Beijing has also made efforts to reduce the emissions that arise from its new building developments. China has minimum mandatary construction and building standards as well as voluntary green building standards, which are denoted with a star rating system. Like LEED, green building standards are certified and verified by an independent third party. In

2017, Beijing achieved its target of 50 percent of the thirteenth five-year plan, with certified green buildings accounting for around 54 percent of the newly built urban buildings in the city.[14] Beijing's newly built government-invested public buildings and large-scale public buildings have been required to implement two-star or above standards.

Promoting Low-Carbon Agriculture

In order to promote low-carbon agriculture, the Beijing Municipal Agricultural Department launched several low-carbon agricultural technology demonstration projects and a quantitative assessment of carbon emissions. For example, in 2018, the Beijing Soil and Fertilizer Station, an agency under the Beijing Bureau of Agriculture and Rural Affairs, monitored the energy consumption, resource efficiency, and manure utilization and treatment of large-scale farms. Data on energy consumption per unit product, feed consumption per unit output, and output per unit land area have been recorded for pigs, dairy cows, and chickens.[15] As another example, the Beijing Bureau of Agriculture and Rural Affairs initiated a project named "The Carbon Emission Assessment and Emission Reduction Technology Demonstration of Dairy Cow Farm."[16]

Increasing the City's Green Carbon Sink Capacity

Beijing has also made achievements in constructing and improving the city's carbon sink capacity, allowing the city to capture carbon that has been emitted. Beijing has sixteen thousand square kilometers of land, though population is mostly located in the central part of the city. From 2015 to 2018, the total forest area in the plain area of Beijing increased year by year, and the per capita green area in Beijing increased year by year. Driven as well by the new round of the One Million-Mu Plain Afforestation Project, Beijing's forestry output value in 2018 increased by nearly 62 percent year on year, accounting for just over 32 percent of the total output value of primary industry. Forest growth has occurred in new and existing urban parks as well as reforestation projects in surrounding mountains.

Encouraging Low-Carbon Consumption and Lifestyles

As Beijing is a mega-city with a large population, the Beijing government has advocated for green living and low-carbon consumption patterns, guided the public to participate in low-carbon actions, and raised public awareness regarding low-carbon activities.

As previously mentioned, Beijing has implemented a block pricing system for residential electricity and water consumption in order to build a resource-saving and environment-friendly society and low-carbon city, meaning that, after using up the first basic-need block quota, the second block price is hiked up significantly. As a result, residents' awareness of electric and water consumption has been increasing, and their living habits have gradually changed. Nearly 90 percent of residents now turn off the lights when they leave the rooms, nearly 75 percent are willing to adjust the air-conditioning temperature to above 26 degrees Celsius in summer, and just under 47 percent reuse water in their daily lives.

Other education and behavior-based programs have also been introduced in the city. In 2017, the Beijing Municipal Commission of Development and Reform, Beijing Environment Exchange, Beijing Energy Conservation and Low-Carbon Engineering Technology Research Institute, and others jointly launched the "I voluntarily drive one more day less per week" platform. By the end of 2018, there were more than 142,000 registered users, and a total of approximately 3.03 million vehicle days were suspended, contributing a cumulative carbon emission reduction of 35,000 tons.[17]

In order to attract enterprises to participate in energy conservation and emission reduction activities, the Beijing Municipal Commerce Bureau offers a subsidy for energy-saving and emission-reducing products, varying from 8 to 20 percent of the sale price (with a maximum subsidy limit of 800 yuan), according to the product's energy-efficiency level.

Overall, Beijing residents appear to be in favor of these efforts. According to a survey conducted by the Beijing Municipal Bureau of Statistics in 2018, over 72 percent of Beijing residents are satisfied overall with the city's promotion and guidance of green and low-carbon life. In addition, the concept of green travel in Beijing has become increasingly popular, with 77 percent of residents saying that they usually choose green travel methods.[18]

Achievements and Challenges in Reducing GHG Emissions in Beijing

The Beijing Municipal Government has recently announced the preliminary results of the carbon reduction efforts taken during thirteenth five-year plan: Beijing's carbon intensity dropped by 23 percent between 2015 and 2020, exceeding the targeted 20 percent drop.[19] By the end of 2020, Beijing's carbon exchange market covered 843 firms in eight sectors, including power, heating, and aviation. In 2020, the total carbon quota trading volume was 4.7 million tons and 245 million yuan. More detailed carbon reduction results are expected to be announced in the coming months in the government's fourteenth five-year plan.

Though Beijing has made progress in moving toward a low-carbon city, there are still many challenges that it must overcome. The most obvious challenge is achieving the energy-consumption level needed for its citizens and supporting its high regional GDP growth, while also maintaining a low-carbon energy mix.

As Beijing enhances its socioeconomic and urban development in the next few years, the energy demand of the people in Beijing will continue to grow, due in large part to several major upcoming mega-projects. For example, Beijing hosted the 2022 Winter Olympics and has plans to construct a new urban subcenter, new metro lines, and a new entertainment facility, similar to Universal Studios. The new Beijing Daxing Airport has recently opened as well, and air transportation is expected to rapidly grow.[20]

Yet another challenge is that the daily energy consumption of Beijing's tertiary industry and its residents has accounted for more than 70 percent of the city's total energy consumption. Though progress has been made so far in conserving energy, the potential for further energy conservation and carbon reduction is becoming more and more difficult.

The Beijing Municipal Government is now developing another carbon reduction strategy within its fourteenth five-year plan, covering the period between 2021 and 2025, based on both its good experiences and lessons learned in the past five years and following China's national strategy of peaking carbon emissions by 2030 and achieving carbon neutrality by 2060. It is very likely that Beijing will plan to achieve its goal of peaking the city's carbon emission within the current plan period, which would be five years earlier than China's national target.

Recommendations

Undoubtedly, Beijing will pave its way to a low-carbon society. In order to develop an even more ambitious carbon reduction strategy within its fourteenth five-year plan, the following recommendations are provided.

The first step that Beijing should take is to continuously optimize the city's industrial structure and promote its low-carbon industrial development. In the future, Beijing will need to continue to promote high-quality economic development, optimize the economic structure, and optimize the internal structure and efficiency of its economy. The primary industry should shift to low-carbon, ecological, and water-saving agriculture. The secondary industry should continue to promote high-tech transformation and strategic emerging industries. Finally, the tertiary industry should continue to increase cultural, tourism, and leisure industries.

The second step is to improve the energy structure and promote low-carbon energy sources. Beijing should continue to implement its zero-coal strategy, strive to achieve 100 percent urban clean energy consumption, increase the proportion of new and renewable energy, and co-control air pollutants and GHG emission reductions.

Third, Beijing should continue to improve its energy efficiency and promote green, energy-saving, and low-carbon demonstrations of major projects. Major projects under construction and new additions should focus on improving energy efficiency and applying energy-saving technologies, including industrial and building energy-saving technologies, as well as new and renewable energy technologies.

Fourth, a comprehensive green and low-carbon transportation system should be built. Beijing should continue to promote the construction of public bus and rail transit and increase the scale of and facilitate the bicycle and pedestrian systems, in order to increase the proportion of green and low-carbon travel; increase the level of electrification and cleanliness of transportation, including public buses, taxies, rental cars, private cars, urban logistics, sanitation, and other vehicles; and strengthen the control of diesel vehicles.

Finally, the scattered service industries, such as the catering, auto repair, and other urban service industries, should be brought into the carbon emission monitoring system. Citizens should better educate themselves to choose low-carbon products, abide by the garbage classification rule, and so on. In addition, Beijing should continue to promote a low-carbon lifestyle with shifts in low-carbon service industries and households.

With each of these steps, Beijing will be on its way to reducing its GHG emissions and achieving a low-carbon society.

NOTES

1. Central People's Government of the People's Republic of China (PRC). 2020. *Speech at the General Debate of the Seventy-Fifth United Nations General Assembly*. www.gov.cn.
2. In 2018, the PRC committed to transition toward an "ecological civilization" and enshrined this concept in the national constitution. The goal of creating an ecological civilization is supposed to guide the country's development trajectory. Hanson, Arthur. 2019. *Ecological Civilization in the People's Republic of China: Values, Action, and Future Needs*. Asian Development Bank, www.adb.org. Beijing Municipal Bureau of Ecological Environment. 2019. "Beijing Holds the 2019 Low Carbon Day Theme Event, Advocates Low-Carbon Action, Defends the Blue Sky and Cheers for Beautiful Beijing." http://sthjj.beijing.gov.cn (website may not be available outside China).
3. Beijing Municipal Government. 2016. *Outline of the 13th Five-Year Plan for Beijing's National Economic and Social Development*. www.beijing.gov.cn.
4. *Beijing Daily*. 2013. "Decision of the Standing Committee of the Beijing Municipal People's Congress on Beijing's Pilot Work for Carbon Emissions Trading under the Premise of Strictly Controlling the Total Amount of Carbon Emissions." Standing Committee of Beijing Municipal People's Congress, December 30, www.bjrd.gov.cn.
5. Center for Climate and Energy Solutions. n.d. "Global Emissions." Accessed July 8, 2021, www.c2es.org.
6. The proportion of the primary industry has dropped from 0.59 percent in 2015 to 0.39 percent in 2018, and the proportion of the secondary industry has dropped from 19.68 percent in 2015 to 18.63 percent in 2018.
7. In 2018, the added value of Beijing's tertiary industry reached 2.46 trillion yuan, accounting for 80.98 percent of the regional GDP, which increased by 7.3 percent over the previous year—0.7 percent higher than the growth rate of regional GDP—and contributed 87.9 percent to economic growth.

8. Shengjie, Zhang. 2020. "Thousand-Year Coal Mining History Draws and End, Beijing's Energy Transition Takes a New Step." *China Energy News*, December 11, http://energy.people.com.cn.
9. *Beijing Daily*. 2019. "Green Development, Efficient Transformation, New Energy Utilization Shines." June 19, http://bjrb.bjd.com.cn.
10. Beijing Municipal Bureau of Statistics. 2019. "Beijing Municipal Statistical Communiqué on the 2018 National Economic and Social Development." Changping District People's Government of Beijing Municipality, April 1, www.bjchp.gov.cn.
11. Beijing Municipal Transportation Commission. 2019. "The Fifth 'Beijing Bicycle Day' Event Was Held." http://jtw.beijing.gov.cn.
12. Sogou Baike. n.d. "Bicycle Highway." Accessed December 2, 2020, https://baike.sogou.com (in Chinese).
13. Beijing Transportation Development Research Institute. 2019. "2019 Beijing Transportation Development Annual Report." www.bjtrc.org.cn.
14. Beijing Housing and Urban-Rural Development Promotion Center. 2018. "The Ministry of Housing and Urban-Rural Development Conducted a Special Inspection on the Implementation of Building Energy Conservation, Green Buildings and Prefabricated Buildings in Beijing in 2017." *Green Building: Beijing in Action* 2: 1–3.
15. Beijing Municipal Bureau of Agriculture and Rural Affairs. 2018. "In 2018, the Low-Carbon Emission Reduction Work of the Aquaculture Industry Was Successfully Completed." http://nyj.beijing.gov.cn.
16. Beijing Municipal Bureau of Agriculture and Rural Affairs. 2018. "Dairy Farm Carbon Emission Assessment and Emission Reduction Technology Demonstration Work Completed." http://nyj.beijing.gov.cn (website may not be available outside China).
17. Developing Beijing. 2019. "Unit GDP Energy Consumption, Carbon Trading: Annual Notes on Energy Saving and Emission Reduction." www.sohu.com.
18. Beijing Municipal Bureau of Statistics. 2019. "Practicing Green Lifestyle, Building and Sharing Beautiful Beijing—2019 Beijing Public Green Lifestyle Survey Report." www.beijing.gov.cn (website may not be available outside China).
19. Chu, Xuanjiao. 2021. "Beijing Is Actively Carrying Out Carbon Peak Assessment and Carbon Emission Reduction Special Plan Research." Beijing Municipal Ecology and Environment Bureau, January 19, http://sthjj.beijing.gov.cn.
20. Beijing Municipal People's Government. 2016. "Beijing's 'Thirteenth Five-Year' Period Energy Saving and Consumption Reduction and Climate Change Planning." www.beijing.gov.cn.

13

Global Sustainable Cities

Berlin Aims at Climate Neutrality

DÖRTE OHLHORST AND MIRANDA A. SCHREURS

The German capital of Berlin has done much to protect the climate in recent decades. It has made considerable progress toward reducing its greenhouse gas emissions, phasing out coal use by 2030, and aims to become climate neutral by 2045. The Berlin Energiewendegesetz (Energy Transition Law), which entered into force in 2016, stipulated that Berlin's CO_2 emissions are to be reduced by at least 40 percent by 2020, 60 percent by 2030, and about 95 percent by 2050, each compared to 1990 emissions levels.[1] In 2021, these targets were amended upward to require a 70 percent reduction by 2030 and at least 95 percent by 2045. Today, Berlin boasts a relatively low per capita CO_2 footprint: 4.6 tons per capita, compared with about 9 tons per capita for Germany as a whole, based on 2017 data.[2]

Berlin is special—partly because of its history, partly because of its future visions. Its actions matter for German climate performance.[3] If the German capital succeeds in significantly reducing its greenhouse gas emissions, it not only will make an important climate contribution to the national climate goals but will also be noticed by people across Europe and the world. Many people look to Germany and its capital city as models for how to transition from a fossil- and nuclear-based energy system to an energy-efficient, renewable-energy-based economy. This is why Berlin's role in advancing climate protection, introducing new climate technologies, and experimenting with approaches to sustainable urban development is of particular importance.

Building on and Breaking with Historical Path Dependencies

To understand Berlin's climate policies, it is helpful to first step back in time and briefly consider how infrastructural path dependencies, industrial legacies, wartime destruction, Cold War division, and eventual reunification shaped energy structures and energy use in this most fascinating of cities. Berlin's goal to become climate neutral will demand changes to energy, building, and transport structures that took root decades or even centuries ago. Today's opportunities and challenges are related to infrastructural decisions made in the past, including the development of a metro system and district heating, as well as the industrial makeup of the city. Achieving climate neutrality will, however, require going even further, as this too means having to break with other path dependencies—such as the city's long-term dependence on coal and continued love of the automobile.

Social scientists often consider how infrastructural, socioeconomic, and political legacies of the past (path dependencies) shape patterns of behavior and activity long into the future unless they are somehow broken, due to a shock or series of disruptive events, new knowledge and ideas, or cultural change.[4] Berlin continues to be influenced by past legacies, but it has also experienced numerous critical junctures that have set it on a new energy trajectory. To understand today's climate change politics, it is helpful to understand these earlier path dependencies and critical moments of change.

When the German state formed in 1871, it brought together numerous kingdoms that had long interacted with each other due to their common German cultural and linguistic heritage. At the time, Berlin was Germany's most important industrial city, with a population of over eight hundred thousand. Just fifty years later, the population had ballooned to over three million.[5]

Already in the latter part of the nineteenth century, Berlin was known as a center of invention and modernity. Firms like Städtische Elektricitäts-Werke (the predecessor to Bewag and later Vattenfall; electricity production), Borsig (locomotives), AEG (electric technology), Siemens & Halske (electric technology), Schering (chemicals), and Deutsche Bank set up in Berlin. The Städtische Elektricitäts-Werke opened the first electricity production facility in Germany in Berlin in 1884. Siemens invented the electric street car in Berlin in 1881 and followed this with the development of a metro system, with the first underground line running between Stralauer Tor (in Friedrichshain) and the city's center, Potsdamer Platz, opening in 1902.[6] Berlin's district heating system, the largest in Europe, was also initiated in the beginning of the 1900s. The city hall (Rathaus) in Charlottenburg was the first building to be heated by steam produced in the nearby electricity plant. This system captures waste heat from electricity production and waste incineration and, through a system of underground pipes, brings the heat to homes and offices throughout the city. It was not initially developed for climate reasons but rather to reduce the potential of fire—a constant threat in cities, where coal stoves in homes were widespread.

With industrialization and the accompanying rapid population growth, many tenement houses were built around the old city center. Farther out, housing associations built large housing projects with four and five stories. Throughout the city, but especially in the outskirts, garden plots were created to provide workers and children with a place to rest from the city and grow food (which was crucial during the war years).[7] Lignite and hard coal were the city's main heat supply.

To fuel the growing city, coal was transported by barge and rail from as far away as England, though the main suppliers of coal were the Ruhr, Silesian, and Lausitz regions in Germany. So important was coal to the city that during the Berlin Air Lift—which kept the city afloat from June 1948 to May 1949, when Soviet forces tried to strangle the city and gain control of it by shutting down all railroads and highways into the city—Allied forces flew supplies to the city, including coal. Airplane deliveries of coal continued for several additional months, even after the blockade ended. Between June 1948 and August 1949 alone, 539,000 tons of coal were flown from Faßberg in the Lüneburg Heath to the Gatow airport, in Spandau near Berlin.[8] Coal and lignite remained the main sources of heating and electricity production in the postwar decades.

Berlin's status as a world city was dealt a dramatic blow by Germany's defeat in World War II. The city was occupied and divided into four zones. With the onset of the Cold War, the city was politically divided. The British, French, and US zones became West Berlin and followed a path of social-democratic capitalism. The Soviet zone became East Berlin and followed a path of communism and economic socialism. Defeat in war became a point of dramatic changes, with important consequences for later climate and energy policies and programs.

Some of the hardest and most destructive fighting at the close of World War II occurred in Berlin. Berlin was heavily bombed and largely destroyed. Hunger, lack of shelter, and unemployment made life in the city a struggle for survival. Various industries abandoned the city (Siemens, for example, left for Bavaria in 1949); others that were located in the Soviet zone were dismantled and sent to Russia as war reparations. Along with the political division of the city, there was an infrastructural division. Electricity production and the metro system were divided. West Berlin became not only a political island surrounded by a different ideological system but also an electricity and public-transportation island.

Over the course of the next decades, building by building, the city was resurrected. In areas where the damage to the buildings was too extreme to allow a reconstruction of the old buildings, new buildings were quickly erected, including with Marshall Plan aid. While this was crucial at the time to provide individuals with housing, these buildings were not built with energy efficiency in mind.

In the postwar years, district heating systems that had been started decades earlier were extended as new housing areas were built. This helped the two halves of the city make more efficient use of the fossil fuels imported into the city. This was especially important in East Berlin, where obtaining adequate energy supply was a challenge. Approximately 50 percent of homes in East Berlin and 14 percent in West Berlin were connected by district heating.[9] In Berlin's efforts to be low carbon, it has benefited from some of these much earlier decisions, made long before climate change was a topic.

Rebuilding after Unification

Unification after the fall of the Berlin Wall in 1989 was a critical juncture of enormous proportions that gave Berlin the opportunity to reinvent itself, albeit at high costs. Many out-of-date industries in the east went bust because they could not compete in the western capitalist marketplace; high levels of unemployment ensued. With the city having little industrial production of its own, Berlin's CO_2 emissions in 1990 (7.8 tons CO_2 equivalent per person) were already well below the national average of 12.3 tons CO_2 equivalent per person.[10] In the years after the fall of the Berlin Wall, Berlin was further deindustrialized, a large factor behind the subsequent rapid decline in the city's greenhouse gas emissions. In

Figure 13.1. Built in the east after the Berlin Wall was erected, the television tower in Alexanderplatz is the city's tallest building. (Photo by Peter Sartig)

the meantime, the city became home to a growing number of service industries and new tech companies.

At the time of the fall of the Berlin Wall in 1989, many of the buildings in the city, and especially in former East Berlin, were in poor shape, with outmoded heating and electric systems; some did not even have running hot water. The modernization of Berlin's infrastructure and buildings was initiated almost immediately upon unification and continues to this day.[11] Although extremely costly, the renovation has helped the city (and other areas in the former East Germany) cut its greenhouse gas emissions even further. The critical juncture provided by reunification proved to be an opportunity for Berlin to begin a process of modernizing and improving its energy infrastructure. This was critical, as the city had fallen into a state of severe infrastructural disrepair.

One remarkable example of the post-unification transformation of Berlin is the Reichstag, the federal parliament building. The heavy stone building was constructed in the late 1800s, burned in 1933, and badly damaged in the last days of battle, when Soviet troops entered Berlin to free it from Hitler's grasp. While the building was sufficiently repaired to be used as a museum during the Cold War years, it needed to be renovated to once again become the seat of parliament of a reunified Germany, and the British architect Norman Foster was commissioned for the work. Not only was the architectural redesign a huge success from an artistic perspective, but the building, together with the neighboring Paul Löbe Haus, where parliamentarians have their offices, was also designed to be a sustainable-energy building. A glass dome, symbolizing the importance of transparency in a democracy, lets sunlight into the plenary hall and collects rainwater to be used to wash the windows. Solar panels were placed on the rooftop. Four diesel engines fueled with biodiesel were specially designed to provide the building with electricity. Heat from the generators is piped through the building in the winter, and absorption coolers help cool the building in the summer. Excess heat is used to warm water taken from the nearby Spree River, which is then stored in a geothermal storage system to be used for additional heating in the winter. Similarly, cool water taken from the Spree helps to keep the building comfortable in the summer.[12] That this most important of symbols of

Figure 13.2. The German Reichstag with its glass dome is a symbol not only of democracy but also of the energy transition. (Photo by Peter Sartig)

Figure 13.3. Symbolizing the unification of West and East across the River Spree, the bridge connects two houses associated with parliament: the Paul Löbe Haus, named after the last democratic president of the Weimar Republic, and the Marie Elisabeth Lüders Haus, named after the women's rights champion and parliamentarian. (Photo by Peter Sartig)

German democracy has also become a hallmark of climate-friendly technological ingenuity speaks to the remarkable transformation of this city.

With the government moving from Bonn to Berlin, the federal government also invested huge sums into the reconstruction of many historic buildings for use by the government. These costly renovations were done to further showcase the potential for energy-efficient and sustainable building renovations.

German Federalism and the State of Berlin

Today, Germany's federal system is composed of sixteen *Länder* (states), three of which are city-states (Berlin, Bremen, and Hamburg). As Berlin is like an island in the middle of the state of Brandenburg, climate and energy issues often require cooperation between these two *Länder*.

Much of the renewable energy that Berlin consumes is imported from Brandenburg and other regions.

The status of Berlin is unique. It is at the same time the capital of all of Germany, a *Land* (a federal state), and a city.[13] As a *Land*, Berlin has its own state parliament that issues climate programs and laws. Within the *Land*, there are twelve districts (*Bezirke*) and ninety-seven communities (*Ortsteile*).[14]

Berlin holds elections for the *Land* parliament once every five years. After the 2021 election, six major parties are represented in the parliament, each designated by a color: Sozialdemokratische Partei Deutschlands (Social Democratic Party, SPD; red, thirty-six seats), Bündnis 90/Die Grünen (The Greens; green, thirty-two seats), Christlich Demokratische Union Deutschlands (Christian Democratic Union, CDU; black, thirty seats), Die Linke (The Left; dark red, twenty-four seats), Freie Demokratische Partei (Free Democratic Party, FDP; yellow, twelve seats), and the far-right party, Alternative für Deutschland (Alternative for Germany, AFD; blue, thirteen seats).

Berlin's Mayer, Franziska Giffey (SPD), is the first woman to hold this post. She heads a coalition among the SPD, Bündnis 90/Die Grüne, and Die Linke. In 2016, Berlin created a new position: minister for environment, traffic, and climate protection. The position is currently held by Regine Günther of Bündnis 90/Die Grünen. She had previously led the Energy Section of WWF Germany and also led its policy and climate programs. The energy transition has been a major theme of her time in office, and she has actively promoted bicycle transport as well. Since 2016, the minister for economy, energy, and business has been Ramona Pop, the first member of Bündnis 90/Die Grünen to ever hold this position. The left-leaning orientation of the Berlin government has provided various opportunities to pursue green politics.

Climate Policy Making in Berlin

Climate policy making has a relatively long history in the city. From 1983 until 1985, the Berlin Parliament hosted an Enquete Commission called "Energy Policy for the Future."[15] The cross-party findings of the Energy Enquete Commission had a strong impact on energy policy in the years that followed. The environmentally friendly, "economical and rational use of energy" concept proposed by the commission became a widely accepted goal. Furthermore, the idea of ecological modernization—integrating an awareness of environmental protection needs and resource constraints into economic decision-making—took hold among representatives elected to the Berlin House of Representatives. One of them, Martin Jänicke, later went on to promote ecological modernization concepts more widely in German academia and politics.

Building on the recommendations of two expert reports and with the goal of achieving an ecological energy transition in Berlin, in 1989, the energy policy spokespersons of Berlin's Red-Green government began to promote what at the time seemed like a quite radical idea: halving Berlin's energy consumption.[16] Their plan was immediately criticized by conservative opponents. In order to avoid a hardening of the lines of conflict and to build up sufficient support for policy change, the Red-Green government established the Berlin Energy Advisory Council. Made up of representatives of the energy supply companies, market specialists, and experts from science, trade unions, and consumer associations, the Energy Advisory Council was tasked with advising the Berlin government on all energy-related issues: industrial, legal, and policy related. For the next five years, the advisory board worked on an energy concept for Berlin. In 1994, it presented its plan. This plan, which was much less ambitious than the Greens had wanted, called for reducing energy-related CO_2 emissions by 25 percent between 1990 and 2010 (a clear deviation from the original goal of halving CO_2 emissions by 2010). Although the plan's development process was seen as quite innovative given various public inputs, the plan lacked concrete and verifiable action targets, did not set specific responsibilities for meeting targets or carrying through with policy implementation, and did not have an adequate financial basis.[17] Climate change, so it seems, was still not so high on the agenda; this was reflected in the many compromises that worked their way into the plan.[18]

Climate policy making gained more traction in the following years. The Berlin Climate Protection Council was set up in 2010 as an independent advisory body to advise the state government on climate protection issues. It was tasked with amending the Berlin Energy Saving Act and developing an energy plan for the city. The council was newly appointed in 2017 and was tasked with ensuring compliance with climate protection goals and with the implementation and updating of the Berlin Energy and Climate Protection Program 2030 (BEK).

In 2014, the Berlin House of Representatives voted to establish an Enquete Commission called "New Energy for Berlin: The Future of Energy-Sector Structures," which prepared the groundwork for Berlin's Act to Implement the Energy Transition and to Promote Climate Protection in Berlin (the Berlin Energy Transition Act) and which came into effect in 2016. The commission's membership included eleven parliamentarians (from the SPD, CDU, Alliance 90/Die Grünen, Die Linke, and the Pirate Party) and five outside experts. The commission met twenty-three times during 2014 and 2015 and issued a report to the Berlin House of Representatives, which noted the cross-party consensus on the need to achieve climate neutrality by 2050 and called for a phase-out of lignite use by 2020, a phase-out of hard coal by 2030, and a focus on energy efficiency and renewable-energy development.[19]

Berlin followed up on these recommendations in its 2016 Energy Transition Law, which was amended to include more stringent and earlier phase-out targets in 2021 due to growing domestic and European pressure to do more to address climate change. The first goal, a 40 percent reduction of 1990 emission levels by 2020, has already been met. In the meantime, the 60 percent cut in CO_2 emissions targeted in the 2016 law for 2030 was raised to 70 percent, and the date for obtaining at least a 95 percent reduction was moved up to 2045 from the earlier target of 2050.[20] In 2016, specific targets for CO_2 emission reductions by 2050 were established for different sectors relative to 2012 levels. These included a reduction of 84 percent for the building and urban development sectors, 77 percent for trade, 90 percent for private households and consumption, and 67 percent for transport.[21] The revised law mandates that solar photovoltaics be installed on all public buildings and that the Berlin administration be climate neutral by 2030.

The Energy Transition Law of 2016 aimed to achieve a secure, low-cost, and climate-compatible energy system that will incorporate a growing share of renewables. It set a target to stop energy generation from lignite in Berlin by December 31, 2017, and from hard coal by the end of 2030 at the latest (Section 15 of the Energy Transition Law). In contrast, it was only in July 2020 that the German Parliament (the Bundestag and the Bundesrat) agreed on a nationwide coal phase-out by 2038 at the latest. Thus, the Berlin Senate moved several years earlier than the federal government on phasing coal out of the energy system. In line with the law, Vattenfall, one of Berlin's main electricity providers, which in 2020 was still running four electricity production facilities with lignite, has announced it plans to stop using coal by 2030 and to be climate neutral by 2050.[22]

Educational measures are also a component of the Energy Transition Law. Section 14 of the law requires climate change and climate protection, as well as energy conservation, to be included in school and preschool education. Section 4 established the Berlin Energy and Climate Protection Program 2030 (Berliner Energie- und Klimaschutzprogramm 2030; BEK), which was launched in 2018 and is developing plans for reducing emission in multiple sectors, including power generation and supply, transportation, households, and business and commercial sectors.[23] It contains around one hundred measures to protect the climate and to adapt to the consequences of climate change. One of the focal points of the BEK 2030 includes measures supporting energy-efficiency and clean-energy improvements (such as a "heating system exchange program"), as well as expert advice provided to city districts on how to reduce energy demand and shift to cleaner energy sources. Multiple Senate departments, under the leadership of the Senate Department for Environment, Transport, and Climate Protection are responsible for achieving progress in greening the energy supply, renovating buildings, rethinking urban development, promoting a circular economy, and making transport more climate friendly. A digital monitoring system (diBEK) has been introduced to monitor progress and assess climate impacts in Berlin.[24]

The Berlin Mobility Act (Mobilitätsgesetz), effective as of July 2018, gives priority to bicycling, walking, and public transit. Cycling infrastructure and public transit, which had already been expanded in previous years, are to be expanded yet further; Berlin already boasts about 620 kilometers of bike lanes.[25] The law also sets a goal for electrification of city buses, and its traffic planning includes a traffic strategy for pedestrians.[26]

Renewable energy has also been a focus of climate policy making in Berlin. In order to expand the use of solar energy, in March 2020, the Berlin Senate initiated a master plan, "Solarcity Berlin."[27] In the summer of 2020, the Coordination Office for the Solarcity Master Plan Berlin was established within the Senate Department for Economic Affairs, Energy, and Operations and became responsible for coordinating the implementation of the plan. Under this plan, solar expansion is to be accelerated, with a goal of achieving 25 percent of Berlin's electricity from solar power. As a first step in this direction, solar panels are to be installed on the roofs of state-owned buildings, including school buildings and public facilities. In January 2021, the first monitoring report for the Solarcity Master Plan was published, which lists some important milestones that have already been achieved. Highlights include the initiation of a support program for energy storage, improved information and legal structures for solar power, and demonstration projects. Plans for 2021 are also discussed.[28] The Berlin Senate is now debating a requirement to include solar panels on new buildings and on older buildings when roofs are replaced.

In December 2019, in response to a successful citizens' initiative calling for the announcement of a climate emergency, Berlin became the first *Land* (and the sixty-eighth German polity) to declare a climate emergency.[29] In doing so, the Senate acknowledged the seriousness of global warming and the need for additional climate protection efforts in Berlin.

Emission Trends

Figures released in December 2020 show that CO_2 emissions in the city were down 40.7 percent compared with 1990 levels. That Berlin met its 2020 emissions target can be credited to the shift away from coal and other measures found in the Berlin Energy and Climate Protection Plan.[30] With this shift away from coal, Berlin was able to save almost one million tons of CO_2 in 2017.[31] Other factors that contributed to the city's reduction in emissions include increases in energy efficiency and more conscious consumers. Emissions certainly also dropped due to the slowdown of the economy and reduced travel associated with the COVID-19 pandemic.

Nationally, Berlin performs very well compared to other major cities in Germany. Berlin's 4.6 tons of CO_2 equivalent per person (2017 data) are much lower than Hamburg's 8.9 tons of CO_2 equivalent per person

(2018 data), which is roughly the same as the national average.[32] Hamburg's higher emission levels are linked to its status as a major harbor city. Nevertheless, Hamburg, like Berlin, has set a goal to achieve climate neutrality by 2050. In December 2019, eight days after Berlin declared a climate emergency, Munich, Germany's third-largest city, also declared a climate emergency. With competition for leadership emerging among German cities, Munich moved up the date it had set to achieve climate neutrality from 2050 (target set in 2017) to 2035 (target set in 2020). The Munich administration further set a goal for itself to achieve climate neutrality by 2030.[33] Munich's greenhouse gas emissions of 5.9 tons CO_2 equivalent per person (2017 data) is, however, still higher than Berlin's. The city's goal of climate neutrality by 2035 means it will need to reduce emissions to 0.3 tons CO_2 equivalent per person per year—a highly ambitious goal.[34]

In 2020, Berlin, like most of Germany, experienced intermittent periods of lockdown and reduced travel as a result of the COVID-19 pandemic. Consequently, it can be expected that CO_2 emissions levels in 2020 and 2021 will be lower than would otherwise have been the case, although perhaps less than in other cities with major industries.

There has been at least one development in which there is a link between the COVID-19 pandemic and Berlin's energy and climate policy: the bicycle infrastructure has been significantly improved in some parts of the city. For example, a lane was set up for cyclists on many busy and wide streets, clearly demarcated from the lane for cars.[35] It is not yet clear, however, whether these cycle lanes will persist in the long term.

Berlin: A City with a Multitude of Climate Actors and Activities

As Berlin is the capital of the country, many federal and international climate-related activities take place here. The very first Conference of the Parties to the United Nations Framework Convention on Climate Change (COP 1) was hosted by then–Environment Minister Angela Merkel in Berlin in 1995. It led to the formation of "the Berlin Mandate," setting the stage for the formation of the Kyoto Protocol of 1997.

Most groups with an interest in lobbying the federal government on issues related to climate change and energy have a presence in the greater Berlin-Brandenburg area. Numerous major think tanks dealing with climate change, energy, environmental, or sustainability issues are either located in the Berlin-Brandenburg region or have a branch office there. These include, for example, the Potsdam Institute for Climate Impact Research, Agora Energiewende, Adelphi, Ecologic, the Wuppertal Institut, the Öko-Institut, and the Institute for Ecological Economy Research. Three of Germany's political foundations have their headquarters in Berlin: the Heinrich Böll Foundation, Friedrich Ebert Foundation, and Konrad Adenauer Foundation. Each holds events, sponsors climate projects in and outside Berlin, as well as internationally, and supports young researchers. Of the political foundations, the Heinrich Böll Stiftung, which is affiliated with the Green Party, is arguably the most active in the climate field, although most of the foundations have become increasingly active on climate issues.

Berlin's civil society is also very active with regard to climate protection and sustainability. Berlin cyclists regularly organize the "Berliner Stadtradeln" campaign, drawing thousands of Berliners who cycle through the city in the evening to draw attention to the importance of bicycles for climate protection and to heighten awareness of the safety needs of cyclists. The "Extinction Rebellion" and "Fridays for Future" movements are also very active in Berlin, drawing attention to the urgency of climate protection with imaginative actions and, in some cases, civil disobedience.

The universities are some of the largest energy users in Berlin. Major steps were taken by the Berlin universities to improve their energy scorecards. The Freie Universität Berlin, a university of approximately thirty-seven thousand students, provides an interesting example. Freie Universität started an energy management process twenty years ago, particularly based on a university-wide energy monitoring system, annual programs to improve the energy efficiency of the university building, and a bonus scheme for energy savings that enables the departments to earn money by saving energy. Those that exceed the two-year averaged baseline consumption have to pay additional payments. Particularly innovative was the idea to have department-by-department expectations with regard to energy use. Departments that save on energy relative to a two-year averaged base line are allowed to keep 50 percent of the cost savings from the central budget. Those that exceed the base line are fined and have to pay 100 percent of the additional energy costs. Building by building, low-hanging fruits were addressed first. Operational technologies like heating systems, including regulation systems and heating pumps, were modernized, ventilation systems were optimized, old inefficient equipment (old computers, refrigerators, and the like) were replaced with high-efficiency models, and energy-use behaviors were addressed with a new communication strategy. Additionally, the university installed four highly efficient block heat and power plants and nine photovoltaic plants on building rooftops. One of these was established together with a student initiative.

Efforts are also being pursued to change individual behavior. One example is in relation to waste. The Senate Department for Environment, Transport, and Climate Protection has further introduced a Zero Waste Strategy and organic waste bins. The focus of Berlin's Zero Waste Strategy is on avoiding waste and recycling that which cannot be avoided. Around half of Berlin's waste is organic—by recycling around three hundred thousand tons of organic waste every year, climate-friendly biogas and useful compost can be produced. The Senate's environmental administration is campaigning to encourage even more households to order an organic waste bin. Berliners can save money with it, because the less rubbish they throw in the residual waste bin, the less waste fees there are to be paid.[36]

Funding Climate Initiatives

With funding from the National Climate Initiative program run by the Federal Ministry of the Environment, since 2008, more than 32,400 projects with a funding volume of around €1.07 billion have been launched to reduce carbon dioxide emissions across the country. From 2008 to 2019, Berlin was awarded 120 projects (compared with 85 for Bremen and 123 for Hamburg, the other two city-states, or *Länder*). Examples of projects include the provision of assistance for small- and medium-sized enterprises to reduce emissions through, for example, digitalization; the reduction of waste in multicultural fast-food institutions; the reduction of energy use in youth facilities run by the Sozialdiakonische Arbeit Berlin GmbH; the renovation of light facilities in a school in Berlin; the reduction of energy use in facilities caring for patients; the development of environmentally friendly delivery services; and so on.[37] The Berlin

Energy Agency (Berliner Energieagentur) has made CO_2 measurement equipment (e.g., heat-sensing cameras; CO_2 temperature and humidity equipment; an electricity price measurement device) available to schools.[38]

The Berlin Senate Department for Economy, Energy, and Businesses also funds climate projects in Berlin. An example is its program supporting projects for the storage of renewable energy (Energiespeicher-PLUS).[39] It also supports the development of public transportation options and bike lanes, energy-efficiency improvements in companies and public buildings, and research into energy-efficient technologies that can be used in Berlin (Berliner Programm für Nachhaltige Entwicklung BENE Klima). BENE Klima ran a campaign called Berlin Saves Energy (Berlin Spart Energie) in the fall of 2020 and has a web portal with best-practice examples.[40] BENE also has been supported by the European regional development fund with a total volume of over €274 million.

The Federal Ministry of Interior, Building, and Community awarded Berlin and thirty-one other cities the designation "Pilot Project Smart City" (Modellprojekt Smart City), making funding for digitalization projects available.[41]

The Climate Protection Partners of Berlin, an alliance of business chambers and trade associations, has awarded a "Climate Protection Partner of the Year" prize for climate-friendly projects every year since 2002. The winners are chosen at the Berlin Energy Days event that takes place every spring.[42]

Challenges for Berlin's Climate Protection Policy and Goals

While Berlin has made substantial progress in reducing its carbon footprint over the years, the next phases of the energy transition toward carbon neutrality will be difficult. Berlin's population continues to grow, and the city is becoming more economically dynamic. This could lead to more traffic, higher levels of consumption, and a concomitant rise in energy and resource demand.[43] The city will need to invest more in renewables in order to further reduce emissions in the energy sector, speed up the rate of building energy retrofits, and electrify the transport sector.

Carbon dioxide emissions can be measured in various ways: as a percentage of primary energy, which includes the energy used to produce energy and lost in transmission, as final energy consumption, or as emitting sectors. With regard to emissions from energy generation and transmission, there were several big changes during the period 1990 through 2019. There were sharp drops in reliance on coal, lignite, and mineral oil, which led to a respective reduction of 75, 99.3, and 39.2 percent in their CO_2 emission contributions. In contrast, the share of CO_2 emissions attributed to natural gas grew by 94 percent during this same period.[44] In 2019, the largest sources of fossil-fuel-related CO_2 emissions were from mineral oil (6.9 million tons, or 44 percent), natural gas (6.5 million tons, or 41 percent), hard coal (2 million tons, or 12.4 percent), lignite (0.04 million tons, or 0.2 percent), and other sources (2.4 percent). Overall, the largest emissions by energy source were from mineral oil and mineral oil products and electricity.[45]

A source-based analysis of CO_2 emissions in 2019 indicates that 36.7 percent of emissions were related to energy generation and energy conversion processes; 28.2 percent to households, businesses, retail, services, and other users; 33.5 percent to transport; and 1.6 percent to mining of stones and earth. Residential and commercial buildings are a major source of emissions; a sector-based analysis of CO_2 emissions in 2017 shows households, businesses, retail, services, and other users as by far the biggest polluting category, accounting for 61.6 percent of emissions (only a slight drop of 3.2 percent from 1990 levels).[46] While many of the buildings in the former East Berlin were renovated and their energy efficiency improved, there is still much that needs to be done throughout the city. Berlin has many monumental and Wilhelminian-style buildings, which add to the city's charm but which present a challenge when it comes to energy renovation. Furthermore, a large proportion of the apartments in Berlin are rented (approximately 86 percent). This complicates the question of who should pay for energy renovations: the owner or the renter? And almost 10 percent of the buildings are under historical preservation status, making building renovation for climate protection more difficult.

Reaching climate neutrality will require cutting heating-related emissions by around 80 percent.[47] To achieve energy-efficiency goals, it is estimated that the city needs to renovate 2.1 percent of its buildings each year through 2025 and increase this to 2.6 percent per year in 2030. Currently, however, only around 1 percent of buildings are renovated per year.[48] The share of renewable energy in the heat supply must increase, and the heat demand of buildings (e.g., with insulation and energy demand control technologies) must be reduced.

To deal with these problems, the Berlin Energy Agency set up the Service Agency for Energy District Development (Servicestelle Energietische Quartiersentwicklung), which is tasked with supporting the creation of energy-efficient district concepts. The service point advises various actors, including the administration, the real estate industry, residents, and trade people on the complex interdisciplinary process of developing energy-related district strategies. An initial focus has been given to districts with significant energy and CO_2 savings potential, as well as willing implementers and decision-makers.[49]

The city is further challenged by a rising number of automobiles.[50] Although Berlin has fewer automobiles per 1,000 residents than other major German cities (Berlin has 390, while Munich has 570, Cologne has 515, and Frankfurt has 512), as the largest of the cities, it also has the largest number of cars (1.4 million, based on 2018 data).[51] The transport sector was responsible for 29.3 percent of 2017 emissions, up from a 1990 share of 17.3 percent. Transport emissions were 5.0 million tons CO_2 equivalent in 1990 and 5.6 million tons in 2017. Within the transport sector, the largest share of emissions is related to street traffic (20.6 percent of total emissions), with a further 5.7 percent from air travel. Emissions increased by 6.7 percent for road transport and 198.2 percent for aviation between 1990 and 2017.[52]

And despite Berlin's image as a relatively climate-friendly city, it has done too little to advance renewables or electric mobility—there is still a very low percentage of renewable energy sources in Berlin's energy mix.[53] In 2017, renewable energy accounted for only 4.2 percent of primary energy in Berlin and 1.7 percent of final energy consumption. Although this was a large increase over 1990 shares and Berlin's performance has improved in the past few years, the city still ranks at the bottom of *Länder* in renewable energy installation, employment of individuals in the renewables sector, and research and development in this field.[54]

The National and International Contexts

Berlin's progress will be tied to national and European climate progress. At the European level, a fundamental paradigm shift is under way with the introduction of the European Green Deal and Climate Protection Act. Europe aims to be climate neutral by 2050 and to reduce carbon dioxide emissions by 55 percent by 2030 compared to 1990 levels. This will mean that deep transitions of economic and energy structures will be taking place across Europe, opening up good opportunities for cross-societal and cross-metropolitan learning. Berlin is a member of several international networks for climate change, including the Covenant of Mayors, C40 Cities Climate Leadership Group, ICLEI—Local Governments for Sustainability, and Climate Alliance (Klimabündnis).[55]

At the national level, an energy-system transformation (Germany's Energiewende) is also under way. Targets set out in the 2016 German climate protection plan included a long-term goal of reducing GHG emissions by 80 to 95 percent by 2050 in comparison to 1990 levels, and the 2019 Climate Protection Law sets a date to achieve a 55 percent CO_2 emission reduction by 2030 and to achieve climate neutrality by 2050.[56] In a historic decision issued on April 29, 2021, however, the German Constitutional Court declared the targets set by the government as inadequate. The court reasoned that the targets placed an unfair burden on younger generations, which would have to shoulder too large a share of the reduction burden, and mandated a revision of the law. In response, the German government revised the Climate Protection Law in the summer of 2021. The carbon dioxide reduction targets was raised to 65 percent by 2030, and the target for climate neutrality was moved forward from 2050 to 2045.

Under the 2019 Climate Protection Law, the Bundestag and Bundesrat decided to price CO_2 from traffic and buildings (with amendments to the Fuel Emissions Trading Act); consumers started to pay higher prices for heating and fuel, beginning in January 2021. This is intended to provide an incentive to switch to more climate-friendly alternatives. In the period between 2021 and 2024, the federal government expects to generate a total income of almost €40 billion from this CO_2 pricing scheme.[57]

Energy justice issues were also considered. The rise in heating and fuel costs due to the new taxes are to be offset in part by a reduction in the electricity price—more specifically, a reduction in the renewable-energy surcharge paid by customers. Earnings from carbon pricing are to be used instead to maintain support for renewables. Commuters traveling longer distances will also be eligible for tax relief. Of particular interest is the provision that increases the housing allowance provided to people on welfare to help cover the increases in energy prices. The German Federal Ministry for Economic Affairs and Energy also recently initiated a funding program to study the socioeconomic impact of the energy transition on individual sectors of society.

The Building Energy Act of August 2020 brings together regulations from several different acts: the Energy Saving Act, the Energy Saving Ordinance, and the Renewable Energies Heat Act. It creates a new, uniform, and coordinated set of rules for more climate protection in the building sector.[58] Critics fear, however, that the funds designated for these reforms are too low.

In another major change, the Federal Ministry of Economics and Energy will be expected to integrate climate protection into economic growth strategies. A paper from the ministry titled "Combating Climate Change and Boosting the Economy" provides the impulse for a win-win situation for the economy and the climate.[59]

For Berlin, it is critical that these changes are under way at the international and national levels. Berlin's efforts at achieving climate neutrality will be interdependent with the pace of change achieved both nationally and internationally.

Conclusion

Berlin's climate policies have been influenced by decisions and structural developments that reach far back into the past. The reunification of Berlin in 1989 offered a chance to modernize industry, improve energy infrastructures, and renovate many buildings that had fallen into states of disrepair. The fact that the city has few major heavy industries, in addition to a good public transport network and district heating system, has contributed to its relatively low CO_2 emission levels. But the city's heavy dependence on hard coal and lignite, citizens' continued love of the automobile, and an old housing stock are some of the major obstacles to achieving yet deeper improvements.

Legislative developments in the past several years, however, suggest that more change can be expected. The coal phase-out is beginning, the push to green the transport sector is strengthening, and there are some signs of progress with regard to the build-out of renewable energy. Thus, Berlin is on the way to becoming climate neutral—but there is still much that needs to be done. With much of the low-hanging fruit already picked, further action will require strong political and economic will. And yet there is reason for hope. Berlin's climate policies are comprehensive and ambitious and, in many ways, can serve as an example from which other metropolitan regions may learn.

NOTES

We thank Lauren Goshen for her assistance with background research for the chapter and editorial contributions, Danielle Spiegel-Feld for helpful comments on content and structure, and Sara Savarani for her detailed editing of the manuscript.

1. Berlin Senate. 2021. "Ziele und Grundlagen der Klimaschutzpolitik in Berlin." *Senatsverwaltung für Umwelt, Verkehr und Klimaschutz*, January 9, www.berlin.de.
2. Amt für Statistik Berlin-Brandenburg. 2019. "Statistischer Bericht EV IV 4—j/17, Energie- und CO_2-Bilanz in Berlin 2017." www.statistik-berlin-brandenburg.de.
3. Reusswig, Fritz, B. Hirschl Bernd, and Wiebke Lass. 2014. *Klimaneutrales Berlin 2050—Ergebnisse der Machbarkeitsstudie*. Berlin: Senatsverwaltung für Stadtentwicklung und Umwelt, 3, www.bgmr.de.
4. *See, e.g.*, Romero-Lankao, Patricia, Daniel M. Gnatz, Olga Wilhelmi, and Mary Hayden. 2016. "Urban Sustainability and Resilience: From Theory to Practice." *Sustainability* 8(12): 1224. https://doi.org/10.3390/su8121224; Sorensen, André. 2015. "Taking Path Dependence Seriously: An Historical Institutionalist Research Agenda in Planning History." *Planning Perspectives* 30(1): 17–38; Stefes, Christoph H. 2014. "Energiewende: Critical Junctures and Path Dependencies since 1990." In *Rapide Politikwechsel in der Bundesrepublik*, edited by Freidbert W. Rüb, 48–71. Berlin: Nomos; Chang Gyu Choi, Sugie Lee, Heungsoon Kim, and Eun Yeong Seong. 2019. "Critical Junctures and Path Dependence in Urban Planning and Housing Policy: A Review of Greenbelts and New Towns in Korea's Seoul Metropolitan Area." *Land Use Policy* 80: 195–204. https://doi.org/10.1016/j.landusepol.2018.09.027.
5. Luisenstädtischer Bildungsverein e.V., 2004. "Bevölkerungsentwicklung in Berlin." https://berlingeschichte.de.
6. Schneider, Stefan, and Martin Carazo Mendez, dirs. "Elend und Fortschritte—Die Geschichte der Industriestädte." Episode 3, *Deutschlands Städte*, ZDF, aired July 19, 2015, www.zdf.de.

7. Senatsverwaltung für Umwelt, Verkehr und Klimaschutz. n.d. "Geschichte des Berliner Stadtgrüns: Kleingärten." Accessed February 12, 2021, www.berlin.de.
8. As part of a school project, students of the Christian-Gymnasium Hermannsburg made a documentary, "Wie die Kohle nach Berlin kam," about the Berlin Air Lift in which interviewees explain that electricity was rationed to just a few hours each day and that meant sitting at home in the dark and in the cold—the situation was dire. Christian-Gymnasium Hermannsburg and Bundeswehr-Fachmedienzentrum Faßberg: Cellesche Zeitung. 2019. "Wie die Kohle nach Berlin kam." YouTube, May 10, www.youtube.com/watch?v=pEPFCOSceq4.
9. Wiedemeier, Juliane. 2011. "Kleine Geschichte der fernen Wärme." *Prenzlauer Berg Nachrichten*, June 7, www.prenzlauerberg-nachrichten.de.
10. Senatsverwaltung für Wirtschaft, Energie und Betriebe, Berlin. n.d. "Berliner Energieverbrauch und CO_2-Bilanz." Accessed January 9, 2021, www.berlin.de.
11. The renovation has been financed in part through a solidarity tax that was initiated in 1991 and only ended (for most taxpayers) in 2021.
12. Deutscher Bundestag. n.d. "Strom, Wärme, Kälte: Das Energiekonzept des Deutschen Bundestages." Accessed January 9, 2021, www.bundestag.de.
13. Berlin was made the capital by the parliamentary decree of 1991 after the reunification of Germany. It had been capital of all of Germany prior to the division of Germany, but with the onset of the Cold War, the capital of West Germany was established in Bonn, while the capital of East Germany was situated in East Berlin.
14. Before 2001, there were twenty-three districts.
15. The Enquete Commission "Zukünftige Energiepolitik" was established on January 28, 1982. The establishment of the commission was the result of a parliamentary motion by the member of parliament Martin Jänicke, which then became an all-party motion. The occasion was the opposition to the Berlin Reuter-West coal-fired power plant, which nevertheless was later built.
16. The two reports were Jänicke, Martin, Lutz Mez, Jürgen Pöschk, Susanne Schön, and Thomas Schwilling. 1987. "Alternative Energiepolitik in der DDR und in West-Berlin. Möglichkeiten einer exemplarischen Kooperation in Mitteleuropa." *Schriftenreihe des Instituts für ökologische Wirtschaftsforschung* 3/87; and Berlin, Senatsverwaltung für Stadtentwicklung und Umweltschutz (SenStadtUm), ed. 1990. *Ziele und Möglichkeiten einer stromspezifischen Energieeinsparpolitik in Berlin (West) unter Berücksichtigung des Stromverbundes mit der Bundesrepublik. Neue Energiepolitik in Berlin, Heft 1*. Berlin.
17. Monstadt, Jochen. 2004. *Die Modernisierung der Stromversorgung. Regionale Energie- und Klimapolitik im Liberalisierungs—und Privatisierungsprozess*. Wiesbaden: VS Verlag für Sozialwissenschaften.
18. Ziesing, Hans-Joachim. 1995. "Berlin nach dem Klimagipfel. Auf dem Weg zu einer Zukunftsfähigen Metropole?" *Stadtforum Journal* 19 (May): 1–2.
19. Miranda Schreurs was a member of the Enquete Commission. An English version of the commission's report is available at Berlin House of Representatives. n.d.

"Conclusions and Recommendations from the Final Report of the Study Commission on New Energy for Berlin—The Future of Energy-Sector Structures." Accessed January 11, 2021, www.parlament-berlin.de.

20. Hirschl, Bernd, and Harnisch, Richard. 2016. "Climate Neutral Berlin 2050—Recommendations for a Berlin Energy and Climate Protection Programme (BEK)." Senate Department for Urban Development and the Environment, Berlin, www.berlin.de; Latz, Christian. 2021. "Koalition beschließt Energiewendegesetzt: Solarpflicht für alle öffentlichen Gebäude in Berlin." *Der Tagesspiegel*, August 9, www.tagesspiegel.de.
21. Berlin Senate. 2017. "Berliner Energie- und Klimaschutzprogramm 2030 (BEK 2030)—Umsetzungszeitraum 2017 bis 2021, Konsolidierte Fassung." Senatsverwaltung für Stadtentwicklung und Umwelt, 36, www.berlin.de.
22. Vattenfall GmbH. n.d. "Ausstieg Kohleenergie." Accessed January 11, 2021, https://group.vattenfall.com.
23. Full text available at Berlin Senate. n.d. "Der Weg zum Berliner Energie- und Klimachutzprogramm (BEK)." Senatsverwaltung für Umwelt, Verkehr und Klimaschutz. Accessed January 11, 2021, www.berlin.de.
24. Berlin Senate. n.d. "Digitales Monitoring- und Informationssystem (diBEK)." Senatsverwaltung für Umwelt, Verkehr und Klimaschutz. Accessed January 11, 2021, www.berlin.de.
25. BerlinMap360. n.d. "Berlin Bike Map." Accessed January 11, 2021, https://berlinmap360.com.
26. Berlin Senate. n.d. "Fußverkehrsstrategie für Berlin." Senatsverwaltung für Umwelt, Verkehr und Klimaschutz. Accessed January 11, 2021, www.berlin.de; Berlin Senate. n.d. "Mobilitätsgesetz: Vorrang für Bus, Bahn, Fahrrad—und Fußgänger*innen." Senatsverwaltung für Umwelt, Verkehr und Klimaschutz. Accessed January 11, 2021, www.berlin.de.
27. Berlin Senate. n.d. "Masterplan Solarcity." Senatsverwaltung für Wirtschaft, Energie und Betriebe. Accessed January 11, 2021, www.berlin.de.
28. Senatsverwaltung für Wirtschaft, Energie und Betriebe. 2020. "Masterplan Solarcity. Monitoring 2020." SolarWende, February 15, www.solarwende-berlin.de.
29. Wikipedia. 2020. "Liste deutscher Orte und Gemeinden, die den Klimanotstand ausgerufen haben." December 12, https://de.wikipedia.org; Reinsch, Melanie. 2019. "Berlin erklärt als erstes Bundesland die 'Klimanotlage.'" *Berliner Zeitung*, December 10, www.berliner-zeitung.de.
30. Berlin Senate. 2020. "Berlin schafft Klimaziel 2020 bereits vorzeitig." Senatsverwaltung für Wirtschaft, Energie und Betriebe, December 10, www.berlin.de.
31. *Der Tagesspiegel* (Berlin). 2019. "CO_2-Emissionen in Berlin gehen zurück." December 18, www.tagesspiegel.de.
32. Behörde für Umwelt, Klima, Energie und Agrarwirtschaft. n.d. "Bilanz des Statistikamtes-Nord für 2018, CO_2-Emissionen in Hamburg." Hamburger Stadtportal. Accessed January 12, 2021, www.hamburg.de; Amt für Statistik Berlin-Brandenburg 2019, 18.

33. Landeshauptstadt München. 2020. "Klimaschutzstrategie der Landeshauptstadt München." Das offizielle Stadtportal, www.muenchen.de.
34. Landeshauptstadt München. 2020. "Neue Zahlen: CO_2-Emissionen in München rückläufig." Das offizielle Stadtportal, March 6, https://ru.muenchen.de.
35. Berlin Senate. n.d. "Temporare Radfahrstreifen." Senatsverwaltung für Umwelt, Verkehr und Klimaschutz. Accessed January 12, 2021, www.berlin.de.
36. Senatsverwaltung für Umwelt, Verkehr und Klimaschutz. n.d. "Abfallstrategien." Accessed January 11, 2021, www.berlin.de.
37. Bundesministerium für Umwelt, Naturschutz und nukleare Sicherheit. 2020. "Projekte." Nationale Klimashutzinitiative, www.klimaschutz.de.
38. Susanne Kramm. n.d. "Klimaschutzprojekte an Berliner Schulen erfolgreich umsetzen—kostenfrei Messgeräte ausleihen." Berliner Energieagentur. Accessed January 11, 2021, www.berliner-e-agentur.de.
39. Berlin Senate. 2019. "EnergiespeicherPLUS Bringt Grünen Strom in Berliner Haushalte." Senatsverwaltung für Wirtschaft, Energie und Betriebe, October18, www.berlin.de.
40. EUMB Pöschk GmBH & Co. KG. 2020. "Video-Content rund um die Aktionswoche 2020." Berlin Spart Energie, www.berlin-spart-energie.de.
41. Bundesministerium des Innern, für Bau und Heimat. 2020. "32 Modellprojekte Smart Cities Ausgewählt." September 8, www.bmi.bund.de.
42. Senate Department for Urban Development and the Environment. 2015. "Working Together for Climate Change Mitigation in Berlin." www.berlin.de.
43. Reusswig et al. 2014, 8; Berlin Senate 2017, 16–17.
44. Amt für Statistik Berlin-Brandenburg. 2020. "Statischer Bericht E IV 5—j19: Energie- und CO2-Daten in Berlin 2019, Vorläufige Ergebnisse." www.statistik-berlin-brandenburg.de, 23.
45. Amt für Statistik Berlin-Brandenburg 2020, 23.
46. Amt für Statistik Berlin-Brandenburg 2020, 25.
47. Berlin Senate. n.d. "Wärmestrategie für das Land Berlin." Senatsverwaltung für Umwelt, Verkehr und Klimaschutz. Accessed January 11, 2021, www.berlin.de.
48. Stamo, Irina. 2018. "Energieeffizienz im Gebäudesektor in Berlin: Interaktion von verschiedenen Schlüsselakteuren. Kopernikus Projekte ENavi." IKEM, July, www.ikem.de.
49. Berlin Energie Agentur. n.d. "Service- und Beratungsstelle für energetische Quartiersentwicklung." Accessed January 11, 2021, www.berliner-e-agentur.de.
50. Schäfer, Andreas. 2019. "Das ist die Berliner Luft." *Der Tagesspiegel*, September 10, www.tagesspiegel.de.
51. *Der Tagesspiegel* (Berlin). 2020. "Berlin hat deutlich weniger Autos als andere Großstädte." January 1, www.tagesspiegel.de.
52. Amt für Statistik Berlin-Brandenburg 2019, 21, 34.
53. Berlin Senate 2017, 17.
54. Amt für Statistik Berlin-Brandenburg 2019, 12, 25; Schill, Wolf-Peter, Jochen Diekmann, and Andreas Püttner. 2019. "Sechster Bundesländervergleich Erneuerbare

Energien: Schleswig-Holstein und Baden-Würtemmberg an der Spitze." *DIW Wochenbericht* 48, www.diw.de.

55. Berlin Senate. n.d. "Internationales Engagement Berlins." Senatsverwaltung für Umwelt, Verkehr und Klimaschutz. Accessed January 11, 2021, www.berlin.de.
56. Bundesministerium für Umwelt, Naturschutz und nukleare Sicherheit. 2016. "Klimaschutzplan 2050. Klimaschutzpolitische Grundsätze und Ziele der Bundesregierung," 6, www.bmu.de.
57. Deutscher Bundestag. 2020. "Einnahmen aus der CO2-Bepreisung." Wirtschaft und Energie/Antwort. 24.08.2020 (hib 866/2020), August 24, www.bundestag.de.
58. Bundesministerium des Innern, für Bau und Heimat. n.d. "Das neue Gebäudeenergiegesetz." Accessed April 24, 2021, www.bmi.bund.de.
59. Federal Ministry for Economic Affairs and Energy Public Relations Division. 2020. "Combatting Climate Change and Boosting the Economy: Proposal for an Alliance of Society, Business and Government for Climate-neutrality and Prosperity." www.bmwi.de.

14

Reducing Greenhouse Gas Emissions from Buildings in New York City

An Evolving Regime

DANIELLE SPIEGEL-FELD

In New York City, buildings are the dominant source of greenhouse gas (GHG) emissions. Whereas across the globe, energy used in the construction and operation of buildings accounts for approximately 39 percent of GHG emissions, in New York City, it accounts for two-thirds.[1] Transportation, which relies heavily on mass transit, has historically made a comparably modest contribution to local GHGs.[2]

Given buildings' outsized contribution to local emissions, as well as the city's limited control over the electricity sector, New York City's climate-mitigation policy has traditionally focused on reducing energy use in buildings.[3] The city began tackling the issue in the first decade of the twenty-first century, during which time the administration of mayor Michael Bloomberg passed a suite of pioneering, but fairly light-touch, building-efficiency regulations. These early regulations were aimed primarily at reducing perceived market barriers to investing in energy-efficiency improvements, rather than mandating that any particular investments be made or any performance standards be met.

More recently, New York City has adopted a stricter approach to regulating buildings. The centerpiece of this new approach is a law known as Local Law 97 of 2019, which sets a cap on the GHG emissions that buildings can release for free.[4] This new law marks a notable departure from past regulations in that it sets a mandatory standard that buildings must meet to avoid being fined. Local Law 97 also differs from the Bloomberg-era regulations in that it measures compliance in terms of the amount of GHG emissions attributable to a given building's energy use instead of the amount of energy used. This is a significant change because it means that the difficulty of meeting the law's caps depends in large part on the GHG intensity of electricity supplied by the grid. And because the state, rather than the city, regulates the electricity sector, state electricity policy will greatly influence the stringency of the city's law.

Local Law 97 was passed toward the end of a long period of prosperity in New York City. Whether the city remains fully committed to the policy independent of state policy in the present era of economic instability that the COVID-19 crisis has wrought remains to be seen. It seems plausible that local leaders will develop implementing regulations for the law that will effectively diminish the costs imposed on the real estate industry for fear of driving businesses—and tax revenue—outside the city's borders.

This chapter reviews the history of New York City's efforts to reduce emissions from the building sector and the major questions with which it is grappling today. It proceeds chronologically, starting with the Bloomberg administration's efforts to promote better information disclosure about building efficiency and then describing the new approach that Local Law 97 set in motion. The chapter concludes with a discussion of the ways in which the COVID-19 crisis may have altered the city's calculations regarding mandatory GHG reductions.

The Beginnings: Assistance and Incentives

In 2009, New York City passed a suite of environmental regulations known as the Greater, Greener Buildings Plan (GGBP). The GGBP, which was championed by Mayor Bloomberg's administration, encompasses four distinct regulations:

1. Local Law 84, which requires buildings to annually report to the city how much energy and water they consumed; the buildings then receive a "benchmarking" score that indicates how their consumption compares to similar properties
2. Local Law 85, which requires buildings to meet the requirements of the most current energy code when they conduct major renovation

3. Local Law 87, which requires buildings to conduct periodic energy audits that identify opportunities for cost-effective retrofits and retro-commission HVAC equipment.[5]
4. Local Law 88, which requires nonresidential buildings to make certain lighting upgrades and separately charge large tenants for their electricity consumption (i.e., "submeter") instead of charging them a fixed percentage of the building's total electricity bill (which had previously been common practice)

The GGBP also established two programs to assist building owners in making energy upgrades: a jobs-training program to bolster the local workforce with the required technical expertise and an energy-efficiency financing corporation known as the New York City Energy Efficiency Corporation (NYCEEC) to provide low-cost funding for energy upgrades.[6]

The GGBP was groundbreaking. At the time it was passed, the buildings that Local Laws 84, 87, and 88 regulated accounted for 45 percent of citywide energy usage.[7] Moreover, Local Law 84 was the first benchmarking law to be implemented in the United States.[8] In the years after it came into being, more than a dozen US cities, including major cities like Atlanta, Boston, Chicago, and Los Angeles, put similar benchmarking laws on their books, covering billons of square feet.[9] Local Laws 84 and 87 also helped create a vast repository of data about building energy use in New York City that has facilitated the subsequent development of nuanced performance standards, including those used in developing Local Law 97.

But the GGBP was also modest. Apart from the requirements to retro-commission HVAC equipment and make lighting upgrades in nonresidential buildings, none of the regulations actually oblige building owners to reduce energy consumption nor do they create a direct financial incentive to do so. Instead, what the laws do is attempt to mitigate the perceived market barriers to energy efficiency that cause owners to forgo cost-beneficial energy improvements (the so-called energy efficiency gap).[10] Local Laws 84 and 87 aim to overcome information deficits among building owners regarding how inefficient their properties may be; Local Law 88 seeks to help tenants internalize the costs of their energy consumption so that they are better incentivized to conserve; and NYCEEC and the jobs-training program aim to mitigate labor and financing shortfalls that could hinder building owners' efforts to implement the upgrades that they want to make.[11] In short, the GGBP sought to empower property owners, not penalize them.

This light-touch approach produced some meaningful results. In fact, one rigorous econometric analysis of Local Law 84 credits the policy with having reduced energy consumption in the regulated buildings by 14 percent in the first four years after it took effect.[12] But impressive as these effects may be, by 2014, when New York City adopted the goal of reducing citywide GHGs to 80 percent below 2005 levels by 2050, it became clear that regulators would need to do much more to achieve deep decarbonization. It was equally clear that buildings would have to lead the way.[13]

The city took a first step in this direction by expanding the disclosure requirements set out in Local Law 84. Whereas initially owners were only required to report their benchmarking scores to the city's Department of Finance, which then posted the scores on a hard-to-find website, the City Council passed a law in 2017 that required benchmarking scores to be translated into letter grades and prominently displayed in buildings' entrances.[14] The idea behind this new law, which real estate interests strongly opposed, was to make the public aware of how efficient (or inefficient) the buildings they occupy are so they could apply pressure on lagging landlords or management boards to do better.[15] To the extent that the law embraces the idea of public shaming to incentivize property owners to make changes they would not otherwise make, it is a departure from the purely facilitative approach that the GGBP set out. And yet the law is still geared toward removing market barriers to energy upgrades—the barrier, in this case, being information asymmetries between property owners and occupants—and does not directly require owners to take any action beyond the disclosure. It is thus a long way off from the approach that the city took next.

The New Regime: Mandates

In September 2017, as international leaders gathered in New York City for the annual Climate Week programs, Mayor Bill de Blasio made a big announcement: New York City would soon become the first city in the United States to set mandatory GHG emissions limits for its buildings. The limits would apply to buildings with more than twenty-five thousand square feet, thus capturing tens of thousands of buildings. Though the mayor offered few details about the plan at that time, he made clear that property owners who failed to meet the limits would face substantial fines, up to millions of dollars per year. "We gave people a very fair amount of time for the private sector to come forward and really agree to voluntary goals that will be sufficient," Mayor de Blasio said. "It [is] time to move to mandates."[16]

The announcement caught many stakeholders off guard. Both the speaker of the City Council and chairman of the Environmental Protection Committee, which needed to pass the relevant legislation, indicated that they had been excluded from deliberations about the proposal and expressly declined to endorse the idea.[17] The real estate industry was also critical, as were affordable-housing advocates who feared that the cost of mandatory energy upgrades would be passed on to tenants via rent increases.[18] Even a number of local environmental groups, which some observers may have guessed had pushed for the law, seemed surprised.[19]

At first blush, the policy appears surprising from a political economy standpoint as well. Greenhouse gases are global pollutants, and New York City produces only a tiny fraction of global emissions. As such, however significantly New York City's large buildings contribute to *local* GHG emissions, they cannot have anything more than a trivial impact on the world's climate. Yet policies that increase housing costs—even market-rate housing costs—can impact the local economy quite significantly by driving residents outside the city's borders. Policies that raise commercial real estate costs or discourage landlords from renting to businesses that use a lot of energy can drive businesses out of the city too. Real estate industry representatives made this threat explicit in their statements opposing emissions caps.[20] And if local governments compete among each other to attract the residents and businesses that build their tax base, as Charles Tiebout famously argued, the last thing a city should want to do is drive people away.[21] From this perspective, the mayor's proposal to put a price on building emissions seems like a lopsided bargain, where New York City would pay all the costs and get only a small fraction of the benefit.

So why did the administration turn its back on the facilitative approach to reducing building emissions and bring out the stick? There are many possible answers to this question.

One possibility is simply that environmentalism is popular among New York City voters. There is some reason to believe that, as the innovation economy became more dominant in certain major US cities like New York, the local costs of environmental regulation declined, and demand for urban environmental policies grew.[22] Air quality regulations, for example, may impose fewer costs or create less political opposition in cities in which heavily polluting industries account for a relatively small share of economic output. But this explanation is unlikely to *fully* account for New York's support for Local Law 97 for two reasons. First, the real estate sector contributes very substantially to New York City's economy; to give one metric, 52 percent of taxes collected in 2018 were real estate related.[23] Second, while Local Law 97 may impose substantial compliance costs on local industry, it is not primarily geared toward producing local benefits.[24] So, to the extent that New Yorkers perceive the law as cost-beneficial, it may be because they assign value to the intangible benefit of expressing their political ideals.

This leads to a second possible explanation for why the city's GHG policy took a more punitive turn: the election of Donald Trump. President Trump was highly unpopular in New York City (he received only 21 percent of the vote in 2016). It seems plausible that New Yorkers' distaste for the former president helped spur a distaste for the real estate industry with which he is associated and a desire for retribution. Along these lines, it may not be a coincidence that the subtitle of a *New York Times* article announcing passage of Local Law 97 specifically called out that the new law would force Trump Tower to reduce its emissions.[25]

A third, related, explanation pertains to Trump's withdrawal from the Paris Agreement. Shortly after Trump withdrew the United States from the Paris Agreement, Mayor de Blasio passed an executive order that committed New York City to adhere to the principles of the agreement. The order also instructed city agencies to develop plans to accelerate achievement of the city's efforts to achieve its "80 by 50" target to align with the Paris Agreement's goal of keeping warming to 1.5 degrees Celsius.[26] These new international commitments may have brought greater attention to the city's climate goals and provided new impetus to make real progress toward them. The mayor may have also sought to demonstrate policy innovation and leadership at a time when global heads of state were gathering in New York to discuss the climate.

Whatever the reason for the mayor's initial embrace of the policy, the administration remained committed to it, and in April 2019, eighteen months after the initial announcement, the City Council finally passed the ordinance that turned the idea of emissions caps into law. The election of Representative Alexandria Ocasio-Cortez, which emboldened the far-left wing of US politics, and the subsequent introduction of the Green New Deal may have helped push the proposal across the finish line in the City Council. Indeed, retrofitting buildings was a key component of the Green New Deal.[27]

To implement the emissions caps, Local Law 97 groups buildings into ten different categories based on their occupancy types and sets different emissions intensity caps (i.e., the maximum amount of carbon emissions per square foot) for each category. Building owners who exceed their cap are liable to pay up to $268 per ton of excess emissions. To calculate a building's annual emissions, owners must multiply the total amount of energy purchased by the carbon intensity coefficient that the city assigns for the relevant type of energy (i.e., electricity procured from the grid, natural gas, fuel oil, etc.). The first compliance period runs from 2024 to 2029, and the caps are supposed to get progressively stricter in five-year increments until 2050.

As a concession to tenants' advocates who worried about rising rents, buildings with one or more rent-regulated unit were excluded from the mandate. These properties were only required to make certain relatively low-cost prescriptive mandates, such as insulating heating and hot-water pipes and fixing leaks in their heating systems. They were not assigned any hard-and-fast performance targets.[28] Notably, the decision to exclude these affordable-housing properties from the caps illustrates a potential tension between the traditional environmental movement, which is chiefly concerned with ameliorating the physical environment, and movements that are concerned with economic or environmental justice, which seek to insulate low-income communities from the cost of environmental regulation.

Building owners got some meaningful concessions too. Of particular note, at least for the first compliance period, building owners will be allowed to claim deductions against their annual reported emissions by purchasing GHG offsets, procuring renewable-energy credits (RECs) for green energy that feeds into the New York City grid, or using energy from a clean distributed energy installation such as rooftop solar.[29] For example, if an owner consumes 10,000 kilowatt hours of energy in a year from a building's rooftop solar installation, not only does it get the benefit of consuming the electricity without incurring a carbon charge, but it can also *subtract* 10,000 kilowatt hours of grid-procured electricity from its annual energy consumption when calculating its annual carbon bill. RECs have the same subtractive effect. These deductions, which do nothing to incentivize energy efficiency, appear primarily aimed at expanding the supply of renewable energy in New York City rather than directly reducing building emissions. This incentive to install distributed renewable energy is part of a broader suite of city initiatives that uses the city's authority to regulate real estate to promote the greening of the electricity-generation mix, which is a matter over which the city otherwise has very limited control.[30]

Impactful as these deductions may be, however, the biggest concession to the real estate industry was arguably the relative laxness of the initial caps themselves. The emissions targets for the first compliance period (running from 2024 to 2029) were set generously enough that the vast majority of building owners will not have to invest in their properties at all to meet their targets. Thus, many owners were given a decade or more to prepare for the changes that the law will eventually require them to make. Subsequent state legislation may have made it easy for many building owners to meet their obligations in later compliance periods as well: two months after Local Law 97 was adopted, New York State adopted a law that aims to entirely decarbonize the state electricity grid by 2040.[31] It has been estimated that if this target is met, a majority of the buildings that Local Law 97 covers will not have to take any action to meet the emissions caps for the second compliance period either.[32]

Intentionally or not, Local Law 97 also gives authority to the city's Department of Buildings (DOB) to effectively loosen the caps via rule making. For instance, DOB gets to decide what kind of projects will qualify as "offsets" and what the designated carbon intensity coefficient will be for different energy sources after 2029. DOB can also set the pace at which emissions targets will decline between 2035 and 2050, which will determine how quickly owners must invest in energy upgrades. The stringency of the rules that DOB develops will ultimately determine the extent to which the policy drives energy reductions. For example, if DOB assigns a high coefficient for natural gas that fully accounts for upstream methane emissions, it may push some building owners toward electrifying heating in order to avoid paying the penalty for exceeding their cap. If, however, DOB assigns a low coefficient for natural gas, many owners may not have an incentive to do anything other than continue business as usual for many years to come. Similarly, if DOB chooses to allow low-cost international offsets to qualify, as opposed to restricting the geographic scope from which offsets are procured, DOB could substantially reduce the incentive to invest in retrofits. DOB's decisions concerning the coefficient for electricity procured from the grid after 2029 could be extremely impactful as well.[33] Notably, the fact that decisions about these issues are highly technical and will be decided via rule-making processes, which typically garner less political attention than legislative processes do, may make industry interests relatively more powerful than they were during the drafting of Local Law 97.

Can the Momentum Be Maintained?

As of this writing, New York City looks like a very different place than it did when Local Law 97 was passed. The COVID-19 crisis reaped tremendous havoc on the local economy. The city's budget for fiscal year 2021 was cut by almost 10 percent, roughly three times as much as it was reduced after the 2009 financial crisis, and the vacancy rate in Manhattan's office buildings is close to 60 percent higher than it was in March 2019.[34] No one knows if, or when, the city's fortunes will change, but few people seem to think that the present challenges will be short-lived; the city's Budget Office has projected budgetary shortfalls through 2024.[35]

It is during these next, probably recessionary, years that DOB will write the implementing rules that determine how much pressure Local Law 97 actually imposes. If the electricity grid does not rapidly decarbonize, as the state has pledged, will the department craft implementing rules that relieve the pressure on property owners to spare them added costs? Some analysts have argued that investments in energy upgrades will actually save property owners money over the long term as their energy bills decline, and there is certainly reason to believe this is the case.[36] But even the staunchest energy-efficiency advocates will generally admit that it takes years for these investments to be paid off. The question is whether, in the current climate, the city will impose these near-term costs on local property owners for the long-term global good.

Publicly, at least, the de Blasio administration appeared steadfast in its commitment to building emissions caps and in the fall of 2020; the City Council passed a bill to extend the number of buildings that would be subject to a cap.[37] Ultimately, however, implementing Local Law 97 will fall to the administration of the newly elected mayor, Eric Adams, as de Blasio's term expired at the end of 2021. Whether the Adams administration decides to continue the fight for emissions caps will say a lot about the durability of New York City's commitment to reducing GHGs.

NOTES

1. UNEP and IEA. 2019. *2019 Global Status Report for Buildings and Construction.* www.unenvironment.org; City of New York. 2017. *Inventory of New York City Greenhouse Gas Emissions in 2016*, 4, www.nyc.gov.
2. City of New York 2017, 4. Note that the share of transportation-related emissions may have increased during the COVID-19 pandemic as subway ridership declined and reliance on private cars appears to have increased. Muoio, Danielle. 2020. "The Coronavirus Comeback No One Wants: New York City Traffic." *Politico*, July 19, www.politico.com.
3. New York State law vests the state, rather than local governments, with near exclusive control to regulate electricity generation, transmission, and distribution. N.Y. Pub. Serv. Law § 5(1). New York City also does not own its electric utility, as some municipalities do.
4. Spiegel-Feld, Danielle. 2019. "Local Law 97: Emissions Trading for Buildings?" *NYU Law Review Online* 94: 148–168.
5. Retro-commissioning essentially requires buildings to tune up the existing HVAC systems. Retro-commissioning does not require owners to invest in new systems.
6. New York City Mayor's Office of Long-Term Planning and Sustainability. 2014. *Overview of the Greater, Greener, Buildings Plan*, 1, www.nyc.gov.
7. New York City Mayor's Office of Long-Term Planning and Sustainability 2014. As initially adopted, Local Law 84 applied to buildings with more than fifty thousand square feet; it was subsequently amended to cover buildings with greater than twenty-five thousand square feet as well.
8. Urban Green. n.d. "Metered New York." Accessed October 23, 2020, https://metered.urbangreencouncil.org.
9. Urban Green, n.d.

10. Gillingham, Kenneth, and Karen Palmer. 2014. "Bridging the Energy Efficiency Gap: Policy Insights from Economic Theory and Empirical Evidence." *Review of Environmental Economics & Policy* 8(1): 18–38.
11. On the ability for submetering to incentivize energy reductions, *see* Gunay, H. Burak, William O'Brien, Ian Beausoleil-Morrison, and Andrea Perna. 2014. "On the Behavioral Effects of Residential Submetering in a Heating Season." *Building & Environment* 81: 396–403. https://doi.org/10.1016/j.buildenv.2014.07.020.
12. Meng, Ting, David Hsu, and Albert Han. 2017. "Estimating Energy Savings from Benchmarking Policies in New York City." *Energy* 133: 415–423. https://doi.org/10.1016/j.energy.2017.05.148.
13. New York City Mayor's Office of Long-Term Planning and Sustainability. 2014. *One City Built to Last*, 6–7, www.nyc.gov.
14. Intro. 1632 of 2017, N.Y.C. Admin. Code. § 28-309.12.
15. Among other objections, the Real Estate Board of New York argued that the new disclosure requirement would have the "unintended consequence [of] endorsing dark, uncomfortable, unproductive spaces, putting the City's sustainability efforts directly at odds with tenant expectations throughout the City." Hum, Carl (Senior Vice President, Management Services and Government Affairs, Real Estate Board of New York). 2017. Testimony before the New York City Council Committee on Environmental Protection on Introduction Nos. 1629, 1632 and 1637. REBNY, June 27, 2017, www.rebny.com. Spiegel-Feld, Danielle. 2016. *Building Demand for Efficient Buildings: Insights from the EU's Energy Disclosure Law*. New York: Guarini Center. https://guarinicenter.org.
16. NYC Press Office. 2017. "Mayor de Blasio: NYC Will Be First City to Mandate That Existing Buildings Dramatically Cut Greenhouse Gas Emissions." September 14, www.nyc.gov.
17. Neuman, William. 2017. "De Blasio Vows to Cut Emissions in New York's Larger Buildings." *New York Times*, September 14, www.nytimes.com.
18. Poon, Linda. 2017. "NYC's Tall Order for Greener Buildings." *Bloomberg*, September 26, www.bloomberg.com.
19. Neuman 2017.
20. *See, e.g.*, Neuman, William. 2019. "Big Buildings Hurt the Climate. New York City Hopes to Change That." *New York Times*, April 17, www.nytimes.com (quoting Carl Hum, general counsel for the Real Estate Board of New York, as saying, "There's a clear business case to be made that having a storage facility is a lot better than having a building that's bustling with businesses and workers and economic activity.").
21. Tiebout, Charles M. 1956. "A Pure Theory of Local Expenditures." *Journal of Political Economy* 64(5): 416–424. https://doi.org/10.1086/257839.
22. Wyman, Katrina, and Danielle Spiegel-Feld. 2019. "Urban Environmental Renaissance." *California Law Review* 108: 326–339.
23. New York City Comptroller. 2018. *The State of the City's Economy and Finances 2018*. http://comptroller.nyc.gov.

24. Notably, a subsequent analysis of the law, published in November 2021, indicated that it might not actually impose very substantial costs on the real estate industry. *See* Spiegel-Feld, Danielle, Mary Jiang, Sara Savarani, Kasparas Spokas, Burçin Ünel, and Katrina Wyman. 2021. *Carbon Trading for New York City's Building Sector: Report of the Local Law 97 Carbon Trading Study Group to the New York City Mayor's Office of Climate & Sustainability*. Guarini Center, www.guarinicenter.org. However, this assessment was not available to the city at the time the law was adopted, and at that time, city officials suggested that the law could, in fact, impose substantial costs. Kim, Elizabeth. 2019. "Here's How NYC Is Trying to Shrink Buildings' Big Carbon Footprints." *Gothamist*, September 19, https://gothamist.com. Salimifard, Parichehr, Johnathan J. Buonocore, Kate Konschnik, Parham Azimi, Marissa VanRy, Jose Guillermo Cedeno Laurent, Diana Hernández, and Joseph G. Allen. 2022. "Climate Policy Impacts on Building Energy Use, Emissions, and Health: New York City Local Law 97." *Energy* 238, pt. C: 121879. https://doi.org/10.1016/j.energy.2021.121879 (noting that Local Law 97 may actually incentivize combustion of biomass, which causes significant adverse local health impacts).
25. Neuman, William. 2019. "Big Buildings Hurt the Climate. New York City Hopes to Change That." *New York Times*, April 17, www.nytimes.com.
26. NYC Press Office 2017.
27. DiChristopher, Tom. 2019. "New York City Embraces Pillar of Green New Deal, Passing Building Emissions Deal." *CNBC*, April 18, www.cnbc.com.
28. In the fall of 2020, the City Council narrowed this exemption for affordable housing such that only properties with 35 percent rent-regulated units or more were exempt from the emissions caps. Council of City of New York Intro. 1947 of 2020, A Local Law to Amend the New York City Charter and Administrative Code of the City of New York in Relation to Rent Regulated Accommodations.
29. N.Y.C. Admin. Code § 28-320.3.6.
30. For example, the city passed another law alongside Local Law 97 that requires certain newly constructed buildings to include either green or solar roofs. Local Law No. 92 (2019) of City of New York.
31. Climate Leadership and Community Protection Act § 4, N.Y. Pub. Serv. Law § 66-p.
32. Spiegel-Feld et al. 2021, 14n16.
33. To some extent, the stringency of the electricity coefficient will be out of DOB's hands as it will have to reasonably align with the state's progress in meeting its grid decarbonization goals; if the city were to assign coefficients for any energy source—electricity or otherwise—that were too far afield from best estimates of the actual state of affairs, the rules could be judicially overturned as arbitrary and capricious. N.Y. C.P.L.R. 7803(3).
34. Stringer, Scott. 2020. "Comments on New York City's Fiscal Year 2021 Adopted Budget." New York City Comptroller, August 3, https://comptroller.nyc.gov. Cushman & Wakefield. n.d. Manhattan Office Market Overall Statistics, March 2019–July 2021. Data on file with author.

35. New York City Independent Budget Office. 2020. "Tumbling Tax Revenues, Shrinking Reserves, Growing Budget Gaps: New York City Faces Substantial Fiscal Challenges in the Weeks and Months Ahead." *Focus On: The Executive Budget*. https://ibo.nyc.ny.us.
36. Spiegel-Feld et al. 2021, 56.
37. Intro. 1947 of 2020.

PART V

Adapting to Climate Change

15

Varieties of Approaches to Climate Adaptation in Cities

Toward a Focus on Equity

ERIC K. CHU, ASIYA N. NATEKAL, HANNE J. VAN DEN BERG, AND CLARE E. B. CANNON

Cities around the world are developing new policies, regulations, and plans to address increasingly severe precipitation, heat, sea-level rise, and other hazardous impacts on their communities, infrastructure, and economies. Over the past two decades, more proactive local government leadership, often partnered with growing scientific research, has informed a suite of best practices to support adaptation and resilience building in the face of climate change.[1] These practices often rely on spatial interventions—such as building protective infrastructure, designing climate-resilient open spaces, and modifying building codes and land use regulations—although many cities have also crafted new institutions to ensure that adaptation actions relate to pertinent socioeconomic priorities, include relevant stakeholders and partners, and respond to the long-term development aspirations of urban residents.[2] Such considerations are particularly poignant given the disproportionate impact of climate change on socially vulnerable communities, as well as the continued exclusion of historically marginalized voices and interests in planning and decision-making processes.[3] Still, the gap between intention and actual on-the-ground implementation remains large—especially in the Global South, where socioeconomic inequalities are relatively high—with capacity and resource deficits hindering adaptation actions both within local governments and in their interactions with regional and national authorities.[4]

This chapter offers a brief overview of different approaches that cities around the world have taken to institutionalize emerging climate adaptation and resilience priorities within governance and planning processes. Drawing on emblematic examples from the Global North and South, we highlight common ways that cities have addressed a wide array of climate impacts, including through strategic or special-purpose actions, mainstreaming into comprehensive or general plans, programmatic interventions that build long-term public financial or human resources capacity, or smaller-scale projects that are time bound and seek incremental changes on the ground. Our objective is not to exhaustively survey adaptation approaches globally; instead, we aim to synthesize notable strategies taken by select cities to build scientific awareness, mobilize policy coalitions, prioritize potential interventions, and—most importantly—pursue actions that further social equity and inclusion. We know from historical experience that development planning has tended to prioritize capital investment, hard infrastructure, and external resource support, which can lead to varying degrees of housing displacement and privatization of formerly publicly supplied urban services.[5] We therefore argue that social equity and inclusion must be central considerations for climate adaptation planning and resilience building in cities.

Recognizing Climate Risks and Vulnerabilities in Cities

Climate change will pose an existential threat to cities around the world, since they house critical—but vulnerable—assets such as infrastructure, manufacturing, financial services, knowledge, and the capacity for innovation. Many cities are located in low-elevation coastal zones, where sea-level rise and extreme storm surges are projected to affect more than 800 million people and cause infrastructure damage of more than US$1 trillion each year by the middle of the century.[6] In this same time frame, nearly two-thirds of the world's population will live in urban areas, with the most rapid growth occurring in cities in the Global South that have large vulnerable populations and low capacity to adapt to climate impacts.[7] Many of these cities have sprawling informal settlements, home to over 880 million residents globally, where residents regularly experience limited access to secure shelter, electricity, clean water, sanitation, and employment opportunities.[8] Climate impacts are likely to worsen access to such services, especially for already-vulnerable populations, including women and girls, children and the elderly, migrants, indigenous populations, and other historically marginalized groups.[9]

Despite clearly documented risks to urban infrastructure, economies, and communities, priorities around vulnerability reduction and adaptation have yet to be fully or systematically incorporated into the wider urban planning and development agenda.[10] Planning actions are often constrained by a lack of capacity and resources. For many cities, adaptation planning encapsulates both "traditional" development goals—such as providing public services or maintaining economic competitiveness—and emerging climate-related development goals like reducing risk, ensuring safety and well-being, and confronting policies that generate more vulnerability. Adaptation priorities can be complicated by a rapidly shifting demography, burgeoning informal economies, political patronage and corruption, a privatization of basic services, and a lack of expertise in local governments. These challenges are particularly profound for secondary cities or smaller localities that are typically resource-scarce, as well as in rapidly urbanizing regions across the Global South—where growth is outstripping their capacity to effectively meet their needs.

Burgeoning research over the past decade has explored the many opportunities and barriers to progress in adaptation planning in cities across different contexts.[11] From this research, we see that many cities cite governance, finance, skills, and capacity constraints as barriers to proactive adaptation action. Political leaders may lack the kind of expertise, incentives, or support required to mobilize climate action, while downscaled climate projections may also be unavailable, inhibiting steps to integrate adaptation priorities into long-term land management, infrastructure development, and asset protection programs.[12] In other cases, cities may lack the authority or autonomy they need to manage the risks they face. For instance, decisions that need to be taken to protect humans and ecosystems in cities may require action at regional, national, and international levels, and this divided authority makes coordination across political jurisdictions especially difficult.[13] Finally, a lack of transparency in decision-making, opportunity for public discourse, or means of seeking accountability can allow elite interests to exclude those who are less politically and economically powerful.[14] Corrupt or inequitable systems, structures, and processes of governance heighten a city's vulnerability to disasters and the impacts of climate change.

However, with increasing political recognition and resource support, a number of cities are also emerging as arenas of climate adaptation planning and policy experimentation.[15] Over the past decade, studies have documented the rise of urban "early adopters" of adaptation, which include larger cities such as Rotterdam and New York City in the Global North and Cape Town, Quito, and Surat in the Global South.[16] Many of these cities have chosen to form strategic or special-purpose bureaucratic units to spearhead adaptation actions—such as in the form of an office of climate resilience at the mayoral level—or have designed cross-sectoral coordination mechanisms to facilitate policies and plans that account for multiple urban development needs, including in the form of regional partnerships, public-private collaboratives, or citywide advisory or working groups.[17] In Rotterdam, the Netherlands, for example, the city's Rotterdam Climate Proof program helped to coalesce adaptation action around key urban spatial development goals and blue-green infrastructure development, as well as delineating adaptation co-benefits with mitigation. There have also been efforts to pilot risk-reduction and capacity-building projects on the ground, often with external donor or private support, to increase resilience of public health services, water and sanitation infrastructure, or housing provision.[18] The goal of many of these efforts is to institutionalize and embed climate adaptation priorities into existing city planning workstreams or to increase policy uptake in sectors by promoting climate awareness and incremental interventions.

Assessments from "early adopter" cities note how robust scientific projections, strong municipal leadership, and relevance to ongoing planning and development agendas are key drivers of adaptation action across the globe. For example, Durban, South Africa, has benefited from strong local leadership—advocating for both proactive climate mitigation and adaptation actions—which has led to green designs of local infrastructure, incorporation of traditional community elders in decision-making, and international political exposure.[19] Similarly, Barcelona, Spain, has recently begun to advocate for a citywide approach that focuses on an ethics of care and takes into account particular needs of women, youth, migrants, and long-term residents in plans to upgrade public spaces.[20]

Other studies have also pointed to cities as nodes of political awareness and arenas of participatory action, especially in places where knowledge, resources, and capacity are limited.[21] The importance of social inclusion and community-based knowledge generation in adaptation plans is highlighted in experiences from Lima, Peru; Santiago, Chile; Hat Yai, Thailand; Saint Louis, Senegal; and Munich, Germany; among others.[22] A networked perspective is important given the rise of transnational municipal networks such as ICLEI–Local Governments for Sustainability, Asian Cities Climate Change Resilience Network (2008–2014), C40 Cities Climate Leadership Group, and the Rockefeller Foundation's 100 Resilient Cities Network (2013–2019) that are helping to facilitate intercity sharing of resources, planning tools, and scientific knowledge.[23]

Although we are seeing some progress in climate adaptation planning in select cities—notably, those that are wealthier or better resourced—the global picture is more fragmented. As downscaled climate models become more accessible and the need for addressing climate risks in cities becomes more apparent, continued poor governance and high levels of inequality are jeopardizing those who are already socioeconomically marginalized and further entrenching their inability to cope with and adapt to projected climate risks. Bengaluru, India, for example, is struggling to address impacts of heat and water scarcity within informal settlements that house populations, many of whom are migrants, who lack secure access to jobs and tenure rights.[24] In other words, climate vulnerabilities and risks stem not only from direct climate impacts but also from a multitude of socioeconomic and cultural conditions that should be better understood and represented. As a result, uncovering and addressing the fundamental and structural drivers of urban inequalities, which contribute to differential vulnerability in the first place, is key to scaling up adaptation action to more cities in ways that are fair, accountable, and inclusive and that lead to prioritization of and benefits for historically disadvantaged communities.

Enabling Climate Adaptation in Cities

Lessons from "early adopter" cities show that cities are promising sites for testing and embedding more innovative adaptation approaches, especially in those with more proactive leadership and resource support to improve public awareness, gather scientific data, and mobilize stakeholder coalitions.[25] However, these opportunities are unevenly distributed both across and within cities, with richer cities taking the lead globally and, within cities, with wealthier residents being prioritized in the delivery of municipal services and protective infrastructure. For example, research from Medellín, Colombia, has shown that even though constructing a green belt can alleviate urban heat-island impacts, the majority of the project's benefits cater to middle-class residents, whose property values increase due to improvements in livability or who are able to use the green belt for leisure while poorer communities are excluded.[26] Similarly, in Miami, New Orleans, and other coastal cities across the United States, efforts to protect against sea-level rise tend to be concentrated in and around high-income neighborhoods with high-value assets and buildings.[27] Finally, in Jakarta, Indonesia, many of the river and canal upgrading projects designed to improve flood management have led to the displacement of informal communities and the exclusion of their voices in urban development decisions.[28] In light of these examples, we reflect on two key lessons and highlight the need to learn from on-the-ground experiences of institutionalizing adaptation, including cases where inequalities were ineffectively addressed, as well as instances when equity was central to planning.

The first key lesson is that there is no one-size-fits-all approach to climate adaptation in cities. From the examples noted already, we see that cities choose to institutionalize adaptation in a variety of ways, including through strategic or special-purpose actions, mainstreaming into comprehensive or general plans, programmatic interventions that build long-term public financial or human resources capacity, or smaller-scale projects that are time bound and seek incremental changes on the ground. The decision on which approach to pursue is political in nature and depends on various internal contextual factors, ranging from the presence of sympathetic politicians to planning regulations that are easily modified to account for climate impacts.[29]

For instance, cities that already have robust planning regulations in place to manage economic development, public transportation, or environmental quality may choose to integrate (or mainstream) climate adaptation priorities into them. Notable examples include New York City's

OneNYC plan or earlier comprehensive strategies from London and Toronto.[30] Recent efforts have also focused on integrating climate adaptation with ecosystem protection priorities toward potential nature-based solutions.[31] In other cases, cities that face high levels of bureaucratic or resource uncertainty—such as the inability to support long-term municipal budgets or a high turnover rate of local government staff—may elect a more piecemeal approach that focuses on distinct local projects with shorter implementation time frames. Pilot projects that support grey-water reuse, neighborhood-level sanitation, and passive cooling in housing developments in cities like Indore in India and Semarang in Indonesia are good examples of this.[32] Other cities have focused on developing internal financial capacities to support longer-term planning. Examples include the Surat Climate Change Trust in India as a clearinghouse for both domestic and international finance, county-level climate adaptation funds in Kenya, and various intergovernmental arrangements, such as Denmark's efforts to bring together municipalities, utilities, and national ministries to invest in wastewater and flood-management infrastructure.[33]

A second key lesson is that clear and strong mandates that enable social equity and inclusion are required to build broad policy coalitions, design accountable decision-making processes, and ensure long-term benefits that recognize historical disadvantage and are not economically exclusionary. Insights from Medellín, Miami, Jakarta, and other cities illustrate how adaptation plans and strategies have tended to prioritize infrastructure investments that primarily benefit those who are able to afford protection against sea-level rise, urban heat, or other extreme climate impacts.[34] This approach to planning and implementing adaptation has yielded patterns of "climate gentrification," where cities have constructed enclaves of economic privilege and climate resilience while marginalizing historically disadvantaged communities of color, migrants, or people without secure employment or shelter.[35] In view of these critiques, cities are only just beginning to recognize the need for equitable distribution of adaptation benefits and the importance of inclusive decision-making processes.[36] However, questions remain around the role of climate adaptation in perpetuating racist planning policies, prioritizing speculative development over public infrastructure and services, and excluding local, traditional, or indigenous knowledge in understanding the implications of compounding and cascading risks on the ground.[37]

As cities continue to confront and rectify historical injustices attributed to different forms of socioeconomic inequality, we are beginning to see a similarly wide array of strategies to address climate inequities through the prisms of race and ethnicity, gender, youth, migration status, and informality. Emerging strategies often attempt to redress maldistributive outcomes of large-scale adaptation infrastructure and investments—such as expensive seawalls that displace communities while providing enclaves of relative protection for the rich—through improving the procedural legitimacy of decision-making and recognizing structural forms of marginalization experienced by historically underserved communities that tend to also be at the front lines of climate change impacts.

For example, climate adaptation efforts in the US cities of Houston, Los Angeles, and Boston, among others, are specifically targeting the intersection of climate and racial inequality, with strategies around racial reconciliation and the provision of housing and employment services in minority neighborhoods.[38] Along similar lines, San Antonio is promoting climate mitigation and adaptation goals that focus on equity and, in response, has developed a pilot equity screening tool to operationalize and track progress of its commitment to equitable climate action. Seattle's adaptation plan also places equity front and center. It calls for the planning process to address underlying causes of disproportionate risk, advocates for improved access to decision-making processes for socially vulnerable communities, and proposes a community-centered planning model. In South Africa, where there have been efforts to promote racial justice in cities since the end of Apartheid, climate adaptation has also taken on these priorities. For instance, Cape Town's efforts have focused on youth empowerment, vocational education, and skills training in townships with a majority of Black residents.[39] In Durban, municipal strategies have prioritized partnerships with indigenous community leaders to better manage land, food, and urban agricultural systems.[40] Inclusion of informal and/or indigenous communities has also been part of urban adaptation strategies in Quito, Ecuador; Rosario, Argentina; and Maputo, Mozambique.[41] Many of these strategies build on efforts to broadly include diverse voices and knowledge in decision-making, while a subset involves targeted projects to build adaptive capacity and reduce risk exposure in underprivileged communities.

The examples noted here show that cities around the world are engaging with social equity and justice in their respective climate adaptation plans, albeit in different ways and to varying extents. There is increasing recognition that socially legitimate, accountable, and inclusive processes must accompany any ongoing efforts to provide housing, economic opportunity, food access, public health, or other urban services. Furthermore, these strategies are often dictated by the presence of different policy coalitions, local contextual factors such as political leadership, or particular opportunities such as a supportive regulatory environment or the availability of external resource support. Compared to previous approaches that rely more on technocratic or speculative forms of infrastructure planning, a focus on social equity ensures that the benefits of adaptation are more widely distributed, especially among historically disadvantaged groups, and that decision-making processes are more inclusive of local, traditional, and indigenous knowledge.

Conclusion

An exploration into the different ways to institutionalize climate adaptation in equitable and inclusive ways is critical given the long-term uncertainties of climate change, the potential for cascading and compounding risks, and the myriad opportunities for elite capture in the provision of public goods and services. The main argument put forward by this chapter is that social equity and inclusion must be central considerations for adaptation and resilience building regardless of the governance and planning approach taken by local authorities. Decisions on whether to pursue mainstreamed, strategic, or project-based approaches are often determined by existing contextual characteristics, such as the presence of strong leadership, active citizenry, clear policy incentives, availability of funds, or historical experience of disaster impacts. Therefore, more normative directions—especially ones based on universal aspirations around democratic participation, fair distribution of losses and benefits, and ameliorating historical injustices—can help frame the overarching goals of adaptation as well as any relevant processes for assessing and learning from adaptation progress in cities.

The examples highlighted in this chapter are meant to serve as illustrations rather than to set out specific metrics and indicators to "measure" progress toward achieving equity and justice in urban adaptation. It is challenging to benchmark adaptation progress given the myriad local contexts found in cities that vary in size, regulatory autonomy, or geographical setting—although emerging research is trying to do this.[42] Instead, we argue that the examples underscore the importance of considering equity and justice in planning action. We also call out specific hurdles and constraints faced by cities wishing to pursue more equitable climate adaptation actions, which include limited finances and bureaucratic capacity, weak local authority, competing development priorities, and political pressure to focus on immediate, rather than long-term, goals.

In summary, given the inevitable variations in governance and planning for climate adaptation in cities, strong social equity and inclusion criteria can inform adaptation tools that prioritize the rights, voices, and interests of urban residents, including those who are historically marginalized, which can then be applied in conjunction with tools that focus on technocratic and spatial responses to climate change. Efforts to realize equity and justice in urban climate adaptation are particularly important given the need to better inform global agendas such as the Sustainable Development Goals (2015), UN Habitat's New Urban Agenda (2016), and the Paris Agreement's Global Stocktake, through which the global community will begin in 2023 to assess its progress toward meeting climate goals and objectives. On top of these policy needs, there are also opportunities to catalyze more proactive adaptation actions in cities that are policy laggards, helping to reframe adaptation from a purely environmental priority to one that speaks to the everyday developmental needs of people and the emerging risks they face in a changing climate.

NOTES

1. Carmin, JoAnn, David Dodman, and Eric Chu. 2013. *Urban Climate Adaptation and Leadership: From Conceptual Understanding to Practical Action*. 2013/26. Paris: Organisation for Economic Co-operation and Development (OECD); Rosenzweig, Cynthia, William D. Solecki, Stephen A. Hammer, and Shagun Mehrotra, eds. 2018. *Climate Change and Cities: Second Assessment Report of the Urban Climate Change Research Network*. Cambridge: Cambridge University Press.

2. Anguelovski, Isabelle, Linda Shi, Eric Chu, Daniel Gallagher, Kian Goh, Zachary Lamb, Kara Reeve, and Hannah Teicher. 2016. "Equity Impacts of Urban Land Use Planning for Climate Adaptation: Critical Perspectives from the Global North and South." *Journal of Planning Education and Research* 36(3): 333–348. https://doi.org/10.1177/0739456X16645166; Anguelovski, Isabelle, and JoAnn Carmin. 2011. "Something Borrowed, Everything New: Innovation and Institutionalization in Urban Climate Governance." *Current Opinion in Environmental Sustainability* 3(3): 169–175. https://doi.org/10.1016/j.cosust.2010.12.017; Lesnikowski, Alexandra, Robbert Biesbroek, James D. Ford, and Lea Berrang-Ford. 2020. "Policy Implementation Styles and Local Governments: The Case of Climate Change Adaptation." *Environmental Politics* 30(5): 753–790. https://doi.org/10.1080/09644016.2020.1814045.
3. Bulkeley, Harriet, Gareth A. S. Edwards, and Sara Fuller. 2014. "Contesting Climate Justice in the City: Examining Politics and Practice in Urban Climate Change Experiments." *Global Environmental Change* 25: 31–40. https://doi.org/10.1016/j.gloenvcha.2014.01.009; Chu, Eric, Isabelle Anguelovski, and JoAnn Carmin. 2016. "Inclusive Approaches to Urban Climate Adaptation Planning and Implementation in the Global South." *Climate Policy* 16(3): 372–392. https://doi.org/10.1080/14693062.2015.1019822.
4. Anguelovski, Isabelle, Eric Chu, and JoAnn Carmin. 2014. "Variations in Approaches to Urban Climate Adaptation: Experiences and Experimentation from the Global South." *Global Environmental Change* 27: 156–167. https://doi.org/10.1016/j.gloenvcha.2014.05.010.
5. Shi, Linda, Eric Chu, Isabelle Anguelovski, Alexander Aylett, Jessica Debats, Kian Goh, Todd Schenk, Karen C. Seto, David Dodman, Debra Roberts, J. Timmons Roberts, and Stacy D. VanDeveer. 2016. "Roadmap towards Justice in Urban Climate Adaptation Research." *Nature Climate Change* 6(2): 131–137. https://doi.org/10.1038/nclimate2841.
6. Chu, Eric, Anna Brown, Kavya Michael, Jillian Du, Shuaib Lwasa, and Anjali Mahendra. 2019. *Unlocking the Potential for Transformative Climate Adaptation in Cities*. Washington, DC, and Rotterdam: World Resources Institute.
7. United Nations. 2019. *World Urbanization Prospects: The 2018 Revision*. New York: Department of Economic and Social Affairs, United Nations.
8. Dodman, David, and David Satterthwaite. 2009. "Institutional Capacity, Climate Change Adaptation and the Urban Poor." *IDS Bulletin* 39(4): 67–74. https://doi.org/10.1111/j.1759-5436.2008.tb00478.x; Satterthwaite, David, Diane Archer, Sarah Colenbrander, David Dodman, Jorgelina Hardoy, Diana Mitlin, and Sheela Patel. 2020. "Building Resilience to Climate Change in Informal Settlements." *One Earth* 2(2): 143–156. https://doi.org/10.1016/j.oneear.2020.02.002.
9. Bartlett, Sheridan, and David Satterthwaite, eds. 2016. *Cities on a Finite Planet: Towards Transformative Responses to Climate Change*. London and New York: Routledge.
10. Woodruff, Sierra C., and Missy Stults. 2016. "Numerous Strategies but Limited Implementation Guidance in U.S. Local Adaptation Plans." *Nature Climate Change* 6(8): 796–802. https://doi.org/10.1038/nclimate3012.

11. *See* Araos, Malcolm, Lea Berrang-Ford, James D. Ford, Stephanie E. Austin, Robbert Biesbroek, and Alexandra Lesnikowski. 2016. "Climate Change Adaptation Planning in Large Cities: A Systematic Global Assessment." *Environmental Science & Policy* 66: 375–382. https://doi.org/10.1016/j.envsci.2016.06.009; Olazabal, Marta, Maria Ruiz de Gopegui, Emma L. Tompkins, Kayin Venner, and Rachel Smith. 2019. "A Cross-Scale Worldwide Analysis of Coastal Adaptation Planning." *Environmental Research Letters* 14(12): 124056. https://doi.org/10.1088/1748-9326/ab5532; Reckien, Diana, Monica Salvia, Oliver Heidrich, Jon Marco Church, Filomena Pietrapertosa, Sonia De Gregorio-Hurtado, Valentina D'Alonzo, Aoife Foley, Sofia G. Simoes, Eliška Krkoška Lorencová, Hans Orru, Kati Orru, Anja Wejs, Johannes Flacke, Marta Olazabal, Davide Geneletti, Efrén Feliu, Sergiu Vasilie, Cristiana Nador, Anna Krook-Riekkola, Marko Matosović, Paris A. Fokaides, Byron I. Ioannou, Alexandros Flamos, Niki-Artemis Spyridaki, Mario V. Balzan, Orsolya Fülöp, Ivan Paspaldzhiev, Stelios Grafakos, and Richard Dawson. 2018. "How Are Cities Planning to Respond to Climate Change? Assessment of Local Climate Plans from 885 Cities in the EU-28." *Journal of Cleaner Production* 191: 207–219. https://doi.org/10.1016/j.jclepro.2018.03.220.
12. Brown, Anna. 2018. "Visionaries, Translators, and Navigators: Facilitating Institutions as Critical Enablers of Urban Climate Change Resilience." In *Climate Change in Cities: Innovations in Multi-Level Governance*, edited by Sara Hughes, Eric K. Chu, and Susan G. Mason, 229–253. Cham, Switzerland: Springer.
13. Nalau, Johanna, Benjamin L. Preston, and Megan C. Maloney. 2015. "Is Adaptation a Local Responsibility?" *Environmental Science & Policy* 48: 89–98. https://doi.org/10.1016/j.envsci.2014.12.011; Shi, Linda. 2019. "Promise and Paradox of Metropolitan Regional Climate Adaptation." *Environmental Science & Policy* 92: 262–274. https://doi.org/10.1016/j.envsci.2018.11.002.
14. Chu et al. 2016.
15. Bulkeley, Harriet, and Vanesa Castán Broto. 2013. "Government by Experiment? Global Cities and the Governing of Climate Change." *Transactions of the Institute of British Geographers* 38(3): 361–375. https://doi.org/10.1111/j.1475-5661.2012.00535.x.
16. *See* Bellinson, Ryan, and Eric Chu. 2019. "Learning Pathways and the Governance of Innovations in Urban Climate Change Resilience and Adaptation." *Journal of Environmental Policy & Planning* 21(1): 76–89. https://doi.org/10.1080/1523908X.2018.1493916; Chu et al. 2016; Huang-Lachmann, Jo-Ting, and Jon C. Lovett. 2016. "How Cities Prepare for Climate Change: Comparing Hamburg and Rotterdam." *Cities* 54: 36–44. https://doi.org/10.1016/j.cities.2015.11.001; Pasquini, Lorena, Gina Ziervogel, Richard M. Cowling, and Clifford Shearing. 2015. "What Enables Local Governments to Mainstream Climate Change Adaptation? Lessons Learned from Two Municipal Case Studies in the Western Cape, South Africa." *Climate and Development* 7(1): 60–70. https://doi.org/10.1080/17565529.2014.886994; Rosenzweig, Cynthia, William D. Solecki, Reginald Blake, Malcolm Bowman, Craig Faris, Vivien Gornitz, Radley Horton, Klaus Jacob, Alice LeBlanc, Robin

Leichenko, Megan Linkin, David Major, Megan O'Grady, Lesley Patrick, Edna Sussman, Gary Yohe, and Rae Zimmerman. 2011. "Developing Coastal Adaptation to Climate Change in the New York City Infrastructure-Shed: Process, Approach, Tools, and Strategies." *Climatic Change* 106(1): 93–127. https://doi.org/10.1007/s10584-010-0002-8.

17. Bauer, Anja, and Reinhard Steurer. 2014. "Multi-Level Governance of Climate Change Adaptation through Regional Partnerships in Canada and England." *Geoforum* 51: 121–129. https://doi.org/10.1016/j.geoforum.2013.10.006.
18. Aylett, Alexander. 2015. "Institutionalizing the Urban Governance of Climate Change Adaptation: Results of an International Survey." *Urban Climate* 14: 4–16. https://doi.org/10.1016/j.uclim.2015.06.005; Carmin et al. 2013; Shi, Linda, Eric Chu, and JoAnn Carmin. 2016. "Global Patterns of Adaptation Planning: Results from a Global Survey." In *The Routledge Handbook of Urbanization and Global Environmental Change*, edited by K. C. Seto, W. D. Solecki, and C. A. Griffith, 336–349. New York: Routledge.
19. Chu, Eric, Isabelle Anguelovski, and Debra Roberts. 2017. "Climate Adaptation as Strategic Urbanism: Assessing Opportunities and Uncertainties for Equity and Inclusive Development in Cities." *Cities* 60: 378–387. https://doi.org/10.1016/j.cities.2016.10.016; Roberts, Debra. 2010. "Prioritizing Climate Change Adaptation and Local Level Resilience in Durban, South Africa." *Environment and Urbanization* 22(2): 397–413. https://doi.org/10.1177/0956247810379948.
20. Zografos, Christos, Kai A. Klause, James J. T. Connolly, and Isabelle Anguelovski. 2020. "The Everyday Politics of Urban Transformational Adaptation: Struggles for Authority and the Barcelona Superblock Project." *Cities* 99: 102613. https://doi.org/10.1016/j.cities.2020.102613.
21. Chu et al. 2016.
22. Miranda Sara, Liliana, and Isa Baud. 2014. "Knowledge-Building in Adaptation Management: Concertacion Processes in Transforming Lima Water and Climate Change Governance." *Environment and Urbanization* 26(2): 505–524. https://doi.org/10.1177/0956247814539231; Barton, Jonathan Richard, Kerstin Krellenberg, and Jordan Michael Harris. 2015. "Collaborative Governance and the Challenges of Participatory Climate Change Adaptation Planning in Santiago de Chile." *Climate and Development* 7(2): 175–184. https://doi.org/10.1080/17565529.2014.934773; Siriporananon, Somporn, and Parichart Visuthismajarn. 2018. "Key Success Factors of Disaster Management Policy: A Case Study of the Asian Cities Climate Change Resilience Network in Hat Yai City, Thailand." *Kasetsart Journal of Social Sciences* 39(2): 269–276. https://doi.org/10.1016/j.kjss.2018.01.005; Vedeld, Trond, Adrien Coly, Ndèye Marème Ndour, and Siri Hellevik. 2016. "Climate Adaptation at What Scale? Multi-Level Governance, Resilience, and Coproduction in Saint Louis, Senegal." *Natural Hazards* 82(S2): 173–199. https://doi.org/10.1007/s11069-015-1875-7; Wamsler, Christine. 2016. "From Risk Governance to City-Citizen Collaboration: Capitalizing on Individual Adaptation to Climate Change." *Environmental Policy and Governance* 26(3): 184–204. https://doi.org/10.1002/eet.1707.

23. Bellinson and Chu 2019; Dzebo, Adis, and Johannes Stripple. 2015. "Transnational Adaptation Governance: An Emerging Fourth Era of Adaptation." *Global Environmental Change* 35: 423–435. https://doi.org/10.1016/j.gloenvcha.2015.10.006; Woodruff, Sierra C. 2018. "City Membership in Climate Change Adaptation Networks." *Environmental Science & Policy* 84: 60–68. https://doi.org/10.1016/j.envsci.2018.03.002.
24. Chu, Eric, and Kavya Michael. 2019. "Recognition in Urban Climate Justice: Marginality and Exclusion of Migrants in Indian Cities." *Environment and Urbanization* 31(1): 139–156. https://doi.org/10.1177/0956247818814449.
25. Hughes, Sara, Eric K. Chu, and Susan G. Mason, eds. 2018. *Climate Change in Cities: Innovations in Multi-Level Governance*. Cham, Switzerland: Springer.
26. Anguelovski, Isabelle, Clara Irazábal-Zurita, and James J. T. Connolly. 2019. "Grabbed Urban Landscapes: Socio-Spatial Tensions in Green Infrastructure Planning in Medellín." *International Journal of Urban and Regional Research* 43(1): 133–156. https://doi.org/10.1111/1468–2427.12725.
27. Aune, Kyle T., Dean Gesch, and Genee S. Smith. 2020. "A Spatial Analysis of Climate Gentrification in Orleans Parish, Louisiana Post-Hurricane Katrina." *Environmental Research* 185: 109384. https://doi.org/10.1016/j.envres.2020.109384; Keenan, Jesse M., Thomas Hill, and Anurag Gumber. 2018. "Climate Gentrification: From Theory to Empiricism in Miami-Dade County, Florida." *Environmental Research Letters* 13(5): 054001. https://doi.org/10.1088/1748-9326/aabb32.
28. Goh, Kian. 2019. "Urban Waterscapes: The Hydro-Politics of Flooding in a Sinking City." *International Journal of Urban and Regional Research* 43(2): 250–272. https://doi.org/10.1111/1468-2427.12756; Salim, Wilmar, Keith Bettinger, and Micah Fisher. 2019. "Maladaptation on the Waterfront: Jakarta's Growth Coalition and the Great Garuda." *Environment and Urbanization ASIA* 10(1): 63–80. https://doi.org/10.1177/0975425318821809.
29. Chu, Eric. 2020. "Urban Resilience and the Politics of Development." In *Climate Urbanism: Towards a Critical Research Agenda*, edited by Vanesa Castán Broto, Enora Robin, and Aidan While, 117–136. Cham, Switzerland: Springer.
30. Fitzgibbons, Joanne, and Carrie L. Mitchell. 2021. "Inclusive Resilience: Examining a Case Study of Equity-Centred Strategic Planning in Toronto, Canada." *Cities* 108 (January): 102997. https://doi.org/10.1016/j.cities.2020.102997; Mees, Heleen-Lydeke P., and Peter P. J. Driessen. 2011. "Adaptation to Climate Change in Urban Areas: Climate-Greening London, Rotterdam, and Toronto." *Climate Law* 2(2): 251–280. https://doi.org/10.3233/CL-2011-036; Rosenzweig, Cynthia, and William Solecki. 2010. "Chapter 1: New York City Adaptation in Context." *Annals of the New York Academy of Sciences* 1196: 19–28. https://doi.org/10.1111/j.1749-6632.2009.05308.x.
31. Kabisch, Nadja, Niki Frantzeskaki, Stephan Pauleit, Sandra Naumann, McKenna Davis, Martina Artmann, Dagmar Haase, Sonja Knapp, Horst Korn, Jutta Stadler, Karin Zaunberger, and Aletta Bonn. 2016. "Nature-Based Solutions to Climate Change Mitigation and Adaptation in Urban Areas: Perspectives on Indicators,

Knowledge Gaps, Barriers, and Opportunities for Action." *Ecology and Society* 21(2): 39. https://doi.org/10.5751/ES-08373-210239.

32. Chu, Eric. 2018. "Urban Climate Adaptation and the Reshaping of State-Society Relations: The Politics of Community Knowledge and Mobilisation in Indore, India." *Urban Studies* 55(8): 1766–1782. https://doi.org/10.1177/0042098016686509; Sari, Aniessa Delima, and Nyoman Prayoga. 2018. "Enhancing Citizen Engagement in the Face of Climate Change Risks: A Case Study of the Flood Early Warning System and Health Information System in Semarang City, Indonesia." In *Climate Change in Cities: Innovations in Multi-Level Governance*, edited by Sara Hughes, Eric K. Chu, and Susan G. Mason. Cham, Switzerland: Springer, 121–138.
33. Cook, Mitchell J., and Eric K. Chu. 2018. "Between Policies, Programs, and Projects: How Local Actors Steer Domestic Urban Climate Adaptation Finance in India." In *Climate Change in Cities: Innovations in Multi-Level Governance*, edited by Sara Hughes, Eric K. Chu, and Susan G. Mason. Cham, Switzerland: Springer; Barrett, Sam. 2015. "Subnational Adaptation Finance Allocation: Comparing Decentralized and Devolved Political Institutions in Kenya." *Global Environmental Politics* 15(3): 118–139. https://doi.org/10.1162/GLEP_a_00314.
34. Anguelovski et al. 2016; Castán Broto, Vanesa, Enora Robin, and Aidan While, eds. 2020. *Climate Urbanism: Towards a Critical Research Agenda*. Cham, Switzerland: Springer; Long, Joshua, and Jennifer L. Rice. 2019. "From Sustainable Urbanism to Climate Urbanism." *Urban Studies* 56(5): 992–1008. https://doi.org/10.1177/0042098018770846.
35. Shokry, Galia, James J. T. Connolly, and Isabelle Anguelovski. 2020. "Understanding Climate Gentrification and Shifting Landscapes of Protection and Vulnerability in Green Resilient Philadelphia." *Urban Climate* 31: 100539. https://doi.org/10.1016/j.uclim.2019.100539.
36. Meerow, Sara, Pani Pajouhesh, and Thaddeus R. Miller. 2019. "Social Equity in Urban Resilience Planning." *Local Environment* 24(9): 793–808. https://doi.org/10.1080/13549839.2019.1645103; Reckien, Diana, Felix Creutzig, Blanca Fernandez, Shuaib Lwasa, Marcela Tovar-Restrepo, Darryn Mcevoy, and David Satterthwaite. 2017. "Climate Change, Equity and the Sustainable Development Goals: An Urban Perspective." *Environment and Urbanization* 29(1): 159–182. https://doi.org/10.1177/0956247816677778.
37. Ranganathan, Malini, and Eve Bratman. 2021. "From Urban Resilience to Abolitionist Climate Justice in Washington, DC." *Antipode* 53(1): 115–137. https://doi.org/10.1111/anti.12555; Olazabal, Marta, Eric Chu, Vanesa Castan Broto, and James Patterson. Forthcoming. "Subaltern Forms of Knowledge Are Required to Boost Local Adaptation." *One Earth*; Robin, Enora, and Vanesa Castán Broto. 2020. "Towards a Postcolonial Perspective on Climate Urbanism." *International Journal of Urban and Regional Research* 45(5): 869–878. https://doi.org/10.1111/1468-2427.12981.
38. Van den Berg, Hanne J., and Jesse M. Keenan. 2019. "Dynamic Vulnerability in the Pursuit of Just Adaptation Processes: A Boston Case Study." *Environmental*

Science & Policy 94: 90–100. https://doi.org/10.1016/j.envsci.2018.12.015; Chu, Eric, and Clare Cannon. 2021. "Equity, Inclusion, and Justice as Criteria for Decision-Making on Climate Adaptation in Cities." *Current Opinion in Environmental Sustainability* (in press).

39. Ziervogel, Gina. 2019. "Building Transformative Capacity for Adaptation Planning and Implementation That Works for the Urban Poor: Insights from South Africa." *Ambio* 48(5): 494–506. https://doi.org/10.1007/s13280-018-1141-9.

40. Roberts, Debra. 2008. "Thinking Globally, Acting Locally: Institutionalizing Climate Change at the Local Government Level in Durban, South Africa." *Environment and Urbanization* 20(2): 521–537. https://doi.org/10.1177/0956247808096126; Roberts, Debra, Richard Boon, Nicci Diederichs, Errol Douwes, Natasha Govender, Alistair Mcinnes, Cameron Mclean, Sean O'Donoghue, and Meggan Spires. 2012. "Exploring Ecosystem-Based Adaptation in Durban, South Africa: 'Learning-by-Doing' at the Local Government Coal Face." *Environment and Urbanization* 24(1): 167–195. https://doi.org/10.1177/0956247811431412.

41. Chu et al. 2016; Hardoy, Jorgelina, and Regina Ruete. 2013. "Incorporating Climate Change Adaptation into Planning for a Liveable City in Rosario, Argentina." *Environment and Urbanization* 25(2): 339–360. https://doi.org/10.1177/0956247813493232; Broto, Vanesa Castán, Emily Boyd, and Jonathan Ensor. 2015. "Participatory Urban Planning for Climate Change Adaptation in Coastal Cities: Lessons from a Pilot Experience in Maputo, Mozambique." *Current Opinion in Environmental Sustainability* 13: 11–18. https://doi.org/10.1016/j.cosust.2014.12.005.

42. *See* Berrang-Ford, Lea, Robbert Biesbroek, James D. Ford, Alexandra Lesnikowski, Andrew Tanabe, Frances M. Wang, Chen Chen, Angel Hsu, Jessica J. Hellmann, Patrick Pringle, Martina Grecequet, J. C. Amado, Saleemul Huq, Shuaib Lwasa, and S. Jody Heymann. 2019. "Tracking Global Climate Change Adaptation among Governments." *Nature Climate Change* 9(6): 440–449. https://doi.org/10.1038/s41558-019-0490-0; Ford, James D., and Lea Berrang-Ford. 2016. "The 4Cs of Adaptation Tracking: Consistency, Comparability, Comprehensiveness, Coherency." *Mitigation and Adaptation Strategies for Global Change* 21(6): 839–859. https://doi.org/10.1007/s11027-014-9627-7; Olazabal et al. 2019; Runhaar, Hens, Bettina Wilk, Åsa Persson, Caroline Uittenbroek, and Christine Wamsler. 2018. "Mainstreaming Climate Adaptation: Taking Stock about 'What Works' from Empirical Research Worldwide." *Regional Environmental Change* 18(4): 1201–1210. https://doi.org/10.1007/s10113-017-1259-5.

16

Shanghai's Strive to Excel in Climate Change Adaptation and Low-Carbon Promises

A Model to Follow?

HARRY DEN HARTOG

The subtitle of Shanghai's latest master plan (2017–2035) is "Striving for an Excellent Global City." According to this plan, Shanghai wants to compete with, and possibly surpass, other global cities such as New York, London, Paris, Singapore, and Tokyo with regard to economy, image, and quality of life. The plan's authors state that "the world has stepped into an era of ecological civilization that puts environmental friendliness and humanistic approach first." Shanghai aims "to play the pioneering role in the reform and opening up into this new era and set the pace for innovation and development."[1]

To achieve these aims, the master plan includes ecological ambitions and promises, such as a 5 percent reduction of total carbon emissions, a halving of particulate matter emissions, a ban on raw-waste landfills, and the development of more than three hundred square kilometers of new green structures, all to be realized before 2035. The plan also puts a cap on Shanghai's total population to a maximum of twenty-four million registered residents and adds red lines around the city to limit its footprint and protect agricultural lands against urban sprawl. Furthermore, the plan commits Shanghai to becoming "a more adaptable and resilient eco-city as well as a benchmark for international megacities in terms of green, low-carbon, and sustainable development by developing pilot spaces and infrastructures."[2] The message is clear: Shanghai wants not only to set a national example for other Chinese cities but also to cross borders to inspire others to become more adaptable and resilient.

How can Shanghai manage to implement large-scale ecological improvements—in the context of high building density, land scarcity, and booming real estate prices—when it took other metropolises, such as New York, Tokyo, and Singapore, many years to realize much less ambitious aims? And how will Shanghai integrate ecological values with its other aims, such as flood defense, place-making, and the preservation of industrial heritage? This chapter considers these questions, highlighting the tension between China's push toward growth and urbanization and the need to safeguard its cities from environmental threats. As is described, this tension has played out in stark terms in Shanghai, where leaders are grappling with how to advance their development objectives, which have historically relied on reclaiming wetlands, while adapting to rising seas and strengthening storms.

Ecological Vulnerability in Urbanizing Deltas

During the past three decades, urbanization across the globe has accelerated dramatically, especially in the world's emerging economies. China is without doubt a frontrunner in this trend. Most of the urbanization in China has occurred in a one-hundred-kilometer zone along its coastline and has been highly concentrated in three main deltaic areas: the Pearl River Delta, the Bohai Rim, and the Yangtze River Delta. These deltas are also where many fertile agricultural lands are situated and where most of China's ecologically important wetlands exist.[3]

Approximately 41 percent of the world's population lives in river deltas.[4] Cities like Shanghai, Rotterdam, Amsterdam, London, Venice, New York, New Orleans, St. Petersburg, Russia, and many others were all at least partially built on wetlands and swamps. Due to their strategic location, deltas are the scene of complex land-use conflicts: urban development, infrastructure, ports, wetlands, and fertile agricultural land all fight for positions on the same land. And due to rapid large-scale urbanization and the prioritization of these other land uses, wetlands are increasingly under threat.

Wetlands are crucial ecosystems. They provide habitat and breeding grounds for almost 40 percent of all plant and animal species, as well as, either directly or indirectly, almost all of the supply of freshwater that is consumed around the world.[5] Wetlands also provide a range of ecosystem services, such as rainwater storage or sponge capacities, water purification, carbon sequestration, biodiversity conservation, and, critically for our perspective, storm surge protection.[6] Wetlands can also provide limited options for urban recreation (limited to protect the wetland), which can be very valuable in cities with scarce open space. Despite all these benefits, nearly 35 percent of the world's wetlands were lost between 1970 and 2015, and this loss has been accelerating since 2000. Moreover, decision-makers often undervalue the importance of wetlands; urban wetland management and policy guidance is lacking around the world, and tensions usually exist between conservation and development.[7]

Since the 1950s, more than half of China's coastal wetlands have disappeared; 53 percent of temperate coastal ecosystems, 73 percent of mangroves, and 80 percent of near-shore coral reefs have vanished.[8] This loss has occurred mainly because "huge economic returns from land reclamation have prompted local governments to 'bypass' regulations issued by the central government."[9] Reclaiming land from the sea is a relatively quick and cheap way to get more land and profits—although land needs four years to firm up and solidify in the Netherlands, for example, construction in China can often start within one or two years. Yet, while development on wetlands may appear economically attractive, it poses serious consequences in light of the rising threats of climate change: weakening the shoreline, increasing the vulnerability to threats of sea-level rise and flooding, and making adaptation more challenging.

Shanghai, China's Economic "Head of the Dragon" in a Vulnerable Yangtze River Delta

Over many centuries, an efficient network of waterways steered the spatial and economic development of the Yangtze Delta in a relatively sustainable manner.[10] This started to change in the middle of the twentieth century, when Chairman Mao ruled the country. Under Mao's leadership, there was a shift toward extreme technocratic engineering: "Man must conquer nature," Mao insisted.[11] Natural capital and landscape values were neglected, and planning practices were accordingly based on a tabula rasa approach. Many natural waterways in the region were transformed into canals, while others were dammed or cleared. More recently, during the past three decades, GDP-oriented motives also started to dominate waterfront planning, with additional collateral damage for ecosystems and livability, fed by extreme urbanization pressure, mass migration to the city from rural areas, and a change of lifestyle in the new urban areas. The combined effects of Maoist disregard for nature and capitalist tendencies to exploit it have wreaked significant damage on Shanghai's coasts and riparian lands.

The origins of Shanghai are inseparable from its location beside the water; the city's name even literally translates as "upon the sea." Along the coastline, there has always been a strip of natural wetlands that grew via sedimentation from the Yangtze River's estuary. This eastward shifting of the coastline largely created the territory of Shanghai. But since the 1950s, this natural process has been greatly accelerated by breakwaters and land reclamations. Approximately 40 percent of the tidal flats around Shanghai have disappeared since 1980, mainly due to land reclamations in the Yangtze estuary zone, Chongming Island, and along the coastline of Pudong. In total, these land reclamations have created about 816.6 square kilometers of new land between 1974 and 2018.[12] Shanghai currently still counts approximately 464,600 hectares of wetlands, mainly spread over Chongming, Pudong, and Qingpu districts.[13]

There is a constant clash between urbanization desires and ecological protection, and new lands are increasingly used for agriculture and urban expansion, including housing, airports, infrastructure, and recreational landscapes. One example of this clash is the Nanhui Coastal Wetland Reserve at the southeast edge of Shanghai's Pudong district, just south of Pudong International Airport, which saw a situation reminiscent of what occurred in Jamaica Bay in New York as that city made way for JFK Airport. In 2002, the huge wetlands reserve, measuring 122.5 square kilometer in area, was a tidal flat; it was reclaimed from the sea one year later. Officials planned to build Lingang New Harbor City there, with an expected eight hundred thousand inhabitants by 2020.[14] This planned city was supposed to accompany the Yangshan Deep-Water Port complex, constructed in 2010—currently the largest container terminal in the world—and adjacent heavy-industry complexes. However, due to the remote and unattractive location, as well as a temporary collapse of world trade and container transport, the growth of the new city stopped. About three-quarters of the planned city has not yet been built and lies fallow.

In the original plan, the city would have been surrounded by lush nature and wetlands. Instead, a large part of the lands reserved for nature development is currently in use for aqua farming and plantations. From an agricultural point of view, wetlands are often seen as wastelands, and thus farming, including aqua farming, is prioritized above nature conservation, especially in times when agricultural grounds are becoming increasingly scarce in the fertile region around Shanghai.[15] However, there has also been pushback against the agricultural use of this land, in favor of conservation. A group of environmentalists, scientists, and nature lovers launched a protest, addressed to Tesla's new Gigafactory that started construction nearby, to raise awareness and counter the threat against the wetlands.[16]

The so-called Long Island project at the northwestern edge of Shanghai's Chongming Eco-Island provides a more scandalous example of the tensions between conservation and urbanization. Chongming Island was appointed as a National Ecological Demonstration Zone in 1996 and was to serve as a pilot project for sustainable urban planning.[17] But there was a loophole in the policy covering the northwest corner of the island, which developers and profit-seeking local governments exploited to their advantage. In short, under the guise of protecting land through conservation, natural wetlands have been reclaimed for the sake of massive speculative real estate.[18] This loophole has since been adjusted by the central government—unfortunately, after the damage was already done.[19]

Besides threats of land reclamation, the wetlands also face threats from sea-level rise and changes in sedimentation due to a decrease in discharge after construction of the Three Gorges Dam in 2003.[20]

Despite each of these threats and challenges, it is encouraging that China, with Shanghai as a forerunner, is attempting to restore the damage that has been done to the environment during the past few decades of extreme massive urbanization. In 2018, the State Council launched a new regulation on land reclamation to protect coastal wetlands.[21] And to compensate for the collateral damage of rapid urbanization, Shanghai is searching for ways to protect the remaining wetlands and stimulate the establishment of new ones.

China's Shift to an Eco-Civilization

China's extreme and hasty shift toward urbanization, accompanied by industrialization and intensified agricultural production, has resulted in prosperity and high living standards for many people. But it has also brought serious environmental pollution, a shortage of resources, social-economic imbalance, and increasing vulnerability to flooding and sea-level rise.

Since the beginning of this century, Chinese national policy has been increasingly searching for a new "green economy," which essentially turns away from the Western idea of industrialization.[22] In fact, since the People's Republic of China's eleventh five-year plan (2006–2010), it has committed itself to achieving a green economy and has specifically pledged to increase the use of renewable energy sources, to reduce carbon emissions drastically, and to increase forest coverage of lands. In the twelfth five-year plan (2011–2015), additional targets were added, including reversing ecological deterioration and enhancing environmental regulatory institutions.[23] China declared a "war on pollution" and started to introduce multiple green policies. It also started to decouple environmental pressure from economic growth and promised that the year 2030 will be a turning point, not only because China promises to react effectively on the Sustainable Development Goals set by the United Nations but also because China is aiming to realize an eco-civilization by that time.[24] On March 11, 2021, the National People's Congress of China voted to pass the resolution on the fourteenth five-year plan and the 2035 long-term goal outline.[25] This plan sets out goals for an 18 percent CO_2 emissions reduction and a 13.5 percent energy-intensity reduction for the coming five years. This goal is significantly higher than what was set in Shanghai's master plan, and in the coming months, it will become clearer what this will mean for local policies. The previous five-year plans showed a trend of overachieving the previously set goals, and according to some researchers, this will happen again in this new period.[26]

The concept behind "ecological civilization" has been gradually integrated into the policies of the ruling Communist Party since the seventeenth National Congress in November 2007.[27] The integration of measures to counterbalance the negative effects of environmental changes became a national strategy under General Secretary Hu Jintao in 2007: "We will adopt fiscal and taxation systems conducive to scientific development and set up sound compensation systems for use of resources and for damage to the ecological environment."[28]

Ecological civilization can be defined as "a dynamic equilibrium state where humans and nature interact and function harmoniously."[29] To realize an ecological civilization means a drastic societal reform with serious consequences for economy, society, and daily life. The concept of ecological civilization has received a lot of skeptical reactions from several international observers.[30] Although some scholars claim that ecological civilization originates from the Western discourses on ecological modernization, it also has deep roots in Marxism and has the potential to challenge or even replace global capitalism.[31] A remarkable aspect of the concept is the call for a new balance between top-down and bottom-up governance approaches and for exploring public-private partnerships and new forms of participation—ideas that are also mentioned in the final chapter of Shanghai's latest master plan.[32] This is still in an elementary phase, and time will tell us how this will work out in practice over the next few years. However, outwardly, at least, General Secretary Xi has strongly endorsed the eco-civilization discourse and called for a more balanced model of economic growth. With his statement that "clear waters and green mountains are as valuable as mountains of gold and silver," he stresses the economic importance of strong environmental action.[33]

In recent years, the concept of an ecological civilization has permeated Chinese urban planning and architecture. For example, in 2009, the National Development and Reform Commission established the "Low-Carbon City Initiative," and in 2015, it launched the "Sponge City" initiative, which aims to create water storage buffers in so-called sponge districts to capture storm water. As part of these initiatives, several experimental pilot projects have also been implemented in Shanghai, with varying results.[34]

Shanghai's Steps toward Ecological Civilization

According to Shanghai's master plan, "citizen happiness" is fundamental to Shanghai's development and a key motivator of officials' efforts to build a prosperous and innovative city. To achieve citizen happiness, officials believe they must engineer "a desirable ecological city," which is formulated as "a beautiful space that meets the demands of the increasing number of citizens, where the water is more blue, and the land is more green, living in harmony with nature to satisfy the citizens yearning for a better life."[35]

A crucial step toward the implementation of Shanghai's ecological civilization and combat against climate change is the promise to create a "green and open eco-network," with at least 60 percent of the municipal land area used for ecological land.[36] According to the master plan, this is an increase of about 10 percent of green lands compared with today—a massive amount, given the high building density and scarcity of land in Shanghai. In dense downtown areas, this green ambition will be realized in part through green roofs and other forms of vertical green infrastructure. To connect the downtown with the surrounding open landscape, a series of green corridors that are more than one thousand meters wide are planned, as well as large new wetlands along the coastline. Several huge new city parks, hundreds of small pocket parks, and small-scale green features on a neighborhood level are planned as well. A showcase of this ambition for ecological restoration is the massive transformation of former industrial waterfronts, explained in more detail in the following section. Many of these green projects have already been implemented over the past few years or are under construction; in some cases, construction plans have been accelerated and prioritized due to the COVID-19 crisis.

Similar to the Green Belt around London and other green buffers, such as the Green Heart of the Randstad metropolis in the Netherlands, a main function of Shanghai's green and open eco-network is to accommodate the leisure needs of the emerging new middle class. The eco-network is intended to bring citizens closer to nature and to reconnect the city with the countryside—literally "introducing the forest to the city." Moreover, this green framework is considered to be a new backbone for urban development (an alternative to a water- or road-based one).

In 2015, the local authorities started constructing the "Overall Plan for Ecological Civilization System Reform," which is an integrated component of the Shanghai master plan (2017–2035). In this plan, the term "ecological space" refers to "land that is used to provide ecosystem services in the city, including green land, forest land, garden land, cultivated land, tidal flat reed land, pond aquaculture water surface, unused lands, etcetera."[37] This broad definition seems to encompass all the land that is not an urban built-up area and not paved. However, the ecological values of these spaces vary greatly. In fact, some built-up spaces can have ecological values—for example, Shanghai has a fast-increasing number of roof gardens—while some unbuilt and unpaved lands have almost zero ecological value. Additionally, a large share of the green or ecologically earmarked spaces are clearly meant for recreational or decorative purposes, to serve human beings; usage by other species to stimulate biodiversity is often a secondary consideration or is entirely absent. It seems that in land-use planning, quantity still prevails over quality in Shanghai.

The Huangpu River Waterfront as a Stage for Innovation and Eco-Civilization

A key project to realize the promise for Shanghai to become an "Excellent Global City" and to fulfill the goals of ecological restoration and eco-civilization is Shanghai's ambitious transformation of the former industrial-dominated waterfront of the Huangpu River and Suzhou Creek. In 2018, the Huangpu waterfront became a "demonstration zone for the development capability of the global city of Shanghai."[38] Besides strengthening the embankments to reduce flood risk, the purposes of the new waterfront were (1) to create a continuous open public space as an "urban living room" and central park to counterbalance the densely populated metropolis, (2) to preserve industrial heritage and emphasize Shanghai's identity as a port city, (3) to create new cultural centers (mainly in vacant industrial heritage buildings) to facilitate the expected needs of a new international-oriented middle class, and (4) to strengthen ecological connections. Combining this effort with urban regeneration or renewal in former industrial waterfronts and adjacent densely built urban areas is a vast challenge, especially in urban economic centers that face land scarcity and additional problems, such as needing flood defense systems. Yet, impressively, more than half of this project has already been completed.[39]

In Shanghai's current master plan, the Huangpu River is regarded as an important ecological corridor and a "green and low-carbon demonstration zone."[40] The Huangpu and Suzhou Creek waterfront transformations are engines to speed up the ecological restoration of former industrial areas. The master plan promises to improve the diversity of green spaces, to benefit from existing eco-system services, and to create a blue and green interconnected ecological network to replace former polluting industries. Attractive greening projects and walking trails have been created along both riversides of the Huangpu River and also along the Suzhou Creek, accompanied by massive real estate projects, thematic office parks aimed at the finance sector, international trade centers, centers for high-tech and artificial intelligence, five-star hotels, and many cultural facilities. However, much-needed housing is lacking at the new waterfronts, especially affordable housing.[41]

Due to the large scale of the project (120 kilometers of waterfront will be transformed by its end), different sections of the Huangpu River's waterfront are in different stages of development and usage; they also belong to different municipal districts and differ in their implementation and maintenance. Today, more than twenty-five kilometers of river length, which means fifty kilometers of waterfront in total, has already been transformed, after many polluting industries were removed. In less than five years' time, an almost continuous and attractive public waterfront with greenery, renovated industrial heritage buildings, cultural facilities, and biking and walking trails emerged here with abundant public recreational space, offices, shopping, and hotels, offering a welcome and pleasant relief from the urban congestion for many people. Plans are under way to relocate the last remaining industries, including Baosteel, the second-largest steel producer in the world. These measures surely benefit the quality of air and water and also add needed green spaces for recreation.

Massive new real estate clusters also emerged along the riverbanks during the past five years. All of the newly built buildings received green labels to match the National Green Building Standard, especially regarding low-carbon emissions, although the application of these standards in practice is questionable.[42] Unfortunately, field surveys and multiple talks with real estate developers and other specialists indicate that a large share of the new buildings are used for speculation purposes and remain mainly empty, even several years after completion. Other office locations, such as those around the Hongqiao Hub, are preferred to the waterfront locations due to lower pricing and better connectivity to elsewhere, according to interviewed leading specialists from the real estate sector.

The extremely dense concentration of buildings on both riverbanks of the Huangpu River, in combination with the almost continuous industrial sites, makes it nearly impossible to create an ecological corridor here that would match the scale of the master plan's ambition. Yet officials have successfully relocated a large share of the polluting industries to outside the edge of the city, even to other provinces—to reduce carbon emissions in the city and to improve the general image and quality of life—and made a place for a scenic landscape crossed by recreational walking and cycling trails, in a period of about five years. Thus, the city has made meaningful, if imperfect, progress toward eco-restoration.

Can Shanghai Become an Excellent Example for Climate Change Adaptation?

The policies and projects that have been launched in Shanghai over the past few years are impressive with regard to scale and speed. Many city leaders and experts from all over China see Shanghai as a model and gateway to the international world, and many trends that have started in Shanghai have since been transplanted all over China. Shanghai's master plan connects convincingly with the discourse and practice of the international community in its language and promises, trying to absorb the ethos of sustainability into its planning approach. Indeed, the plan has the ambition to exceed international best practices with regard to speed, scale, and quality. Moreover, many of the promising words in Shanghai's master plan have already been translated into specific plans, and a large part have already been implemented, thanks to the decisive centrally managed government, in possession of money flows and land positions. There have been some impressive accomplishments as well: the new Huangpu waterfronts are breathtaking, and it is mind-boggling that they were realized in such a short time span. The amount of greening integrated in a new eco-network, partly already under construction, is also unprecedented.

Yet these facts, supplemented with impressive numbers of square kilometers, distract us from some substantial deficiencies. In many cases, such as with the new waterfronts in downtown Shanghai, the aim still seems to be to improve the public image and status of the city, attracting foreign investment, or to create a comfortable living environment for a selective upper middle class.[43] And the plan to cap Shanghai's total population at a maximum of twenty-four million registered residents, meant to limit the urban pressure, is also causing social-economic tensions. Real estate values are booming, and Shanghai is increasingly becoming the domain for the happy few; the affordable-housing crunch is especially acute in the new waterfronts, where low-income housing neighborhoods have quite literally been erased. Moreover, the municipality of Shanghai is already home to several million more inhabitants than the desired cap of twenty-four million, if unregistered residents are included in the calculation.[44] What will happen with these unregistered migrant workers? Will they return to their rural villages? And what about informal street markets? Systematically, they are disappearing. The disappearance of the street markets has taken place at an accelerated pace following the COVID-19 crisis, as the government has demolished a number of such markets and traditional low-rise, working-class neighborhoods. Relatedly, unemployment will increase, and the gap between rich and poor will widen further.

Although public awareness is increasing, many voices are not being considered during implementation of the master plan. In recent government policy documents and communications, the focus on unbridled GDP growth seems to have decreased and been replaced by terms like "ecosystem services," "eco-civilization," "ecological restoration," "Green GDP," "harmony with nature," and so on. Yet, as we have seen, even when governments have very good environmental intentions, projects or policies can easily fail when public participation is lacking. And although Shanghai has made some efforts to include the public in the development and implementation of the master plan—the fact that the master plan is largely available online in several languages is unprecedented and a step toward real openness—there is still further to go.

Surprisingly, China's constitution has recognized the importance of public participation and consultation since the time of Mao, but the government has generally neglected the public's role in the years since. Several scholars have called for the government to reprioritize the public's role in China's environmental policy making.[45] The realization of a true eco-civilization is necessarily a process of gradual adjustment and understanding, which cannot be implemented from the top down at once; rather, it needs more involvement, consultation, and incentives.[46] If this participatory approach can be incorporated into the Chinese pilot or demonstration projects, it might lead to more effective and sustainable outcomes.

Conclusion

Although China is aware of its environmental challenges, including the need to adapt to climate change, and is willing to play a leading role in a green transition, it must surmount several obstacles before it can realize its ecological objectives. There are discrepancies in the definition, appreciation, and valuation of ecological assets such as wetlands. Greening is frequently used as a means of beautifying real estate projects. Terms like "eco-civilization" and "green eco-network" sound promising and create high expectations but seem primarily aimed at creating benefits for people, such as making cities more desirable places for people to live. There also seems to be a tendency to undervalue eco-system services—those benefits that people receive from healthy ecosystems—and the use and protection of wetlands. Going forward, Shanghai and other Chinese cities would do well to create clearer definitions for assets like wetlands and terms like "ecological restoration" or "eco-system services" and to communicate these definitions with all stakeholders.

China, with Shanghai leading the way, is shifting from a production economy toward a consumption society. In a process of trial and error, there is a search for a new balance. Although the newly implemented public spaces along the Huangpu River are visually attractive, there are still shortcomings in their daily-life functionality, as well as in their functionality as an ecological corridor.[47] Eco-civilizations need to serve people, of course, but also other species, if they are to effectively combat climate change and restore ecosystems.

China is also in a different phase of development than many established Western cities are and must therefore deal with a different audience of stakeholders and end users, usually with different educational backgrounds, experiences, and lifestyles. Environmental pressure rose quickly during recent decades due to extreme urbanization, and consequently, there are other priorities and expectations that must be considered in urban planning and societal transformation.

Despite all these challenges, Shanghai has commenced a journey toward an eco-civilization that will make a big impact on the daily life of its inhabitants. Hopefully, more thoughtful experiments will follow in Shanghai to establish this metropolis further as a world-leading lab for sustainable transition and urban innovation. Shanghai, and China as a whole, can play a leading role in sustainable transitions and be a role model for other cities and countries, such as in developing countries in the Global South but perhaps also in developed countries in the Global North.

NOTES

1. Shanghai Urban Planning and Land Resource Administration Bureau. 2016. "Shanghai Master Plan 2017–2035: Striving for the Excellent Global City."
2. Shanghai Urban Planning and Land Resource Administration Bureau 2016.
3. King, Franklin Hiram. 1911. *Farmers of Forty Centuries; or, Permanent Agriculture in China, Korea and Japan*. Madison, WI: Democrat.
4. Edmonds, Douglas A., Rebecca L. Caldwell, Eduardo S. Brondizio, and Sacha M. O. Siani. 2020. "Coastal Flooding Will Disproportionately Impact People on River Deltas." *Nature Communications* 11: 4741. https://doi.org/10.1038/s41467-020-18531-4.
5. UN Climate Change News. 2018. "Wetlands Disappearing Three Times Faster than Forests." UNFCCC, October 1, https://unfccc.int.
6. Sutton-Grier, Ariana, and Jennifer Howard. 2018. "Coastal Wetlands Are the Best Marine Carbon Sink for Climate Mitigation." *Frontiers in Ecology and the Environment* 16(2): 73–74; Möller, Iris, Matthias Kudella, Franziska Rupprecht, Tom Spencer, Maike Paul, Bregje Van Wesenbeeck, Guido Wolters, Jensen, Kai Jensen, Tjeerd J Bouma, Martin Miranda-Lange, and Stefan Schimmels. 2014. "Wave Attenuation over Coastal Salt Marshes under Storm Surge Conditions." *National Geoscience* 7: 727–731. https://doi.org/10.1038/ngeo2251.
7. Ramsar Convention. 2018. "Wetlands—World's Valuable Ecosystem—Disappearing Three Times Faster than Forests, Warns New Report." September 27, www.ramsar.org.
8. Paulson Institute. 2016. "Blueprint of Coastal Wetland Conservation and Management in China." www.paulsoninstitute.org.
9. Larson, Christina. 2015. "China's Vanishing Coastal Wetlands Are Nearing Critical Red Line." *ScienceMag*, October 23, www.sciencemag.org.
10. Ball, Philip. 2017. *The Water Kingdom: A Secret History of China*. Chicago: University of Chicago Press; King 1911.
11. Shapiro, Judith. 2001. *Mao's War against Nature: Politics and the Environment in Revolutionary China*. Cambridge: Cambridge University Press, 148.
12. Li, Xing, Xin Zhang, Chuanyin Qui, Yuanqiang Duan, Shu'an Liu, Dan Chen, Lianpeng Zhang, and Changming Zhu. 2020. "Rapid Loss of Tidal Flats in the Yangtze River Delta since 1974." *International Journal of Environmental Research*

and Public Health 175: 1636. https://doi.org/10.3390/ijerph17051636; Tian, Bo, Yun-Xuan Zhou, Ronald M. Thom, Heida L. Diefenderfer, and Qing Yuan. 2015. "Detecting Wetland Changes in Shanghai, China Using FORMOSAT and Landsat TM Imagery." *Journal of Hydrology* 529: 1–10.

13. National Bureau of Statistics of China. 2020. "China Statistical Yearbook: 2020." *China Statistics Press*, December 1, www.stats.gov.cn.
14. Den Hartog, Harry. 2010. *Shanghai New Towns: Searching for Community and Identity in a Sprawling Metropolis*. Rotterdam: 010.
15. Li, Xue. 2019. "The Coastal Wetlands in the 'Great Protection of the Yangtze River' Are Indispensable, and It Is Not Feasible to Focus on Construction but Despise Protection." Research in Nanhui Dongtan, China Biodiversity Conservation and Green Development Foundation. www.cbcgdf.org (in Chinese); Li, You. 2020. "The Fierce Debate over Shanghai's New Forest." *Sixth Tone*, May 28, www.sixth tone.com.
16. Brelsford, Craig. 2019. "Letter to Elon Musk." *Shanghai Birding*, December 1, www .shanghaibirding.com.
17. Ma, Xin, Martin de Jong, and Harry den Hartog. 2017. "Assessing the Implementation of the Chongming Eco Island Policy: What a Broad Planning Evaluation Framework Tells More than Technocratic Indicator Systems." *Journal of Cleaner Production*. https://doi.org/10.1016/j.jclepro.2017.10.133.
18. Den Hartog, Harry. 2017. "Rural to Urban Transitions at Shanghai's Fringes, Explaining Spatial Transformation in the Backyard of a Chinese Mega-City with the Help of the Layers-Approach." *International Review for Spatial Planning and Sustainable Development* 5(4): 54–72.
19. Den Hartog, Harry. 2019. "Re-defining the Appreciation and Usability of Urban Watersides in the Urban Center and Peri-urban fringes of Shanghai." *European Journal of Creative Practices in Cities and Landscapes* 2(1): 37–64.
20. Yang, Shilun L., Jianmin Zhang, Jacob Zhu, Joseph P. Smith, S. B. Dai., A. Gao, and P. Li. 2005. "Impact of Dams on Yangtze River Sediment Supply to the Sea and Delta Intertidal Wetland Response." *Journal of Geophysical Research* 110: F03006. https://doi.org/10.1029/2004JF000271.
21. State Council. 2018. "State Development (2018) No. 4 Notice of the State Council on Strengthening the Protection of Coastal Wetlands and Strictly Controlling Enclosed and Reclamation." July 25, www.gov.cn.
22. Linster, Myriam, and Chan Yang. 2018. *China's Progress towards Green Growth: An International Perspective*. OECD Green Growth Papers 2018/05. Paris: OECD. In this regard, President Hu Jintao was greatly influenced by the philosophy of the American architect William McDonough and chemist Michael Braungart, particularly their idea of cradle-to-cradle design and a circular economy. McDonough, William, and Michael Braungart. 2002. *Cradle to Cradle: Remaking the Way We Make Things*. New York: North Point.
23. In 2012, former president Hu called to "actively respond to global climate change." Eighteenth National Congress of the Communist Party of China. 2012. "Report

of Hu Jintao to the 18th CPC National Congress, VIII. Making Great Efforts to Promote Ecological Progress." November 16, www.china.org.cn; Henderson, Geoffrey, and Paul Joffe. 2016. "China's Climate Action Since the 12th Five-Year Plan: A Look Back at China's Path towards Renewable Energy and Its Shift Away from Coal." March 17, https://chinadialogue.net.

24. Henderson and Joffe 2016; United Nations Department of Economic and Social Affairs: Sustainable Development. n.d. "The 17 Goals." Accessed February 23, 2022, https://sdgs.un.org; UN Environment Programme. 2019. "New UN Decade on Ecosystem Restoration Offers Unparalleled Opportunity for Job Creation, Food Security and Addressing Climate Change." March 1, www.unenvironment.org; Hansen, Mette H., Hongtao Li, and Rune Svarverud. 2018. "Ecological Civilization: Interpreting the Chinese Past, Projecting the Global Future." *Global Environmental Change* 53: 195–203. https://doi.org/10.1016/j.gloenvcha.2018.09.014.
25. Xinhua News Agency. 2021. "The Fourth Session of the Thirteenth National People's Congress Voted to Pass the Resolution on the '14th Five-Year Plan' and the 2035 Long-Term Goal Outline." March 11, www.gov.cn.
26. Liu, Hongqiao, Jianqiang Liu, and Xiaoying You. 2021. "What Does China's 14th 'Five Year Plan' Mean for Climate Change?" China Policy, CarbonBrief, www.carbonbrief.org.
27. Marinelli, Maurizio. 2018. "How to Build a 'Beautiful China' in the Anthropocene: The Political Discourse and the Intellectual Debate on Ecological Civilization." *Journal of Chinese Political Science* 23: 365–386. https://doi.org/10.1007/s11366-018-9538-7.
28. Hu, Jintao. 2016. "Hu Jintao's Report at the 17th Party Congress." *Qiushi Journal*, August 3, www.cscc.it.
29. Frazier, Amy E., Brett A. Bryan, Alexander Buyantuev. 2019. "Ecological Civilization: Perspectives from Landscape Ecology and Landscape Sustainability Science." *Landscape Ecology* 34: 1–8. https://doi.org/10.1007/s10980-019-00772-4.
30. Hansen, Mette H., and Zhaohui Z. Liu. 2017. *Air Pollution and Grassroots Echoes of "Ecological Civilization" in Rural China*. Cambridge: Cambridge University Press; Wang, Zhihe, Huili He, and Meijun Fan. 2014. "The Ecological Civilization Debate in China: The Role of Ecological Marxism and Constructive Postmodernism—Beyond the Predicament of Legislation." *Monthly Review*, November 1, http://monthlyreview.org; Wang-Kaeding, Heidi. 2018. "What Does Xi Jinping's New Phrase 'Ecological Civilization' Mean? An Investigation of the Phrase Is Pressing." *Diplomat*, March 6, https://thediplomat.com.
31. Zhang, Lei, Arthur P. J. Mol, and David A. Sonnenfeld. 2007. "The Interpretation of Ecological Modernization in China." *Environmental Politics* 16: 659–668. https://doi.org/10.1080/09644010701419170; Gare, Arran. 2020. "The Eco-Socialist Roots of Ecological Civilization, Capitalism Nature Socialism." *Capitalism Nature Socialism* 32(2): 1–19. https://doi.org/p10.1080/10455752.2020.1751223.
32. Shanghai Urban Planning and Land Resource Administration Bureau 2016.

33. Rudd, Kevin. 2020. "The New Geopolitics of China's Climate Leadership." *China Dialogue*, December 11, https://chinadialogue.net.
34. A discussion on "Low-Carbon City Initiative" projects in Shanghai can be found in den Hartog, Harry, Frans Sengers, Ye Xu, Linjun Xie, Ping Jiang, and Martin De Jong. 2018. "Low-Carbon Promises and Realities: Lessons from Three Socio-technical Experiments in Shanghai." *Journal of Cleaner Production* 181: 692–702. A discussion about the "Sponge City" project in Nanhui (Lingang) can be found in Roxburgh, Helen. 2017. "China's 'Sponge Cities' Are Turning Streets Green to Combat Flooding." *The Guardian*, December 27, www.theguardian.com.
35. Shanghai Urban Planning and Land Resource Administration Bureau 2016.
36. Shanghai Urban Planning and Land Resource Administration Bureau 2016.
37. Shanghai Urban Planning and Land Resource Administration Bureau 2016.
38. Shanghai Urban Planning and Land Resource Administration Bureau 2016.
39. Shanghai Urban Planning and Land Resource Administration Bureau. 2018. "Striving for a World-Class Waterfront Area: Shanghai Huangpu River and Suzhou River Planning." August 23, https://mp.weixin.qq.com.
40. Shanghai Urban Planning and Land Resource Administration Bureau 2018.
41. Den Hartog 2019.
42. Den Hartog et al. 2018.
43. Den Hartog 2019, 37–64; Li, Yifou, and Xiaohua Zhong. 2021. "'For the People' without 'By the People': People and Plans in Shanghai's Waterfront Development." *International Journal of Urban and Regional Research* 45(5): 835–847. https://doi.org/10.1111/1468-2427.12964.
44. World Population Review. 2020. "Shanghai's Population in 2020." https://worldpopulationreview.com.
45. Li, Yifou, and Judith Shapiro. 2020. *China Goes Green: Coercive Environmentalism for a Troubled Planet*. Cambridge, UK: Polity.
46. Xie, Linjun, Christof Mauch, May Tan-Mullins, and Ali Cheshmehzangi. 2020. "Disappearing Reeds on Chongming Island: An Environmental Micro History of Chinese Eco-Development." *Nature and Space*, December 2. https://doi.org/10.1177/2514848620974375.
47. Den Hartog 2019, 37–64; Li and Zhong 2021.

17

Adapting New York City

How the Largest US City Is Addressing the Impacts of Climate Change on Its Coastal Communities

ADALENE MINELLI

Cities across the world have been behind some of the most ambitious attempts to curtail greenhouse gas emissions in an effort to stabilize the concentration of greenhouse gases in the atmosphere before the planet reaches a critical tipping point.[1] Global temperatures have already risen to alarming levels, and for many New Yorkers, the reality of a warming climate has already set in. In 2019, the New York City Council adopted a resolution declaring a climate emergency in the city, joining hundreds of other municipalities across the world and dozens in the United States.[2] Acknowledging the "current, expected, and potential effects of climate change," local policy makers committed to accelerating the race to achieving net-zero emissions. Though nonbinding and largely symbolic, the declaration is evidence of the mounting pressure being placed on governments to take stronger actions to reduce carbon emissions.[3]

Despite these efforts, the city is having to come to grips with a stark reality: while there is still a chance that the impacts of climate change can be mitigated to some extent, many climate risks simply cannot be eliminated. The most likely climate scenario suggests that local emissions reductions will not be enough to stop or reverse the catastrophic effects of a rapidly changing climate. The reality is that even if New York City were to reach net-zero emissions in the short term, it would still not be immune from the consequences of emissions occurring outside its borders.[4] And even if the *world* were to reach net-zero emissions in the short term, New York City would still not be immune from the residual effects of past emissions.[5] Indeed, New York City is already reeling with the impacts of climate change today. This means that local climate policy must consider not only strategies to reduce or prevent the emission of greenhouse gasses but also how to adapt and build resilience to a rapidly changing climate.

This chapter examines New York City's response to the particular climate threats it faces as a coastal city: more frequent and severe flooding, rising seas, and intensifying storms. New York's local government is empowered to act in significant ways toward securing the city's coastlines to these threats, and the city is taking some important steps toward doing so. There are, however, questions as to whether the steps the city is taking are adequately tracking the magnitude of the risks it faces over time to sufficiently protect its coastal communities. Moreover, it is unlikely that the city will be able to fully address these risks in the long term absent assistance from higher levels of governments.

Climate Risks as a Coastal City

The largest and most densely populated city in the United States, New York City owes a lot of its success to what is now one of its biggest threats: its coastal geography. Situated on the Hudson River and the Atlantic Ocean in a naturally sheltered harbor with hundreds of miles of protected coastline, the area quickly became a hub for commercial activity in its early history. Indeed, by the early twentieth century, New York had become the busiest port in the world.[6] Building on this early success as a port city, it has since cemented its role as a commercial, cultural, and financial center of the world. However, its vicinity to major water bodies has made it particularly vulnerable to flooding and major storm events.

In 2012, Hurricane Sandy made landfall as one of the strongest and most destructive storms to hit the New York City region. Over the course of two days, Sandy crippled the city, damaging critical public and private infrastructure and inflicting an estimated $19 billion in damages and lost economic activity across the city.[7] But the humanitarian crisis that ensued in the aftermath was the most significant. The storm took the lives of forty-three people, destroyed nearly three hundred homes, and damaged more than sixty-nine thousand residential units, displacing thousands of New Yorkers. With this, hundreds of thousands of residents were left without power and with limited access to food, drinking water, health care, and other critical services.[8] The city's immediate preparation and response to Hurricane Sandy was one of the largest mobilizations of public services in its history. But for many people, Sandy was a wake-up call to the threats of climate change and quickly became a touchpoint for adaptation efforts in New York.

By many measures, the most destructive aspect of Hurricane Sandy was not the wind or rain but the storm surge.[9] The storm surge, coupled with coincidentally high tides, generated record-breaking storm tides, reaching fourteen feet at the southern tip of Manhattan, flooding roads, tunnels, and subway lines, and causing power outages in and around the city.[10] In total, about 17 percent of the city's total land mass—around fifty-one square miles—was under water.[11] Earlier that year, scientists issued a grim warning about the threat that storm surges posed to New York City but estimated that a storm surge the size of Sandy's was something that might only occur once every five hundred years.[12] Almost a decade later, scientists now estimate that the likelihood of a similar flood event has fallen to once every twenty-five years and could occur every five years within the next three decades.[13]

Sea-level rise has played a major role in this trend.[14] Since the start of the twentieth century, oceans have risen about foot in New York Harbor (a foot and a half since the mid-1800s), driven in part by melting glaciers and thermal expansion of the oceans.[15] Rising sea levels, combined with growing storm surges, increases the risk of coastal flooding from storm events—even moderate ones.[16] In fact, sea-level rise is believed to have extended the reach of Sandy by twenty-seven square miles, bringing an additional eighty-three thousand people into its grip.[17] But the flooding New York City experienced during Sandy also provided a dark glimpse into the city's future.[18] Scientists have estimated that the city could see as much as two and a half feet of additional sea-level rise by the end of 2050, and as much as nine and a half feet by the end of the century.[19] By comparison, much of the sea wall protecting lower Manhattan is only about five feet above the mean sea level.[20] Many low-lying coastal areas across the city are already experiencing increased tidal flooding—today, nearly six thousand residential properties are at risk from tidal flooding, and that number is expected to increase to more than eight thousand by 2033.[21] Absent defensive measures, large swathes of the city may very well be completely submerged in the next hundred years.[22]

Historically overburdened communities have been among the most vulnerable to the effects of New York's climate crisis. Indeed, low-income households and seniors made up a higher-than-average share of the population in the neighborhoods within Hurricane Sandy's surge area, and Black households (and low-income Black households in particular) were more likely to live in these flooded areas.[23] The hurricane damaged nearly 20 percent of affordable-housing units owned by the New York City Housing Authority, more units than the entire stock of any other public housing authority in the country.[24] These communities already face persistent disparities, including social, economic, and health inequities, that are exacerbated by the impacts of climate change. For example, low-income households that have been displaced by storms are at a higher risk of being unable to find new housing that is affordable to them, and seniors are at a higher risk of being unable to properly evacuate during times of emergency.[25] Flooding, which has become increasingly routine in many vulnerable low-lying communities, also aggravates existing environments risks, such as spreading contamination from nearby industrial sites into residential areas.[26]

Strategizing Resiliency

Disaster response—while important—is a highly inadequate framework for addressing the worsening impacts of climate change. One way or another, New York will need to adapt to the full spectrum of future challenges posed by the climate crisis.

One crisis that New York City *has* managed to avert, however, is a crisis of information. In one of the most comprehensive adaptation efforts by a municipal government, the city launched the New York City Climate Change Adaptation Task Force in 2008 to advise the city on preparing its infrastructure for the projected impacts of climate change. To support these goals, the city convened the New York City Panel on Climate Change (NPCC), an independent advisory body of experts, including climatologists, engineers, and practitioners from the insurance and legal sectors, modeled on the Intergovernmental Panel on Climate Change. The NPCC was specifically charged with reviewing the scientific data on climate change and producing climate projections specific to New York City that the local government could use to develop adaptation strategies.[27] Shortly after its inception, the NPCC published its first set of findings on climate risks facing New York City, as well as recommendations for managing those risks.[28]

While the city had already begun to contemplate its adaptation needs, the damage caused by Hurricane Sandy in 2012 served as a major impetus for policy makers and residents to take coastal resiliency efforts more seriously. In 2012, not even a month after Sandy made landfall, New York City Council passed Local Law 42 amending the New York City Charter to institutionalize the NPCC as a permanent body.[29] Not too long after, in early 2013, the NPCC convened for a second time to provide the latest scientific information and analyses on climate risks for use in a Special Initiative on Rebuilding and Resiliency (SIRR). Later that same year, the NPCC developed new climate projections, including maps outlining future coastal flood risks.[30]

With a better understanding of the city's climate risks, it developed plans for rebuilding communities that had been destroyed by the storm and began to rethink its efforts to strengthen the resiliency of its systems and infrastructure. Armed with this new data, the SIRR released a report outlining the city's strategy for preparing for and protecting against a range of climate threats, which was incorporated into the mayor's citywide sustainability plan, *PlaNYC*.[31] In 2015, a new mayoral administration incorporated this work into its a successor document, *OneNYC*, outlining its own strategic plan for responding and building resiliency to coastal climate risks.[32]

Achieving climate resiliency in a city as large as New York is a particularly onerous task. Developing viable solutions can be technically complex because of the city's vast infrastructure (including a massive network of underground railways) and its unique location and proximity to other large coastal cities.[33] Adaptation measures are also expensive to implement. The Mayor's Office of Climate Policy and Programs team recently announced that it was investing $20 billion to adapt the city to climate threats, with around three-quarters of that amount coming from the federal government in response to Hurricane Sandy.[34] Hard, protective infrastructure will cost magnitudes more—a proposal for a single six-mile-long sea wall, for example, was estimated to cost around $120 billion and take over two decades to build.[35]

Given the high cost of adaptation, New York's toolbox of options is limited absent the support of the higher levels of government; however, relying on funding from higher levels of government poses several challenges. To begin with, these funds typically come with various contingencies on how they may be spent and how that spending must be reported, and city agencies must constantly "navigate reams of red tape imposed by federal [or state] agencies," which can make developing and administering resiliency projects difficult.[36] There is another problem with relying on funding from higher levels of government too: this funding has typically flowed *reactively* after a disaster (as it did after Sandy), while climate adaptation inherently requires that the city secure financing for taking measures *proactively*.[37]

Adding to the difficulty of securing funding is the need to secure the political consensus for action. Although we are certain that climate change is real and happening, there is a level of uncertainty in predicting the impacts that will fall on New York in several years' time, let alone several decades or centuries into the future. Cities may struggle to justify investing billions of dollars into adaptation measures without the certainty of knowing whether they will be adequate or sufficient in the long term. As a result, there has frequently been a lack of consensus among both policy makers and stakeholders in New York around which adaptation measures the city should pursue.[38]

There is also a social debate about adaptation that is adding to this political paralysis—that is, how much should New York be expending to rebuild and protect climate-vulnerable neighborhoods in the face of inevitable disaster?[39] The city spent considerable resources rebuilding communities after Sandy, and many people have questioned whether those resources might have been better spent on developing coastal defenses given the high risk of disaster and displacement. Others question whether the city should be expending resources to protect coastal properties at all and whether they should instead be developing plans for retreat.[40] Responding to these questions, which are inherently value-laden, will require the city to make difficult social choices.

Implementing Resiliency

New York City's ability to innovate, experiment, and generate tangible solutions for the climate crisis through law and policy is largely a function of the authority granted to it by the state of New York.[41] The state's constitution grants local governments various "rights, powers, privileges and immunities," including the authority to manage their own affairs through lawmaking.[42] Among other things, this grant empowers local governments to adopt or amend local laws relating to their "property, affairs or government" and the "protection, order, conduct, safety, health and well-being of persons or property therein," so long as these laws are not inconsistent with state law.[43]

New York City's regulatory authority makes it well positioned to tackle various aspects of climate adaptation. Among its lawmaking powers, New York City is principally responsible for determining how land within its borders is used. It does this by enacting and amending zoning laws, which affect the types and scale of structures throughout the city, as well as building code laws, which determine design and construction standards for those structures. Altogether, New York City's one-hundred-year floodplain covers approximately 15 percent of the city's land area and contains more than eighty thousand buildings that vary greatly in use and construction.[44] Addressing the climate risk to buildings in these vulnerable areas means not only setting better standards for new construction to protect them from flooding, wind, and prolonged power outage but also offering ways to retrofit the city's existing buildings to protect the residents and businesses that inhabit them.

Managing climate risks has also required the city to rethink the types of buildings it permits to be placed in certain areas. In 2013, shortly after Hurricane Sandy, the Federal Emergency Management Agency (FEMA) issued new preliminary flood maps for New York City, which included higher flood elevations and a larger one-hundred-year flood zone containing nearly twice as many buildings as the previous maps.[45] Following FEMA's issuance of the new maps, the Department of Buildings issued an emergency rule modifying the city's building code to incorporate new flood-resistant construction standards, including a requirement that buildings be protected to a level around one or two feet higher than the FEMA-designated flood elevation, depending on the type of building.[46] As a result of new flood maps and building code rules, a substantial number of additional buildings became subject to new standards for flood protection.[47] Later that year, New York City Council passed several pieces of building code legislation aimed at improving building resiliency, including Local Law 96, which required all new construction projects to comply with the more restrictive of available FEMA maps.[48]

Zoning laws were also amended. In 2013, Mayor Bloomberg, using his emergency powers, issued Executive Order No. 230, which temporarily suspended zoning height and other restrictions to allow buildings to be rebuilt to meet newer flood-resistant standards based on the updated flood maps from FEMA. The executive order was codified later that year when City Council adopted the Flood Resilience Text Amendment, which amended the city's zoning laws to facilitate reconstruction of storm-damaged buildings and ensure that new and existing buildings within the one-hundred-year floodplain would meet new building resiliency standards.[49] Then in 2015, City Council adopted the Special Regulations for Neighborhood Recovery, which established new rules allowing storm-damaged residential units located on small lots to be reconstructed.[50]

Both sets of zoning law amendments were adopted on an emergency, temporary basis, and in 2021, the city enacted a package of permanent zoning laws aimed at protecting buildings in the city's floodplain from major disasters and sea-level rise, known as Zoning for Coastal Flood Resiliency (ZCFR).[51] While the 2013 and 2015 zoning law amendments were largely an effort to encourage the reconstruction of properties damaged by Sandy and speed up postdisaster recovery efforts, ZCFR is largely focused on facilitating long-term resiliency improvements in buildings. Importantly, the new zoning laws expanded on their predecessors' reach by covering buildings located wholly or partially within both the city's one-hundred-year *and* five-hundred-year floodplains, representing an additional 44,600 buildings.[52]

Notably, apart from new prohibitions on the construction or expansion of nursing homes, the law does little to limit development in these areas.[53] In fact, one of the city's stated goals in enacting the ZCFR was to "maintain prevailing land uses and the planned density in neighborhoods across the floodplain."[54] In the past few decades, many of the neighborhoods in the city's floodplains, including Long Island City, Williamsburg, Greenpoint, and Hudson Yards, have become hotbeds for new large-scale development following aggressive rezoning initiatives aimed at developing the city's waterfront.[55] It is perhaps unsurprising, then, that the city has "no official strategy" on utilizing retreat as an adaptation response.[56] The city has taken some steps to restrict densification in highly vulnerable areas. For example, in 2017, the city established Special Coastal Risk Districts in Broad Channel and Hamilton Beach, Queens, to limit future density in these areas due to their exceptional vulnerability to coastal climate risks. However, in planning for resilience, the city's broader objective has been to rebuild, protect, and in some cases, *promote* waterfront development, albeit with stronger resiliency standards for buildings.[57]

Given how vast and populous the city's floodplains are, it may not be feasible for the city to plan for retreat in the immediate term. It might also not be the best use of resources given current climate risk projections for the immediate term, particularly when these risks can be managed through other adaption measures, like improved building resiliency standards. After decades of public and private investment in the waterfront, the city's reluctance to plan for retreat may also reflect a hesitance to stifle development in areas that have become important to the city. But not everyone agrees with the city's waterfront agenda, and experts and residents alike have called for the city to consider retreat more seriously.

Indeed, some owners of properties in vulnerable areas have indicated that they could be motivated to relocate with the right incentives. Following Hurricane Sandy, several storm-devasted communities mobilized to petition the government for buyouts.[58] In 2013, the state government initiated a voluntary buyout program to purchase the properties of homeowners whose homes were substantially damaged or destroyed by storms.[59] The program generated a great amount of interest from affected homeowners, though it was ultimately limited to a small number of properties in Staten Island.[60] Under the state buyout program, land purchased by the government can never be redeveloped and is instead turned into public space that doubles as natural protective buffers to safeguard against climate impacts like sea-level rise, coastal erosion, and extreme weather events.[61]

Buyouts, however, received little support from city officials. In contrast to the state program, the city's Build It Back program instead focused on rebuilding and acquisition. If homeowners wanted to relocate, properties acquired by the city were resold to people who were willing to assume the costs of rebuilding to meet higher flood-resistant construction standards.[62] Cost was probably a major factor in the city's reluctance to pursue buyouts after Sandy.[63] In fact, even in the rare instance when the city has adopted retreat as a localized strategy, it has identified a need to secure funding for buyouts.[64] But the city's post-Sandy acquisition program raised additional concerns about gentrification as developers moved in to scoop up waterfront real estate in low-income communities and communities of color.[65]

With the city showing no signs of departing from its long-standing waterfront development agenda, it has instead been thinking about how to fortify its coast, raising questions about whether the city's push to continue to develop its waterfront, even with improved resiliency standards, is compatible with its long-term adaptation needs.[66] Many of these options come at a high cost and have required coordination among the city, state, and federal governments. In 2021, the city begun work on its $1.45 billion East Side Coastal Resiliency Project, which is jointly funded by New York City and the federal government.[67] Stretching across 2.4 miles of Manhattan's East River shoreline, the project will use a combination of elevated public spaces, floodwalls, and floodgates in an effort to improve flood resiliency in lower Manhattan.[68] In 2019, New York State partnered with the US Army Corps of Engineers (USACE) to secure $400 million in federal funding for a 5.1-mile combined seawall and walkway on Staten Island in an area that was severely damaged by Hurricane Sandy, officially called the Staten Island Multi-Use Elevated Promenade.[69] Using a combination of interconnected levees, berms, and seawalls, the project is expected to reduce flood-related damages on the island by $30 million each year.[70]

As the city discovered, however, land-based adaptation measures are not always technically feasible. In these cases, the city has explored other coastal defense options. In the low-lying Seaport and Financial District, where land is highly constrained, the city developed a plan to extend the shoreline into the East River to create a new strip of land that would sit about twenty feet above sea level.[71] A variety of offshore coastal defense options are also being considered, including, perhaps most notably, a $119 billion offshore barrier wall that would stretch from the Rockaways in Queens to the coast of New Jersey, south of Staten Island, which was one of five options being studied by USACE in consultation with New York City to protect the New York area from storms.[72] These proposals not only would be extremely costly to implement but would also require extensive coordination across various levels of jurisdictions—New York City, New York State, and New Jersey would each have to approve any barrier and collectively cover 35 percent of the cost, and Congress would have to agree to fund the remaining 65 percent of the cost.[73]

There is also a lack of certainty as to how effective hard protective infrastructure will be in the long run—opponents of USACE's proposed barriers argue that the defenses could be obsolete within decades because of future sea-level rise.[74] Others have been concerned about the impact of hard coastal protection measures on the city's coastal ecosystems. One innovative project, however, is seeking to use offshore ecological systems to protect New York City's coastlines. The Billion Oyster Project, which will cost $60 million in federal funding, aims to introduce vast quantities of oysters into New York City's waterways with the expectation that they will eventually form dense reefs that can slow the movement of water and mitigate the impact of storm surges in vulnerable areas.[75]

It is clear that no single adaptation approach can sufficiently mitigate all the risks that New York City faces from climate change. Rather, it appears that the city will need to pursue a combination of approaches. While the city has made notable advances in gathering knowledge and data about the climate risks it faces, and while it has taken some steps to protect its coastal communities and infrastructure from current climate hazards, it must also be prepared to take proactive measures to adapt to future risks. Given the time needed to secure funding and implement these measures, it is important that the city is able to anticipate and act far enough in advance to address climate risks as they grow over time. Full adaptation will not be cheap, and it will require the city to make difficult policy choices. But it also offers the city an opportunity to do what it does best: innovate.

NOTES

1. Leading climate experts have determined that limiting global warming to below 1.5 degrees Celsius (2.7 degrees Fahrenheit) above preindustrial temperatures will lower risk of the most serious negative impacts. Intergovernmental Panel on Climate Change (IPCC). 2018. "Special Report on Global Warming of 1.5°C." October 8, www.ipcc.ch. Achieving sufficient global emissions reductions in time to prevent global mean temperature from rising more than 1.5 degrees Celsius

from preindustrial levels is an ambitious goal, but many people believe that it is still possible. *See* International Energy Agency. 2021. *Net Zero by 2050: A Roadmap for the Global Energy Sector*.

2. NYC Council, Res. 0864-2019 (declaring a climate emergency and calling for an immediate emergency mobilization to restore a safe climate) (adopted June 26, 2019). New York City's climate emergency declaration came just days after the state legislature adopted the sweeping climate change law known as the Climate Leadership and Community Protection Act. N.Y.S. Senate Bill S6599 (adopted June 19, 2019; signed July 18, 2019). Climate Mobilization Project. "Climate Emergency Declaration Tracking Data." Accessed August 20, 2021, www.theclimatemobilization.org (on file with author) (citing International Climate Emergency Forum. "Government's Emergency Declaration Spreadsheet" (on file with author)).
3. Barnard, Anne. 2019. "A 'Climate Emergency' Was Declared in New York City. Will That Change Anything?" *New York Times*, July 5, www.nytimes.com.
4. Broadly speaking, emissions from any local source mix uniformly into the atmosphere and are distributed globally. Thus, local efforts to reduce or eliminate emissions would not have much bearing on the localized impacts of global climate change absent concurrent worldwide efforts. Simply put, carbon emissions do not respect political boundaries.
5. Hausfather, Zeke. 2021. "Explainer: Will Global Warming 'Stop' as Soon as Net-Zero Emissions Are Reached?" Carbon Brief, April 29, www.carbonbrief.org.
6. Hood, Clifton. 2016. *In Pursuit of Privilege: A History of New York City's Upper Class and the Making of a Metropolis*. New York: Columbia University Press, 176.
7. New York City. 2013. *PlaNYC: A Stronger, More Resilient New York*, 13.
8. New York City 2013, 12–18.
9. New York City 2013, 11–13.
10. NOAA defines a "storm surge" as "an abnormal rise of water generated by a storm, over and above the predicted astronomical tides." "Storm surge should not be confused with storm tide, which is defined as the water level rise due to the combination of storm surge and the astronomical tide." National Hurricane Center and Central Pacific Hurricane Center. n.d. "Storm Surge Overview." Accessed August 20, 2021, www.nhc.noaa.gov. Blake, Eric S., Todd B. Kimberlain, Robert J. Berg, John P. Cangialosi, and John L. Beven II. 2013. *Tropical Cyclone Report: Hurricane Sandy*. National Hurricane Center, 8, www.nhc.noaa.gov ("A storm surge of . . . 9.40 ft was reported at the Battery on the southern tip of Manhattan [and] the storm tide reached 14.06 ft above Mean Lower Low Water.").
11. New York City 2013, 13.
12. Lin, Ning, Kerry Emanuel, Michael Oppenheimer, and Erik Vanmarcke. 2012. "Physically Based Assessment of Hurricane Surge Threat under Climate Change." *Nature Climate Change* 2: 464. https://doi.org/10.1038/nclimate1389 ("The estimated present 50-yr storm surge is about 1.24 m, the 100-yr surge is about 1.74m and the 500-yr surge is about 2.78 m.").

13. Garner, Andra J., Michael E. Mann, Kerry A. Emanuel, Robert E. Kopp, Ning Lin, Richard B. Alley, Benjamin P. Horton, Robert M. DeConto, Jeffrey P. Donnelly, and David Pollard. 2017. "Impact of Climate Change on New York City's Coastal Flood Hazard: Increasing Flood Heights from the Preindustrial to 2300 CE." *PNAS* 114(45): 11861–11866. https://doi.org/10.1073/pnas.1703568114.
14. Garner et al. 2017, 11861 ("projected sea-level rise leads to large increases in future overall flood heights associated with tropical cyclones in New York City"); Miller, Kenneth G., Robert E. Kopp, Benjamin P. Horton, James V. Browning, and Andrew C. Kemp. 2013. "A Geological Perspective on Sea-Level Rise and Its Impacts along the U.S. Mid-Atlantic Coast." *Earth's Future* 1(1): 3–18. https://doi.org/10.1002/2013EF000135.
15. There are other factors that are contributing to sea-level rise in New York City that are not attributable to climate change, including land subsidence. Gornitz, Vivien, Michael Oppenheimer, Robert Kopp, Philip Orton, Maya Buchanan, Ning Lin, Radley Horton, and Daniel Bader. 2019. "New York City Panel on Climate Change 2019 Report, Chapter 3: Sea Level Rise." *Annals of the New York Academy of Sciences* 1439(1): 71–94. https://doi.org/10.1111/nyas.14006. The combination of these various factors mean that New York City is likely to experience "higher than average local sea level rise." Gornitz et al. 2019, 72.
16. Orton, Philip, Ning Lin, Vivien Gornitz, Brian Colle, James Booth, Kairui Feng, Maya Buchanan, Michael Oppenheimer, and Lesley Patrick. 2019. "New York City Panel on Climate Change 2019 Report, Chapter 4: Coastal Flooding." *Annals of the New York Academy of Sciences* 1439(1): 95–114. https://doi.org/10.1111/nyas.14011. Notably, by the time Hurricane Sandy made landfall in New York City, it had been downgraded to a tropical storm.
17. Milman, Oliver. 2017. "Hurricane Sandy, Five Years Later: 'No One Was Ready for What Happened After.'" *The Guardian*, October 28, www.theguardian.com.
18. Lin et al. 2012, 466 ("Some climate models predict the increase of the surge level due to the change of storm climatology to be comparable to the projected [sea level rise] for New York City.").
19. Gornitz et al. 2019. These projections take into account a scenario in which there is a substantial destabilization of the Antarctic Ice Sheet. Studies indicate that "this destabilization is highly unlikely to emerge until the second half of the century, and only then under high emission scenarios." Gornitz et al. 2019, 83. Should this destabilization occur, it is likely that sea level rise would rapidly accelerate after 2050. In the event that this destabilization does not occur, the city could nonetheless see as much as 4.17 feet by the end of the century. Gornitz et al. 2019, table 3.2. Efforts to cut emissions consistent with current global goals are not likely to eliminate this threat. IPCC 2018 (reporting, with "high confidence," that "sea level rise will continue beyond 2100 even if global warming is limited to 1.5°C in the 21st century").
20. Lin et al. 2012, 465 (citing Colle, Brian A., Frank Buonaiuto, Malcolm J. Bowman, Robert E. Wilson, Roger Flood, Robert Hunter, Alexander Mintz, and Douglas

Hill. 2008. "New York City's Vulnerability to Coastal Flooding." *American Meteorological Society* 89: 829–841).

21. SeaLevelRise.org. n.d. "New York's Sea Level Has Risen 9″ since 1950." Accessed August 20, 2021, https://sealevelrise.org. "In the long term, sea level rise could eventually raise tidal flooding to levels even more severe than those that occurred during Hurricane Sandy." Orton et al. 2019, 104.
22. While New York City is not at an immediate risk of extensive land inundation, low-lying areas that experience regular tidal flooding are likely to be among the first areas to experience permanent land loss due to sea-level rise. Areas that are permanently inundated by the 2080s could include "portions of Rockaway Peninsula, Howard Beach, Coney Island, Red Hook, and Staten Island, as well as edges of lower Manhattan waterfront, the Gowanus Canal in Brooklyn and Newtown Creek in Brooklyn and Queens, and Pelham Bay in the Bronx." Patrick, Lesley, William Solecki, Vivien Gornitz, Philip Orton, and Alan Blumberg. 2019. "New York City Panel on Climate Change 2019 Report, Chapter 5: Mapping Climate Risk." *Annals of the New York Academy of Sciences* 1439(1): 121–122. https://doi.org/10.1111/nyas.14015. *See also* Gornitz, Vivien, Michael Oppenheimer, Robert Kopp, Radley Horton, Daniel A. Bader, Philip Orton, and Cynthia Rosenzweig. 2018. *Enhancing New York City's Resilience to Sea Level Rise and Increased Coastal Flooding*. NASA Technical Reports Server, https://ntrs.nasa.gov.
23. Furman Center. 2013. *Sandy's Effects on Housing in New York City*. https://furmancenter.org; Faber, Jacob. 2015. "Superstorm Sandy and the Demographics of Flood Risk in New York City." *Human Ecology* 43(3): 363–378.
24. Furman Center 2013.
25. Furman Center 2013.
26. Foster, Sheila, Robin Leichenko, Khai Hoan Nguyen, Reginald Blake, Howard Kunreuther, Malgosia Madajewicz, Elisaveta P. Petkova, Rae Zimmerman, Cecil Corbin-Mark, Elizabeth Yeampierre, Angela Tovar, Cynthia Herrera, and Daron Ravenborg. 2019. "New York City Panel on Climate Change 2019 Report, Chapter 6: Community-Based Assessments of Adaptation and Equity." *Annals of the New York Academy of Sciences* 1439(1): 126–173. https://doi.org/10.1111/nyas.14009.
27. NYC Special Initiative for Rebuilding and Resiliency. n.d. "What Could Happen in the Future?" Accessed August 20, 2021, www.nyc.gov.
28. New York City Panel on Climate Change. 2009. *Climate Risk Information: New York City Panel on Climate Change*. www.nyc.gov; Bloomberg, Michael R., Jeffrey D. Sachs, and Gillian M. Small. 2010. "Climate Change Adaptation in New York City: Building a Risk Management Response." *Annals of the New York Academy of Sciences* 1196(1): 1–354. https://doi.org/10.1111/j.1749-6632.2009.05415.x.
29. NYC Int 0834-2012.
30. NPCC. 2013. *Climate Risk Information 2013: Observations, Climate Change Projections, and Maps*. www.nyc.gov. New York City relies on Flood Insurance Rate Maps (FIRMs) produced by the Federal Emergency Management Agency (FEMA) for its building code. NYC Department of City Planning. n.d.-a. "NYC

Flood Hazard Mapper." Accessed August 20, 2021, https://dcp.maps.arcgis.com. Notably, however, FEMA's maps had not been updated for several decades prior to Sandy and severely underestimated the storm's risk to coastal properties in many parts of the city. NYC Special Initiative for Rebuilding and Resiliency, n.d. (noting that "approximately 1/2 of all impacted residential units were outside 100-year floodplain [and m]ore than 1/2 of all impacted buildings were outside 100-year floodplain"). NYC Flood Maps. n.d. "New York City Launches Residential Flood Insurance Affordability Study." Accessed August 20, 2021, www1.nyc.gov ("FEMA's FIRMs have not been significantly updated since 1983, and the City's maps are currently being updated by FEMA."). NPCC's 2013 maps incorporated projections of sea-level rise onto FEMA's flood maps to estimate potential impacts of sea-level rise on one-hundred- and five-hundred-year flood zones. NPCC 2013.

31. New York City 2013.
32. New York City. 2015. *OneNYC, Vol. 7: A Livable Climate.* https://onenyc.cityofnewyork.us.
33. SeaLevelRise.org, n.d.
34. Orcutt, Mike. 2019. "New York City Has Big Plans—and $20 Billion—to Save Itself from Climate Change." *MIT Technology Review*, September 19, www.technologyreview.com.
35. Barnard, Anne. 2020. "The $119 Billion Sea Wall That Could Defend New York . . . or Not." *New York Times*, January 17, www.nytimes.com.
36. Stringer, Scott. 2019. *Safeguarding Our Shores: Protecting New York City's Coastal Communities from Climate Change.* https://comptroller.nyc.gov. Despite this, the report notes that "every federal grant dollar dedicated towards flood mitigation can save $6 in future disaster costs."
37. Stringer 2019.
38. Cohen, Stephen. 2020. "The Politics and Cost of Adapting to Climate Change in New York City." *State of the Planet*, January 21, https://news.climate.columbia.edu.
39. Milman 2017.
40. Nonko, Emily. 2020. "NYC's Coastline Could be Underwater by 2100. Why Are We Still Building There?" *Curbed*, January 2, http://ny.curbed.com; Cohen, Ilana. 2020. "In New York City, 'Managed Retreat' Has Become a Grim Reality." *Inside Climate News*, July 4, http://insideclimatenews.org; Pierre-Louis, Kendra. 2019. "How to Rebound after a Disaster: Move, Don't Rebuild, Research Suggests." *New York Times*, April 22, www.nytimes.com (citing Sliders, A. R., Miyuki Hino, and Katharine J. Mach. 2019. "The Case for Strategic and Managed Climate Retreat." *Science* 365(6455): 761–763. https://doi.org/10.1126/science.aax8346).
41. New York City's various strategies, plans, and emergency declarations, despite their good intentions, have no legal teeth. They are nonbinding and unenforceable, though yearly progress reports under *PlaNYC* and *OneNYC* toward certain resiliency targets have provided some accountability through transparency.
42. N.Y. Constitution, art. IX, § 1; Committee on the New York State Constitution. 2016. "Home Rule Report." New York State Bar Association, April 2, 2, citing

Ward, Robert B. 2006. *New York State Government* 545, 2nd ed. ("New York's constitutional and statutory provisions regarding home rule are more extensive than those in many states.").

43. N.Y. Constitution, art. IX, § 2(c)(i), § 2(c)(ii)(1)–(10).
44. This statistic is based on those areas "currently designated on FEMA's FIRMs and Preliminary FIRMs" as having 1 percent annual chance of flood. NYC Department of City Planning. n.d.-b. *Zoning for Coastal Flood Resiliency: Project Description*, 2. Accessed August 20, 2021, www1.nyc.gov.
45. In 2016, the city won an appeal of FEMA's Preliminary Flood Insurance Maps (PFIRMs) on the basis that they overestimated the size of the one-hundred-year floodplain and the height of base flood elevations. Mayor's Office of Recovery and Resiliency. 2015. *Appeal of FEMA's Preliminary Flood Insurance Rate Maps for New York City*. FEMA has agreed to revise the flood maps, but until the new maps are issued, New York City continues to apply the PFIRMs for its building code, except where the older "effective" FIRMS are more restrictive. NYC Flood Hazard Mapper. 2017. "About Flood Hazards Today." https://dcp.maps.arcgis.com.
46. NYC Building Code 2014, Appendix G Flood-Resistant Construction. The requirement applied to new buildings, buildings that were destroyed or substantially damaged during Hurricane Sandy, and buildings undergoing substantial improvements. Other building owners could voluntarily choose to make their buildings meet the new standards. City Planning Commission. 2013. *In the Matter of an Application by the Department of City Planning Pursuant to Section 201 of the New York City Charter for an Amendment of the Zoning Resolution of the City of New York, Pertaining to Enabling Flood Resilient Construction within Flood Zones.* Resolution N130331(A)ZRY.
47. City Planning Commission 2013.
48. Other laws passed include Local Law 83, which required buildings in certain flood-prone areas to be fitted with backwater values to prevent backflow of sewage; Local Law 99, which removed regulatory hurdles around the elevation of certain building systems in flood-prone areas (like telecommunications cabling and fuel storage tanks); and Local Law 101, which clarified wind-resistance specifications for facade elements. NYC Buildings. n.d. "Resiliency Legislation." Accessed August 20, 2021, www1.nyc.gov.
49. NYC Council. 2013. *Flood Resilience Text Amendment*. October 9, www1.nyc.gov. *See also* NYC Department of City Planning. 2013. *Flood Resilience Zoning Text Amendment*. www1.nyc.gov.
50. NYC Planning Zoning Resolution. 2021. "Chapter 4: Special Regulations Applying in Flood Zones." May 12, https://zr.planning.nyc.gov.
51. NYC Department of City Planning. n.d.-c. "Overview, Zoning for Coastal Flood Resiliency." Accessed August 20, 2021, www1.nyc.gov; NYC Council, Zoning for Coastal Flood Resiliency (enacted April 27, 2021), www1.nyc.gov. *See also* NYC Office of the Mayor. 2021. *Press Release: Recovery for All of Us: New York City*

Adopts New Zoning Rules to Protect Coastal Areas from Climate Change. May 12, www1.nyc.gov.

52. NYC Department of City Planning, n.d.-b, 4.
53. The ZCFR "[prohibits] the development of new nursing homes and [restricts] the enlargement of existing facilities within the 1% annual chance floodplain and other selected geographies likely to have limited vehicular access because of the storm event." NYC Department of City Planning, n.d.-b, 23.
54. NYC Department of City Planning, n.d.-b, 4, 7.
55. Santora, Marc. 2010. "New York's Next Frontier: The Waterfront." *New York Times*, November 5, www.nytimes.com.
56. Nonko 2020.
57. Pierre-Louis 2019 ("Until now, much of the focus has been on disaster response, with very little discussion of orderly, strategic retreat from areas at risk. After Hurricane Sandy in 2012, for example, the New York State governor's office ran a public service announcement that made clear the focus would be on rebuilding.").
58. "Buyouts are a form of managed retreat, the process of relocating buildings, infrastructure, and populations away from areas vulnerable to effects of climate change." Koslov, Lisa. 2014. "Fighting for Retreat after Sandy: The Ocean Breeze Buyout Tent on Staten Island." *Metro Politics*, April 23, http://metropolitics.org.
59. Governor's Office of Storm Recovery, New York State. n.d. "Buyout & Acquisition Programs." Accessed August 20, 2021, https://stormrecovery.ny.gov.
60. Nonko 2020.
61. Koslov 2014.
62. Koslov 2014.
63. Notably, both the state and city programs were funded through federal disaster assistance. NYC Housing Recovery. n.d. "Welcome to NYC Housing Recovery." Accessed August 20, 2021, www1.nyc.gov.
64. *See* NYC Department of Housing Preservation and Development. 2017. *Resilient Edgemere Community Plan*. www1.nyc.gov ("Identify funding for a long-term buyout program to relocate current homeowners away from flooding and coastal storm hazards.").
65. Khafagy, Amir. 2021. "She Survived Hurricane Sandy. Then Climate Gentrification Hit." *The Guardian*, April 21, www.theguardian.com; Gould, Kenneth A., and Tammy L. Lewis. 2018. "From Green Gentrification to Resilience Gentrification: An Example from Brooklyn." *City & Community* 17(1): 12–15. https://doi.org/10.1111/cico.12283.
66. Stevens, Mark R., Yan Song, and Philip R. Berke. 2010. "New Urbanist Developments in Flood-Prone Areas: Safe Development, or Safe Development Paradox?" *Natural Hazards* 53: 605–629.
67. New York City. n.d. "East Side Coastal Resiliency Project." Accessed August 20, 2021, www1.nyc.gov.
68. Zhao, Kellie. 2021. "The East Side Coastal Resiliency Project Breaks Ground, but Opponents Aren't Backing Down." *Architect's Newspaper*, May 28, www.archpaper.com.

69. Whiteman, Hilary. 2019. "Staten Island Seawall: Designing for Climate Change." *CNN*, July 14, www.cnn.com.
70. Plitt, Amy. 2019. "Staten Island's Five-Mile Seawall Will Move Forward with Federal Funding." *Curbed*, February 19, https://ny.curbed.com. Notably, the project is currently delayed due to radiation contamination. Michel, Clifford. 2021. "A Five-Mile Seawall Was Supposed to Protect Staten Island by 2021. A Fight over Radiation Cleanup Stands in the Way." *The City*, February 15, www.thecity.nyc.
71. New York City 2015, vol. 7.
72. Some advocates have also argued that "the use of locally tailored, onshore solutions alone, like berms, wetlands restoration and raised parks, would likely benefit wealthy areas first, not the low-income communities that suffered disproportionately from Sandy in 2012." Barnard 2020. Federal funding for the study was halted in 2019. The study is included in the president's budget request to Congress for 2022, and USACE is also seeking other federal funding options. Work on the study will resume when federal funding is provided. US Army Corps of Engineers. n.d. "NY & NJ Harbor & Tributaries Focus Area Feasibility Study (HATS)." Accessed August 20, 2021, www.nan.usace.army.mil.
73. Barnard 2020.
74. Barnard 2020.
75. Klinenberg, Eric. 2021. "The Seas Are Rising. Could Oysters Help?" *New Yorker*, August 2, www.newyorker.com.

18

Climate Change Adaptation in Abu Dhabi

KATIE ZAVADSKI

Most of the United Arab Emirates' oil wealth is concentrated in Abu Dhabi.[1] The largest emirate by land mass, it is home to upward of 90 percent of the country's total oil reserves. That "black gold" funded Abu Dhabi's development into an international research hub and sovereign philanthropist and remains a significant source of its GDP to this day. Yet Abu Dhabi has, over the past twenty years, emerged as a prominent player on the international environmental stage. The emirate funds environmental programs abroad and makes bold commitments about its own environmental benchmarks, like greenhouse gas emissions and environmental conservation.

Today, the Emirate of Abu Dhabi is a paradox: the most oil-rich emirate wants to be a leader against climate change on the world stage. Acknowledging that rising temperatures and sea levels threaten its very existence, the emirate says it wants to transition to green energy and is investing millions in adaptation and mitigation efforts.[2] And given that each of the seven component emirates—some barely more than one city—retain much independence in formulating their own initiatives on climate change, Abu Dhabi has substantial latitude to pursue its adaptation goals.

Abu Dhabi couches its commitment in an internal mythology about the longtime ruler of the emirate who created the confederate nation and served as its president for more than three decades.[3] The origin story that Emiratis present to the world says that man, Sheikh Zayed bin Sultan al-Nahyan, "tamed the desert" before adaptation became a buzzword on the international stage.[4] Today, many of the initiatives that began as cultural touchpoints are being recast through an environmental lens as the emirate confronts increasing climate vulnerability. The successors of Sheikh Zayed, who died in 2004, say they are following in his footsteps, even as a rapidly changing climate and the challenges posed by an ever-growing population complicate the task.

The dangers facing the rest of the world are especially significant in this coastal desert emirate. Temperatures, already unbearable at their peak, are projected to rise 4 to 6 degrees Fahrenheit annually within the next fifty years, coupled with a 10 percent increase in humidity.[5] Some parts of the emirates may experience up to a 100 percent increase in rainfall while sea temperatures and salinity increase, causing greater coastal flooding.[6]

Yet even as Abu Dhabi gears up to combat climate change, its mitigation efforts seem more developed than attempts at adaptation. Many of the existing adaptation efforts require extravagant resource expenditures or themselves worsen carbon emissions, and the UAE's authoritarian political structure makes independent measurements of success difficult to obtain.

Background

The United Arab Emirates was established in 1971, when seven existing emirates joined forces to create a joint federal government.[7] The unique nature of this union meant that each emirate had preexisting governance structures, rulers, and priorities, and the structure of the new government provided that the ruler of one of the individual emirates would also serve as president of the whole. Sheikh Zayed, for instance, led the Emirate of Abu Dhabi for five years before the confederation formed and continued in that role while serving as president of the UAE for more than three decades.[8]

This system of government allows for some powers to be reserved for the federal level but also allows for a great deal of independence in determining how to implement policy, and what policies to implement, among each of the component emirates. While the federal level has a Ministry of Climate Change and Environment (MOCCAE), many climate-change-related laws and policies are implemented by the emirates themselves. The Emirate of Abu Dhabi has used this power to issue laws, resolutions, and decrees that address various environmental matters, including managing waste, regulating groundwater, setting up an urban planning council, and establishing a comprehensive environment, health, and safety management system.[9] It has also established Estidama (Arabic for "sustainability"), a construction sustainability evaluation system similar to LEED that awards developments "Pearls" on a tiered system.[10]

Each emirate also has its own administrative agencies, known as departments, which administer and implement policies. Abu Dhabi's environment department (EAD) heads up environmental protection, as well as adaptation and mitigation initiatives, and coordinates with other departments.[11] One of EAD's key tasks is ensuring the proper implementation of federal and emirate laws within Abu Dhabi and providing guidance on how to facilitate their implementation.[12] For example, federal law states that the competent authority is responsible for waste management within each emirate and grants it the right to "undertake the preparation of plans, programs and measures necessary for waste management to improve environmentally sound practices."[13] In turn, the Emirate of Abu Dhabi has introduced its own laws on the matter and has stated that EAD is designated as the competent authority for its territory to implement the law.[14] EAD may thus develop and introduce its own policies regarding waste management—as demonstrated when, in 2020, EAD issued a single-use plastic policy that banned single-use plastic bags in the emirate starting in 2021.[15] Abu Dhabi has a Department of Energy as well, which regulates the water, wastewater, and electricity sector within the emirate.[16] This chapter focuses on adaptation efforts on the emirate level, as governmental authority is centered with the emirate, and the City of Abu Dhabi and other municipalities are administered by departments within the emirate itself.

The emirates have seen tremendous population growth and have become popular tourist destinations over the past four decades, and the Emirate of Abu Dhabi is no exception.[17] Its population has topped two million people, many of them expats and migrant workers, from less than one hundred thousand when the confederation was formed. More than 85 percent of people in the emirates are non-Emirati, and many of those work in construction or the service industry.[18] This population growth was driven in part by the profits generated by the discovery of oil fields under the emirate, which by itself is home to 94 percent of the UAE's oil reserves.[19] Though the Emirate of Abu Dhabi is the least oil-rent-dependent member of the confederation, oil revenues still play a major role in its economy.[20] A push toward economic diversification means that the World Bank estimated that oil revenues accounted for only 31 percent of its GDP in 2006. Fifteen years later, the UAE still indicates 30 percent as a top-line number for the combined economy.[21] Yet independent assessments suggest that the Emirate of Abu Dhabi is more heavily reliant on oil revenues than it publicly acknowledges. A Standard & Poor's assessment from 2019 determined that 50 percent of the emirate's GDP comes from oil revenues and as much as 90 percent comes from the hydrocarbon sector overall. The same report projected that the emirate's non-oil economic sector will grow at a slower rate than the oil sector.[22]

Despite this continuing reliance on oil revenues and continued investment in oil production, Abu Dhabi aspires to be a global leader in clean energy.[23] It pledged that a quarter of its energy use would come from clean energy sources by 2021 and expanded that pledge to 50 percent clean energy target by 2050.[24] Yet, as of July 2021, it was unclear whether the emirate would meet its 2021 target. Nonetheless, emirate-owned oil and gas companies are said to be growing their non-oil investment portfolios in appreciation of the climate challenges facing the emirate.[25] Even so, Abu Dhabi's economic model elides the fact that oil and gas exports, even in the face of an internal shift to clean energy, will contribute to the environmental threats facing the emirate.

Critical Areas

The UAE's national plan for 2017 to 2050 highlights some of the national priorities of the country. The emirates pledged to "manage" greenhouse gas emissions while continuing to promote economic and population growth and to implement a variety of climate change mitigation measures. But importantly, they have also said they want to "increase climate resilience" by adopting adaptive measures. In this vein, the country's future is pinned on making its economy and infrastructure "climate proof."[26] The need for adaptive measures is clear perhaps nowhere as much as in the Emirate of Abu Dhabi. Qais Bader Al Suwaidi, an official with the Climate Change Department of the MOCCAE, identified heat, water, and infrastructure adaptation measures as top priorities for Abu Dhabi.[27]

Heat is of particular concern to the emirate and one that the government has already made attempts to address. In Abu Dhabi, for example, outside labor is already prohibited between 12:00 and 3:00 p.m. during the summer months because of heat.[28] This prohibition puts pressure on the construction industry, which may soon have to switch to nighttime operations because of increasing temperatures.[29] Those who disobey the law are subject to a fine of AED 5,000 (roughly $1,350) per worker. But the ban is not absolute, and employers can argue that they fit one of the exemptions that allow for continued operation for "technical" reasons.[30]

At the same time, the heat means that the emirate must employ more and more cooling measures, driving up carbon emissions and energy use. In summer months, cooling mechanisms make up between 50 and 70 percent of the emirate's energy bill.[31] Some estimates suggest that energy consumption, largely for cooling, may increase up to 20 percent by 2040 in commercial buildings.[32] The emirate has also set up district cooling systems, which are said to be a more energy-efficient way to provide cooling for large networks of buildings. District cooling works by aggregating the cooling needs and linking a group of buildings, potentially lowering carbon emissions by as much as 40 percent.[33] Abu Dhabi issued its first district cooling license to the Saadiyat Cooling Company in 2021.[34] Other initiatives within the emirate employ heat mapping to explore the micro-climates within the region.[35]

Water is another critical area (see chapter 4 in this volume), particularly for human consumption. Groundwater accounts for 65 percent of the emirate's water consumption, but increasing demand and rising salinity levels threaten that resource.[36] In Abu Dhabi, residents consume more than double the world average of 180–200 liters a day.[37] Desalination, a popular choice for addressing water needs beyond groundwater capacity, is an expensive and energy-intensive process.[38] The emirate now seeks a water reserve to last for ninety days, instead of just forty-eight hours, in case of emergency.[39] To address these water concerns, the emirate recently developed a ten-year water management plan that outlines steps the sector must take by 2030 to prepare for water scarcity. Under the plan, Abu Dhabi will look at ways to address the freshwater shortage through reverse-osmosis desalination, harnessing solar energy to prevent water dam evaporation, and developing new technologies for water harvesting. It also plans to look at ways to conserve agricultural water usage through innovative irrigation techniques.[40]

Finally, the Emirate of Abu Dhabi is facing infrastructure issues relating to shoreline vulnerability and coastal flooding. The UAE's energy and climate change report from 2017 suggested that, by the end of the century, the country may lose 6 percent of its coastline to rising sea levels. Rising sea levels also threaten to flood parts of Abu Dhabi, including tourist attractions that draw many visitors to the emirate.[41] Additionally, many of the emirate's residents live in densely populated coastal areas—the same areas where the emirate is investing in much new construction, all barely above sea level. Some studies suggest that as many as 528 square kilometers may be affected.[42] One of the emirate's longest-running adaptation projects seeks to address this threat: it has planted millions of mangroves to shore up the coastal ecosystems.[43]

As the risks of climate change multiply, the emirate's resource allocation indicates areas of prioritization for the government. One key area of concern for human rights advocates is the emirate's large immigrant labor population, which is especially vulnerable in fields like construction.[44] Despite the emirate's changes in labor policies, advocates say that migrant workforces are still at risk and that the existing measures are insufficient to address the dangers of high temperatures and prioritize the emirate's continuing economic advancement over human life. More than half of Indian nationals who died in the UAE in 2018 died of heart attacks, for instance, which critics of the UAE's labor policies say is an indicator of cardiac stress from working in extreme heat.[45] Business interests in the emirate maintain that daytime work breaks are a high-efficacy, low-investment solution, but the emirate's political structure and the disadvantaged status of the affected workers make it difficult to ascertain the impact and efficacy rates.[46]

Attempts at Innovation

Two of the Emirate of Abu Dhabi's biggest adaptation projects have been the Masdar City Initiative and the mangrove-planting project. The projects are notable for their difference: one, a futuristic city with self-driving personal cars, and the other, a low-tech attempt to harness the region's historical defenses to stave off an unprecedented challenge.

Masdar

The Masdar City Initiative, started in 2006, is one of the emirate's most ambitious projects. The idea was a model city that, once fully realized, would be home to fifty thousand people, a hub for clean-energy entrepreneurship, and an example of what a carbon-neutral city can be. Located a short distance from Abu Dhabi City, it would be a Middle Eastern Silicon Valley that would allow workers to commute from the city center while also developing an internal city life. In addition to meeting leading environmental standards, Masdar City was to be a place where the next generations of green technology could be developed and tested.[47]

Masdar City stands out as a project of the emirate-level government and was billed as a model for how cities can adapt to a changing climate. Its chief funder was the Mubadala Development Company, an Abu Dhabi sovereign wealth fund and investment vehicle.[48] Mubadala boasts that the development is "the UAE's natural home for sustainability."[49] The project brought in a British architectural firm to build the city and promoted it to world leaders, including US President Joseph Biden.[50]

Environmental actors cautiously supported the development. While groups like Greenpeace emphasized that the priority should be retrofitting existing cities to be more energy efficient, they praised the Masdar initiative as a potential alternate model.[51] The US Department of Energy (USDOE) also signed a memorandum of understanding with the initiative for collaboration on clean-energy technologies in 2010 and entered into a joint testing agreement with the initiative the following year. Daniel Poneman, a deputy secretary at the USDOE, praised the two countries' "shared vision" on clean-energy innovation in the announcement.[52]

But the city has fallen behind schedule. Though initially planned for fifty thousand residents, fifteen years into its construction, Masdar City is home to just two thousand.[53] A lack of public transit access means city workers and tourists alike must rely on personal vehicles for the trip from Abu Dhabi City. The road between the cities is lined with greenery: "date palms, green grass, and at times wildflowers": "It was widely believed that the Abu Dhabi government expended an extraordinary amount of resources to keep these roads verdant, nearly as much as the military budget for the emirate."[54] The reality is a long way from the car-free city imagined at the start.[55]

Another concern is replicability. The project costs upward of $20 billion for a two-and-a-half-square-mile area—roughly a tenth of the area of Manhattan.[56] Few other municipalities have the resources to undertake a project of this cost. A replication of the Masdar City initiative requires the confluence of tremendous government wealth, long-term commitment, and a fortunate real estate location in relative proximity to other big cities and airports. This may mean that, even if resources were not an issue, few major cities would have the requisite ingredients to replicate this experiment.

But EAD says that the Masdar development has adapted to changes in the surrounding environment since it was first conceived. Development and expansion of Abu Dhabi City has minimized the distance between it and Masdar, and while developments there still meet the original plan's mission for sustainability, they are now less considered in isolation and more responsive to market needs.[57] Further, Emirati officials say that Masdar served a powerful public-opinion purpose in recentering Emirati ideas about sustainability and became a hub for testing the feasibility of regulations, like EAD's Estidama rating system, that would later be applied elsewhere. Even now, developments in Masdar are held to a higher Estidama Pearl rating than those elsewhere in Abu Dhabi and must be developed to sustainability levels higher than those required by LEED's Gold standards.[58]

Mangroves

Loss of wetlands, coastal erosion, and flash floods are three of the most pressing climate challenges facing Abu Dhabi, and the government has placed a long-term bet on mangrove reforestation helping to combat these scourges.[59] While many indigenous mangrove trees were destroyed during the UAE's rapid economic growth period, there are now 150 square kilometers of mangroves along Abu Dhabi's coast. The reforestation campaign began in Abu Dhabi before the unification of the UAE and continued on a local level, with national expansion, during the subsequent decades.[60]

The trees are known for their expansive root structures, which expand above and below the water and allow them to survive in harsh conditions. In Abu Dhabi's case, mangroves can thrive in hypersalinated water due to internal systems that keep salt at bay. This makes them particularly useful for coastal areas where climate change is projected to further raise salinity levels, where they can help secure shorelines and guard against erosion. An added benefit of mangrove reforestation is that the trees and their roots maintain important habitats for oysters, birds, fish, and other animals native to the region.[61] The mangrove-planting effort re-creates habitats that are endangered by or have been destroyed through desertification. Shaikha Al Dhaheri, the head of Abu Dhabi's EAD, called the trees a "first line of defense for any coastal city."[62]

The United Arab Emirates indicates that it planted more than three million mangroves in the decade preceding 2019.[63] In Abu Dhabi, the emirate government directed the EAD to implement sustainable management for the emirate's forests. The emirate boasts more than seventy square kilometers of forest with twenty million trees and places a special emphasis on developing wildlife habitats.[64] But as many as 20 percent of Abu Dhabi's mangrove forests are in ill health, raising concerns about the sustainability of the reforestation initiative.[65] The continued development of coastal areas in Abu Dhabi—the very areas the mangroves are said to protect—may threaten the planted forests by creating conditions that are inhospitable to their survival.[66] However, some scientists believe that increasing urbanization and freshwater runoff from everyday urban activities contributes to the mangroves' growth by diluting the salinity of the seawater and providing them with a more habitable environment.[67]

Emirati officials also tout mangroves as a natural way to combat CO_2 emissions. "One hectare of mangrove forest can store about 3,754 tons of carbon, according to a study done by Abu Dhabi's government. That's the equivalent of taking about 2,651 cars off the road for a year," a CNN article on the subject declared.[68] The trees are known as "natural carbon sinks," and the Emirati government has funded multiple studies about their efficacy.[69] A study commissioned by the Abu Dhabi Global Environmental Data Initiative and the MOCCAE, for example, analyzed carbon sequestration by the mangrove forests. It found "meaningful rates for soil carbon sequestration given that they continue year after year as long as the mangrove remains healthy" and concluded that the mangroves are strong contributors to efforts to combat climate change.[70]

Yet the mangroves alone will probably be insufficient to combat the threat of rising sea levels. While they can help prevent coastal erosion, they are not impervious to the threat of rising sea levels.[71] And in heavily developed coastal areas, like those in Abu Dhabi City, additional structural impediments to coastal flooding are necessary to protect residents and developments.[72]

Nevertheless, Abu Dhabi continues to invest in its "Blue Carbon" initiatives. The term refers to carbon captured and stored by coastal and aquatic ecosystems, including mangroves.[73] The Blue Carbon Project began in 2012 with a study on the impact of carbon sequestration and has since continued in various iterations.[74] In 2020, the emirate began to use drones to plant mangrove seeds around the Mirfa power plant, relying on the drones' technology to map out where to disperse the seeds and, later, to monitor their progression.[75] The plant, which also includes a desalination facility to create more potable water for the emirate, relies on fossil fuels.[76]

Looking Forward

One of the complications of Abu Dhabi's adaptation efforts is that many of the measures it must take to adapt to climate change can actually exacerbate climate change. Some of the most energy-consumptive adaptation efforts include cooling and nighttime operations, as well as desalinating ocean water for consumption.[77] Furthermore, as Masdar City illustrates the challenges of building a carbon-neutral city from scratch, further adaptation efforts will be needed to bring Abu Dhabi's existing infrastructure into a carbon-neutral era.

And while Abu Dhabi and the UAE's adaptation initiatives are often entwined with mitigation objectives, the latter are more developed than the former. For instance, the national plan highlights benchmarks for mitigation, deadlines for clean-energy targets, greenhouse gas emission management systems, and other programs. In contrast, adaptation benchmarks are described as assessments, policy planning, and monitoring, without clear guidance on what each of those steps entails and how success will be measured.[78]

And yet Abu Dhabi's model of relying on ancient local customs and investing in local ecosystems offers a promising path forward. These include Abu Dhabi's investments in clean energy, the Blue Carbon initiative, and its development of coastal ecosystems. Abu Dhabi's investments in local ecosystems includes programs to promote sustainable aquaculture, and the emirate is also tracking and addressing the presence of invasive alien species in its environs.[79] Yet Abu Dhabi's unique circumstances, including a strong emirate-level government and vast wealth, make its experience in adaptation difficult to replicate.

Conclusion

Abu Dhabi's posture as a global leader in the climate change movement is somewhat tempered by its continued reliance on oil for economic growth. The emirate has some of the highest carbon emissions per capita globally and is one of the world's largest oil producers.[80] While it sets ambitious goals for domestic shifts to clean energy, this research, development, and adaptation is essentially funded by revenues from global oil sales. Its reliance on oil sales for its GDP is still significant, and despite claims of diversification, as much as 90 percent of the emirate's GDP is linked to the oil industry, according to Standard & Poor's.[81] Because climate change is a global phenomenon, the emirate's oil exports will continue to threaten its long-term survival, even as it attempts to implement green policies domestically. Domestic efforts to switch to green energy, furthermore, are difficult to evaluate, and it is not always clear whether, or how, the government is meeting its ambitious goals.

NOTES

1. Most chapters in this book address climate policies on a municipal level. The governmental structure of the UAE, however, means local governance powers are centralized on the emirate level. This chapter, therefore, focuses on the adaptation measures undertaken by the Emirate of Abu Dhabi.
2. Without intervention, experts say Abu Dhabi may be uninhabitable within a century. Pal, Jeremy S., and Elfatih A. B. Eltahir. 2016. "Future Temperature in Southwest Asia Projected to Exceed a Threshold for Human Adaptability." *Nature Climate Change* 6(2): 197–200. ("By the end of the century, annual [maximum temperature] in Abu Dhabi, Dubai, Doha, Dhahran and Bandar Abbas exceeds 35°C several times in the 30 years, and the present-day 95th percentile summer

(July, August, and September; JAS) event becomes approximately a normal summer day.").

3. Abdul Kader, Binsal. 2018. "Zayed: The Man Who Tamed the Desert." *Gulf News*, November 7, http://gulfnews.com.
4. Salloum, Habeeb. 1997. "How Sheikh Zayed Turned the Desert Green." *Christian Science Monitor*, May 27, www.csmonitor.com.
5. Indeed, the Middle East and North Africa region as a whole is expected to be particularly vulnerable to the effects of climate change, and some scholars have questioned whether "the negative impacts will be so great that no amount of resilience will suffice to revive the MENA region." Thompson, William R., and Leila Zakhirova. 2021. *Climate Change in the Middle East and North Africa: 15,000 Years of Crises, Setbacks, and Adaptation*, 138; UAE Ministry of Climate Change and Environment (MOCCAE). 2017. *National Climate Change Plan of the United Arab Emirates, 2017–2050*. www.moccae.gov.ae.
6. MOCCAE. 2021. *The UAE State of Climate Report: A Review of the Arabian Gulf's Changing Climate & Its Impacts*.
7. UAE. n.d. "History." Accessed June 12, 2021, http://u.ae.
8. Salloum 1997.
9. Law No. (21) of 2005 (requiring waste generators to reduce waste); Law No. (17) of 2008 (establishing a Center for Waste Management in the emirate); Law No. (5) of 2016; Law No. (23) of 2007. The council has since been merged with the department now known as the Department of Municipalities and Transport. Decree No. (42) of 2009.
10. Alobaidi, Khaled Ali, Abdrahman Bin Abdul Rahim, Abdelgadir Mohammed, and Shadiya Baqutayan. 2015. "Sustainability Achievement and Estidama Green Building Regulations in Abu Dhabi Vision 2030." *Mediterranean Journal of Social Sciences* 6(4): 509–518.
11. Environment Agency, Abu Dhabi. n.d. "About Us." Accessed June 14, 2021, www.ead.gov.ae.
12. Environment Agency, Abu Dhabi. n.d. "What We Do." Accessed June 14, 2021, www.ead.gov.ae.
13. Abu Dhabi Emirate. 2020. *Single Use Plastic Policy*. www.ead.gov.ae. Federal Law No. (12) of 2018 gives the responsibility for waste management to the competent authority.
14. Under Law No. (21) of 2005 and Law No. (17) of 2008, Abu Dhabi has stated that EAD is the competent authority to implement the waste management law.
15. Abu Dhabi Emirate 2020.
16. Department of Energy. n.d. "What We Do." Accessed June 14, 2021, http://rsb.gov.ae.
17. While Dubai remains the emirates' most popular tourist destination, Abu Dhabi's beaches, mosques, and museums—including a branch of the Louvre—bring in over five million tourists annually. Department of Culture and Tourism Abu Dhabi. 2019. *Tourism: Annual Report 2019 Volume 2*. https://tcaabudhabi.ae.

18. US Central Intelligence Agency (CIA). 2021. "United Arab Emirates." *CIA World Factbook*, August 23, www.cia.gov.
19. CIA 2021; United Arab Emirates Country Commercial Guide. 2019. "United Arab Emirates—Oil and Gas Field Machinery and Services." *Export.gov*, August 7, www.export.gov.
20. Luomi, Mari. 2016. *The Gulf Monarchies and Climate Change: Abu Dhabi and Qatar in an Era of Natural Unsustainability*. Oxford: Oxford University Press.
21. UAE Embassy in Washington, DC. n.d. "About the UAE—UAE Economy." Accessed June 19, 2021, www.uae-embassy.org.
22. Fattah, Zainab. 2019. "Abu Dhabi GDP to Grow an Average of 2.5% through 2022, S&P Says." *Bloomberg*, June 2, www.bloomberg.com.
23. Di Paola, Anthony. 2020. "Abu Dhabi Plans to Spend $122 Billion on Oil in Next Five Years." *Bloomberg*, November 23, www.bloomberg.com.
24. UAE. 2021. "The UAE's Response to Climate Change." August 27, http://u.ae; MacMillan, Arthur. 2021. "UAE Announces 50 Percent Clean Energy Target by 2050 at UN." *The National*, July 5, www.thenationalnews.com.
25. *Reuters*. 2020. "UAE's ADNOC to Explore Clean Energy Expansion, CEO Says." October 14, www.reuters.com.
26. MOCCAE 2017.
27. Al-Suwaidi, Qais Bader. 2020. Presentation at Global Sustainable Cities Conference, NYU Abu Dhabi Saadiyat Campus, November 5.
28. Abueish, Tamara. 2020. "UAE Bans Outdoor Work Ahead of Summer Season, Issues up to $13,000 Fine for Violators." *Al Arabiya English*, June 14, http://english.alarabiya.net.
29. Al-Suwaidi 2020.
30. Abueish 2020.
31. Pamuk, Humeyra. 2011. "UAE's Mission Impossible: Cooling the Desert." *Reuters*, July 13, www.reuters.com.
32. Shanks, Kirk, and Elmira Nezamifar. 2013. "Impacts of Climate Change on Building Cooling Demands in the UAE." Paper presented at SB13 Dubai: Advancing the Green Agenda Technology, Practices and Policies, Dubai, United Arab Emirates, December 8–12.
33. Godinho, Varun. 2021. "Abu Dhabi Issues First District Cooling License to Saadiyat Cooling." *Gulf Business*, April 18, http://gulfbusiness.com.
34. Günel, Gökçe. 2016. "The Infinity of Water: Climate Change Adaptation in the Arabian Peninsula." *Public Culture* 28(2): 296, 298. "Desalination, an energy-intensive process, costs the UAE about $18 million each day," and "the Abu Dhabi Water Resources Master Plan . . . notes that 'overall fossil fuel use in the cogeneration plants is around 21 million tons equivalent of CO_2 per year and the share attributed to water production and use lies between 20 and 45%.' 'Thus water use probably contributes between 4 and 9 million tons of CO_2 equivalent per year' and promotes ways of thinking past these issues, such as research and

development of newer, less resource intensive desalination technologies." *See also* Savarani, chapter 4 in this volume.

35. Ghedira, Hosni. 2019. "Lessons from Abu Dhabi's Urban Heat Island Effect." Khalifa University, December 10, www.ku.ac.ae.
36. EAD. 2019. *First Groundwater Atlas for Abu Dhabi to Promote Efficient Water Resource Management in Emirate*. January 16, www.ead.gov.ae.
37. Gökçe 2016, 295.
38. Gökçe 2016, 295. *See also* Thompson, William R., and Leila Zakhairova. 2021. *Climate Change in the Middle East and North Africa*, 192 ("Taking salt out of seawater is an extremely energy-intensive process. Desalination uses about 15,000 kilowatt-hours of power for every million gallons of fresh water produced. This amount of energy is enough to power an average American household for a year and a half.") (internal citations omitted).
39. Dawoud, Mohamed A. 2018. "A Strategic Water Reserve in Abu Dhabi." Groundwater Solutions Initiative for Policy and Practice, November 5, http://gripp.iwmi.org.
40. MEED. 2021. "Abu Dhabi Launches Water Management Plan." *Water Technology*, February 26, www.water-technology.net.
41. Shabeeh, Rabiya. 2017. "How Climate Change Can Impact UAE Industries." *Khaleej Times*, April 26, www.khaleejtimes.com.
42. Ksiksi, Taoufik Saleh, and Tarek Youssef. 2012. "Sea Level Rise and Abu Dhabi Coastlines: An Initial Assessment of the Impact on Land and Mangrove Areas." *Journal of Ecosystem & Ecography* 2(4): 1–5.
43. Burke, Louise. 2020. "Why the UAE'S Mangroves Are So Important—And How to Save Them." *The National*, January 16, www.thenationalnews.com.
44. Van Esveld, Bill. 2020. "'The Island of Happiness:' Exploitation of Migrant Workers on Saadiyat Island, Abu Dhabi." Human Rights Watch, November 13, www.hrw.org.
45. Eapen, Nikhil. 2020. "The Invisible Migrant Workers Dying of 'Natural Causes' in the Arab Gulf." *Caravan*, March 31, http://caravanmagazine.in.
46. Jwoubert, Darren, Jens Thomsen, and Oliver Harrison. 2011. "Safety in the Heat: A Comprehensive Program for Prevention of Heat Illness among Workers in Abu Dhabi, United Arab Emirates." *American Journal of Public Health* 101(3): 395–398.
47. Kingsley, Patrick. 2013. "Masdar: The Shifting Goalposts of Abu Dhabi's Ambitious Eco-City." *Wired UK*, December 17, www.wired.co.uk.
48. Günel, Gökçe. 2019. *Spaceship in the Desert: Energy, Climate Change, and Urban Design in Abu Dhabi*. Durham, NC: Duke University Press.
49. Mubadala Investment Company. 2021. "Masdar City." April 26, www.mubadala.com.
50. Flint, Anthony. 2020. "What Abu Dhabi's City of the Future Looks Like Now." *Bloomberg City Lab*, February 14, www.bloomberg.com.
51. Cottrill, Jacob. 2015. "What We Can Learn from Masdar City." US Chamber of Commerce Foundation, September 9, www.uschamberfoundation.org.

52. US Department of Energy. 2011. "U.S. Department of Energy and Masdar Collaborate in Testing Cutting-Edge Solar PV Coating Technologies." February 28, www.energy.gov.
53. Gammon, Katherine. 2016. "Lessons from a City Built without Light Switches and Water Taps." *TakePart*, September 19, www.takepart.com.
54. Günel 2019, 3.
55. WWF. 2008. "WWF, Abu Dhabi Unveil Plan for World's First Carbon-Neutral, Waste-Free, Car-Free City." January 13, https://wwf.panda.org.
56. Jeppesen, Helle. 2010. "Built on Sand: Masdar City to Become Eco-City of the Future." *Deutsche Welle*, April 21, www.dw.com.
57. Baselaib, Yousef. 2019. "Sustainable Development Is Key to Urban Future." MEED, May 2, www.meed.com.
58. Griffiths, Steven, and Benjamin K. Sovacool. 2020. "Rethinking the Future Low-Carbon City: Carbon Neutrality, Green Design, And Sustainability Tensions in the Making of Masdar City." *Energy Research & Social Science* 62: 101368. https://doi.org/10.1016/j.erss.2019.101368.
59. Paleologos, Evan, B. A. Welling, M. E. Amrousi, and Huda Masalmeh. 2019. "Coastal Development and Mangroves in Abu Dhabi, UAE." *IOP Conference Series: Earth and Environmental Science* 344: 012020. https://doi.org/10.1088/1755-1315/344/1/012020.
60. Burke 2020.
61. Burke 2020.
62. Tremblay, Sophie. 2019. "Abu Dhabi Is Replanting Mangroves in the Fight against Climate Change." *CNN*, July 15, www.cnn.com.
63. Tremblay 2019.
64. EAD. n.d. "Mangroves: Avicennia Marina." Accessed June 23, 2021, www.ead.gov.ae.
65. Zacharias, Anna. 2018. "A Fifth of Abu Dhabi's Mangroves in Moderate or Poor Health, Study Shows." *The National*, October 18, www.thenationalnews.com.
66. Paleologos et al. 2019.
67. Burt, John, Mary Killilea, and Anne Rademacher. 2021. "Unexpected Nature? Proliferating Mangroves on the Coast of Abu Dhabi." In *Urban Environments as Spaces of Living in Transformation*. Urban Environments Initiative. https://urbanenv.org.
68. Tremblay 2019.
69. Ray, Raghab, and Tapan Kumar Jana. 2017. "Carbon Sequestration by Mangrove Forest: One Approach for Managing Carbon Dioxide Emission from Coal-Based Power Plant." *Atmospheric Environment* 171: 149–154. https://doi.org/ 10.1016/j.atmosenv.2017.10.019.
70. Crooks, Stephen, Katrina Poppe, Angela Rubilla, and John Rybczyk. 2019. *Mangrove Soil Carbon Sequestration of the United Arab Emirates: Trial Application*. Abu Dhabi Global Environmental Data Initiative/Environment Agency Abu Dhabi, Silvestrum Climate Associates, and Western Washington University.

71. Ksiksi, Taoufik, Essam Abdel-Mawla, and Tarek Youssef. "Sea Level Rise and Abu Dhabi Coastlines: An Initial Assessment of the Impact on Land and Mangrove Areas." *Journal of Ecosystem & Ecography* 2(4): 1–5. https://doi.org/10.4172/2157-7625.1000115.
72. Zaman, Samihah. 2018. "Abu Dhabi Looks at Drain Systems to Prevent Flooding." *Gulf News*, May 3, https://gulfnews.com.
73. GRID. 2013. "The Abu Dhabi Blue Carbon Demonstration Project." February 13, www.grida.no.
74. Abu Dhabi Global Environment Data Initiative. n.d. "AGEDI: Abu Dhabi Blue Carbon Demonstration Project." Accessed June 23, 2021, http://agedi.org.
75. Siegel, Fern, and Matthew B. Hall. 2020. "Abu Dhabi Plants Mangrove Forest to Reduce Carbon Footprint." *Tennessee Tribune*, November 24, https://tntribune.com.
76. International Trade Administration. n.d. "United Arab Emirates—Country Commercial Guide." Accessed June 19, 2021, www.trade.gov.
77. Novo, Cristina. 2020. "Desalination Enhances Abu Dhabi's Water Security." *Smart Water Magazine*, July 27, https://smartwatermagazine.com.
78. MOCCAE 2017, 26. On the mitigation front, the UAE established a clean-energy target for 2021, developed a "GHG emissions management system" by 2020, and strives for a "high level of eco-efficiency" by 2030. In contrast, the adaptation priorities state that, by 2020, a "climate risk and vulnerability assessment [will be] performed, [and] immediate measures put in place;" by 2025, "adaptation planning [will be] mainstreamed in development policy"; and the following decades will bring "continuous monitoring and evaluation to ensure evidence-based adaptation measures."
79. EAD. 2019. *Sustainable Aquaculture Policy*. www.ead.gov.ae; EAD. 2018. *Taking Action on Terrestrial and Freshwater Alien Species in Abu Dhabi: From Prevention to Control*. www.ead.gov.ae.
80. "Total energy consumption per capita is very high (the 6th highest worldwide in 2019) and represents 8.2 toe in 2019. Per capita electricity consumption is also very high at 12.7 MWh in 2019 (8th highest in the world)." Enerdata. n.d. "United Arab Emirates Energy Information." Accessed July 2, 2021, www.enerdata.net; Carpenter, J. William. 2021. "The Biggest Oil Producers in the Middle East." *Investopedia*, May 19, www.investopedia.com.
81. Saadi, Dania. 2020. "UAE's Abu Dhabi GDP to Contract 7.5% in 2020 on Oil Price Crash, Virus: S&P Global Ratings." S&P Global Platts, May 31, www.spglobal.com.

ACKNOWLEDGMENTS

First and foremost, we would like to express our deep appreciation to Sara Savarani for her tireless assistance in putting this book together. We would also like to thank NYU Abu Dhabi Institute, the D'Agostino Fund, and all those who provided commentary, either during the workshop or on prior written drafts, including Yishai Blank, Mohamed Abdel Hamyd Dawoud, Anel Du Plessis, R. Andreas Kraemer, Yifei Li, Jolene Lin, Dane McQueen, Cynthia Rosenzqeig, Qais Bader Al Suwaldi, and Johanna Wolff.

ABOUT THE EDITORS

DANIELLE SPIEGEL-FELD is Executive Director at the Guarini Center on Environmental, Energy and Land Use Law, where her research focuses on urban environmental law and sustainable building policy. She is also an adjunct professor of law at NYU School of Law and Bard College. Previously, Spiegel-Feld served as a research fellow at the University of Copenhagen's Faculty of Law, where she studied interactions between international trade law and domestic environmental policy. She has published widely on topics in environmental law and policy in scholarly publications and media outlets including the *New York Times*, *U.S. News*, *Crain's*, and others. She received her JD from NYU and her BA, summa cum laude, from the University of Pennsylvania.

KATRINA MIRIAM WYMAN is Sarah Herring Sorin Professor of Law at NYU School of Law. Her research interests include urban environmental law, property, and natural resources. She coteaches a seminar on urban environmental law with Danielle Spiegel-Feld.

JOHN J. COUGHLIN serves as Global Distinguished Professor of Religious Studies and Law at NYU, Abu Dhabi, and affiliated faculty at the NYU Law School. He holds a JCD, JCL, from Gregorian University, Rome; a JD from Harvard Law School; a ThM from Princeton Seminary, an MA from Columbia University, and a BA from Niagara University. He has taught at the Gregorian University, Notre Dame Law School, St. John's Law School, St. Joseph's Seminary, and St. Bonaventure University. The author of books and articles, he is the founding Program Head of Legal Studies at NYU, Abu Dhabi.

ABOUT THE CONTRIBUTORS

BARBARA ANTON is Senior Coordinator at the ICLEI European Secretariat. Her thematic focus is on urban water management and governance, including blue-green solutions and climate change adaptation. As part of her work, she has provided policy advice and contributed to stakeholder consultations at the European and global levels.

TIRTHA BISWAS is a policy analyst working on the development of sustainable and competitive pathways for Indian industry to support its low-carbon growth aspirations. At the Council on Energy, Environment and Water (CEEW), his research involves mineral resource security, clean energy solutions, and energy efficiency of the domestic industrial sector in India. He holds a dual degree from the Indian Institute of Technology (Indian School of Mines), Dhanbad, with an undergraduate degree in mineral engineering and a master's degree in mineral resource management.

CLARE E. B. CANNON is Assistant Professor in the Department of Human Ecology at the University of California, Davis, and Research Fellow in the Department of Social Work at the University of the Free State, South Africa. She researches intersections of social inequality, health disparities, and environmental injustice in urban, rural, and disaster contexts.

ERIC K. CHU is Assistant Professor in the Department of Human Ecology and Co-Director of the Climate Adaptation Research Center at the University of California, Davis. He is interested in how local governments and communities around the world adapt to climate change impacts and reduce risks in ways that are resilient, inclusive, and equitable. He is Lead Author in the Working Group 2 contribution to the Sixth Assessment Report to the Intergovernmental Panel on Climate Change and Chapter Lead in the Fifth US National Climate Assessment.

PATRYCJA DŁUGOSZ-STROETGES is Officer in the Sustainable Resources, Climate, and Resilience team at the ICLEI European Secretariat. Her work focuses on holistic urban environmental management. She currently works on the conceptual development and implementation of the European Commission's Green City Accord.

HARRY DEN HARTOG is a Dutch urban designer, researcher, and critic. Since 2008, he has been based in Shanghai, where he runs his think tank studio, Urban Language (www.urbanlanguage.org). Since 2012, he has been a faculty member at Tongji University in Shanghai, College of Architecture and Urban Planning. During these years, he has done a lot of research, especially on new town developments, on urban-rural transitions, and on waterfronts transformations.

TANUSHREE GANGULY is an air quality researcher who is working on developing a data-driven approach toward clean-air policy making in India. At the Council on Energy, Environment and Water (CEEW), her work focuses on assessing the potential of alternative methods of monitoring air quality and understanding and addressing the current regulatory hurdles in effective implementation of clean-air policies. Prior to joining the Council, she worked with the Centre for Science and Environment, where she helped develop clean-air action plans for the nonattainment cities of Andhra Pradesh. During her brief professional stint as an air quality consultant for an environmental consulting firm in California, she estimated the potential health risk stemming from construction and operation of over twenty proposed land-use development projects. She has a master's in environmental engineering from Georgia Institute of Technology and is a certified engineer-in-training under California law.

HE FENG serves as a senior consultant of the E20 Environment Platform. He has been involved in drafting a low-carbon development report for Beijing and other low-carbon city research projects.

HU TAO is President of Lakestone Academy of Marine Science and Technology, Shenzhen, China; Director of Lakestone Institute for Sustainable Development, USA; Chairman of the Board of the Professional

Association for China's Environment; Vice President of the Union of Chinese American Professional Organizations; Member of China Carbon Neutrality 50 Experts Forum; and Member of UN PRME China Academic Committee. Hu is also Senior Research Fellow of the Center for Energy Security and National Development, Peking University, and Senior Consultant for the Asian Development Bank and World Bank. Hu's research and expertise cover environmental economics, policy and governance, trade, investment and environment, globalization, co-control of air pollutants and GHGs, negative emission technologies, 3e (energy, environment, and economics) policy, and the like.

Frank J. Kelly holds the Chair in Community Health and Policy at Imperial College London, where he is Director of the Environmental Research Group, Director of the NIHR Health Protection Research Unit on Environmental Exposures, and Health and Deputy Director of the MRC-PHE Centre for Environment & Health. Kelly has published over four hundred peer-reviewed papers as well as many conference papers and books (as author or editor) on the toxicology and health effects of air pollution. In addition to his academic work, Kelly has provided policy support to the Greater London Authority and WHO. He chaired COMEAP, the UK's Department of Health and Social Care Expert Committee on the Medical Effects of Air Pollutants, for nine years, and he is a member of the Health Effects Institute Review committee.

Neeraj Kuldeep co-leads the Renewables team at the Council on Energy, Environment and Water (CEEW) to accelerate the grid-interactive distributed renewable-energy deployments in India. He has worked and published extensively on renewable-energy markets, jobs and skills, and power markets. He is an IIT-Bombay alumnus, having graduated with a bachelor's in energy science engineering, and also holds an MTech in energy systems.

Yael R. Lifshitz is Lecturer (Associate Professor) in Property Law at King's College London. She researches and teaches on issues of energy law, natural resources, and property law and theory. She is a member of the Climate Law and Governance Centre at King's College London and a fellow with the Guarini Center on Environmental, Energy and Land Use Law at New York University School of Law.

ALVIN LIN is Climate and Energy Director at the Natural Resources Defense Council (NRDC) office in Beijing, China. Lin's work focuses on analyzing China's climate and clean-energy policies and advocating for their continual improvement. His areas of expertise include the environmental impacts of coal and shale gas development, energy-efficiency technologies, and air pollution law and policy. Prior to joining NRDC, Lin worked as a litigator and a judicial clerk in New York City. He holds a bachelor's degree from Yale University, a master's from the Chinese University of Hong Kong, and a JD from New York University School of Law.

MAO XIANQIANG is Professor at the School of Environment, Beijing Normal University. His research interests range from low-carbon city, low-carbon economy, and co-control for GHGs and local pollutants to trade and environment problems.

ADALENE MINELLI is Senior Fellow at the Guarini Center on Environmental, Energy and Land Use Law at New York University School of Law, where she researches and writes on matters of local environmental law. Her areas of expertise include environmental markets, environmental review, and local climate policy.

ASIYA N. NATEKAL is Post-Doctoral Researcher in the Center for Regional Change at the University of California, Davis. Her research interests lie at the intersection of the built environment and sustainability. Prior to joining the center, she worked as an adjunct professor at the University of California, Irvine, and California State Polytechnic University, Pomona. She holds a bachelor of architecture from Mumbai University, master of urban and environmental planning from Arizona State University, and a PhD in planning, policy, and design from the University of California, Irvine.

DÖRTE OHLHORST is Lecturer at the Bavarian School of Public Policy at the Technical University of Munich. Her research focuses on energy, environmental and innovation policy, governance in multilevel systems, sustainability, and participative decision-making processes. She is currently coordinating projects dealing with citizen acceptance of hydrogen technologies and renewable energy technologies. From 2012 to 2016, she led the project "ENERGY TRANS" at the Environmental Policy Research Center, Freie Universität Berlin, where she investigated the governance of the energy transition in Germany and Europe. From 2009 to 2012, she was research associate at the German Advisory Council for the Environment and Head of the Climate and Energy Department of the Center for Technology and Society of the Technical University of Berlin. In 2011, she founded the Institute for Sustainable Use of Energy and Resources together with colleagues.

Sara Savarani is Senior Fellow at the Guarini Center of Environmental, Energy and Land Use Law at New York University School of Law. Her research spans from urban environmental law and sustainability to global ocean governance and plastics. She received her BS in environmental science from UCLA and her JD from New York University School of Law.

Miranda A. Schreurs is Professor of Climate and Environmental Policy at the Bavarian School of Public Policy, Technical University of Munich. She investigates environmental movements, green politics, and climate policy making both comparatively and internationally. In 2011, Schreurs was appointed by Chancellor Angela Merkel as a member of the Ethics Committee for a Secure Energy Supply. She was a member of the German Council on the Environment (2008–2016), the Berlin Enquete Commission: "New Energy Berlin" (2014–2015), and the Berlin Climate Advisory Council (2012–2014) and is currently Vice Chair of the European Advisory Council on Environment and Sustainable Development and Co-Chair of the Citizens' Committee Accompanying the Search for a High Level Radioactive Waste Site.

Hanne J. van den Berg currently works at the European Environment Agency as a climate change adaptation expert specialized in justice and equity in climate change adaptation. She has close to fifteen years of international experience analyzing and managing consultancy and applied research projects as well as policy development on climate change adaptation, flood risk management, nature-based solutions, international development, governance, (participatory) urban planning, and decision-support tools from a multisectoral perspective. She holds a doctoral degree from Harvard University's Graduate School of Design. She contributed to this publication in a personal capacity.

Josephine van Zeben is Professor of Law and Chair of the Law Group at Wageningen University, the Netherlands. Her work focuses on interdisciplinary scholarship related to the regulation of environmental impacts by public and private actors. Much of her work relates to developments in the European Union, at supranational, national, and local levels.

Vidya Vijayaraghavan is an environmental lawyer from India, with special interests in biodiversity, climate change, and sustainability issues. She is a fellow at the Guarini Center of Environmental, Energy and Land Use Law at New York University School of Law. She graduated as a Vanderbilt scholar with a master's in environmental law at the New York University School of Law. Previously, she advised the UNDP, India, the Indian Ministry of Environment and worked with IDLO and the governments of Namibia and Vietnam on environmental legal and policy issues.

Xing Youkai is a senior engineer at the Transport Planning and Research Institute, Ministry of Transport. His research interest focuses on co-control for GHGs and local pollutants, low-carbon transportation, and the like.

Gao Yubing is Research Associate at the School of Environment, Beijing Normal University. Her research area focuses on co-control for GHGs and local pollutants, low-carbon cities, and the like.

Katie Zavadski is a JD candidate at the New York University School of Law, where she is a staff editor on the *Law Review*. She holds an AB from Harvard University and an MA from NYU. She was previously the research editor at ProPublica and taught journalism at the CUNY Newmark Graduate School of Journalism.

INDEX

Page numbers in italics indicate photos and tables

Printed in the United States
by Baker & Taylor Publisher Services